CELLULAR AND MOLECULAR BIOLOGY OF AUTISM SPECTRUM DISORDERS

Anna Strunecka

Department of Physiology, Faculty of Science, Charles University in Prague, Prague, Czech Republic

&

Institute of Medical Biochemistry, Laboratory of Neuropharmacology, 1st Faculty of Medicine, Charles University in Prague, Prague, Czech Republic

Russell L. Blaylock

Institute for Theoretical Neuroscience, LLC & Belhaven University, Ridgeland, MS, USA

Mark A. Hyman

Institute for Functional Medicine & The UltraWellness Center, Lenox, MA, USA

Ivo Paclt

Psychiatric Department, 1st Faculty of Medicine, Charles University in Prague, Prague, Czech Republic

&

Institute for Postgraduate Medical Education, Prague, Czech Republic

Cover Design

Authors of figures   : Anna Strunecka and Hana Kruzikova

Graphics and cover   : Hana Kruzikova, E-mail: hana.kruzikova@tiscali.cz

Picture for cover   : Faiths of Hope, 2003, oil painting with apophyllites, 60x46.

PhDr. Ing. Zdenek Hajny, PhD. Gallery Cesty ke Svetlu (Ways to Light), Zakouřilova 955/9, Prague - Chodov, Czech Republic. www.cestykesvetlu.cz

eBooks End User License Agreement

CONTENTS

Foreword   *i*

Preface   *ii*

List of Contributors   *iii*

CHAPTERS

1. **Autism Spectrum Disorders: Clinical Aspects**   1
   I. Paclt, A. Strunecka

2. **The Cerebellum in ASD**   17
   R. L. Blaylock

3. **Dysregulation of Glutamatergic Neurotransmission in ASD**   32
   A. Strunecka

4. **Immunoexcitotoxicity as a Central Mechanism of ASD**   47
   R. L. Blaylock

5. **Immune Dysfunction in ASD**   73
   R. L. Blaylock

6. **Gastrointestinal Disorders and ASD: A Causal Link or a Secondary Consequence?**   82
   A. Strunecka

7. **Biochemical Changes in ASD**   100
   A. Strunecka

8. **Searching the Role of Mercury in ASD**   121
   A. Strunecka, R. L. Blaylock

9. **Fluoride and Aluminum: Possible Risk Factors in Etiopathogenesis of ASD**   148
   A. Strunecka, R. L. Blaylock

10. **The Role of Melatonin in Etiopathogenesis and Therapy of ASD**   162
    A. Strunecka

11. **The Search for Plausible Role of Oxytocin in Etiology and Therapy of ASD**   173
    A. Strunecka

12. **Regulation of Cortisol Levels in Autistic Individuals and their Mothers**   186
    A. Strunecka

13. **Reproductive Hormones and ASD**   199
    R. L. Blaylock

14. **Addendum. Autism: Is It All in the Head?**   206
    M. A. Hyman

Index   217

FOREWORD

The history of the biomedical movement in autism treatment began in the 1960's when Dr. Bernard Rimland, founder of the Autism Research Institute (ARI), took a monumental step forward by declaring that autism was due to a physiological abnormality rather than a result of poor nurturing by uncaring parents. Soon after his 1964 book, *Infantile Autism*, was published, he was besieged with letters and telephone calls from parents worldwide who praised his writings and shared their own personal journeys with autism.

Shortly after the publication of this seminal book, Dr. Rimland was astonished at the number of parents who reported observing significant improvements in their children soon after giving them a nutritional supplement. He conducted several formal and informal studies, and concluded that vitamin B6 with magnesium might help up to 50% of the autism population. To date, there are 11 placebo-controlled studies supporting the efficacy of vitamin B6 and magnesium as a treatment for autism. Another biomedical-related intervention, reported by Dr. Rimland in the 1970s, was the importance of restricted--and healthy diets.

Over the years, parents as well as clinicians continued to write to Dr. Rimland about their experiences; in turn, ARI would share this information with research scientists and clinicians around the world.

The year 1995 was a turning point in the biomedical field; Dr. Rimland, along with two of his close colleagues, Drs. Sidney Baker and Jon Pangborn, convened the first international think tank on autism. Over 30 researchers and clinicians were invited to meet for two-and-a-half days. Toward the end of the meeting, they agreed on the importance of investigating gastrointestinal (GI) and immune system problems more deeply, to better understand and treat individuals on the autism spectrum.

2010 has also been an important one for the biomedical field. The American Academy of Pediatrics' journal, *Pediatrics*, published a consensus report on the state-of-the-art research on GI problems associated with autism. A few months later, a large-scale multi-center survey involving 1,185 children and teenagers on the autism spectrum showed that nearly half (45%) had one or more forms of GI problem.

Viewed from a broader perspective regarding the treatment of autism, one of the problems in the field is a clash with clinicians and researchers who favor other effective forms of treatment, such as Applied Behavior Analysis (ABA) and sensory interventions. The viewpoint taken by many people in the biomedical field, including those at ARI, is that many, but not all, individuals on the autism spectrum suffer from some type of medical problem, such as GI and/or immune system dysfunction, and these problems can lead to discomfort or pain, cause sensory dysfunction, impede executive functioning, and more. Once the person's health improves, many of their sensory problems are reduced or are eliminated, and they will be primed to attend, and thus to learn in an educational setting.

Clinicians and researchers worldwide are striving for a global standard of care for individuals on the autism spectrum. We need to recognize their various needs or problems--medical, sensory, and behavioral--but we also need to be cognizant of individual differences in this population. Through networking, communicating, and discouraging politically-oriented science, it will be made possible for individuals on the autism spectrum to reach their true potential, and their quality of life will improve significantly.

Stephen M. Edelson,
Autism Research Institute
USA

PREFACE

Autism spectrum disorders (ASD) are a group of neurodevelopmental disorders characterized by abnormalities in social interaction, language function and communication, and abnormalities in the realm of behavior. Over the past several decades the incidence of ASD has increased dramatically, with much of the increase not being explained by improved diagnosis. The etiology of ASD remains an unsolved puzzle to scientists, physicians, pediatricians, psychiatrists, and pharmacologists. Of great concern is that no central mechanism has been proposed to explain the various clinical presentations of the ASD and no evidence-based therapy has been offered. The advantage of this eBook is to discuss the state of knowledge regarding the pathophysiology, cellular and molecular biology of these disorders.

A great number of biochemical and pathological changes have characterized ASD, adding confusion to discovering a common etiology. A recent review of the genetic links to ASD found that the most common genes suspected were operate glutamate receptors (GluRs), either ionic or metabotropic. A considerable amount of evidence suggests a role for a dysfunctional immune system in the ASD. The crosstalk between GluRs and cytokine receptors leads to neurodegeneration, abnormal neuronal migration patterns, seizure generation, and dysfunctional brain connectivity. When combined with the finding of elevated glutamate in a number of autistic children, this indicates a possible hyperactivity of GluRs in those at greatest risk.

Our eBook explains, for the first time, the central role of immunoexcitotoxicity in the etiopathogenesis of the broad spectrum of autistic disorders. Based on our hypothesis of immunoexcitotoxity, we integrate various findings in ASD with this hypothesis. A careful review of known environmental and pathological links to ASD indicates that most, if not all, are connected to the immunoexcitotoxic process. Our eBook also offers treatment proposals that address each of these mechanisms. It explains how previous, often successful treatment methods, may indeed operate through the immunoexcitotoxic mechanism.

The tremendous research of individuals with ASD shows most explicitly that ASD is neither a disease of one gene, neurotransmitter or hormone, nor a disease of a single isolated second messenger disturbance. The enormous increase of autism during last decade inevitably requires an integrative approach, which brings together not only specialized scientific knowledge, but also knowledge about the homeostatic mechanisms of the whole human being. We simultaneously realize that the living system does not behave as a static jigsaw puzzle. The behavior of a whole cannot be predicted by knowing the separated parts. We hope that the integration of specialized knowledge about molecular and cellular mechanisms could lead to understanding why new generations suffer with an epidemic of autism.

Our eBook reviews the studies of scientists from the broad area of neurosciences and neuropharmacology, cognitive and affective developmental neuroscience; researchers from immunology, pathophysiology, and developmental biology; researchers from the developmental psychopathology and applied behavioral analysis; practicing physicians, pediatricians, psychiatrists, and psychologists; but also parents and care-givers, who are in daily contacts with children and adults with autism.

Author's thanks belong to Hana Kruzikova for her excellent cooperation in preparing diagrams and figures and to PhDr. Ing. Zdenek Hajny, PhD., for providing his picture for the cover.

Anna Strunecka
Charles University in Prague
Czech Republic

and **Russell L. Blaylock**
Institute for Theoretical Neuroscience
USA

LIST OF CONTRIBUTORS

Russell L. Blaylock, MD

Neurosurgeon. Institute for Theoretical Neuroscience, LLC and Visiting Professor of Biology, Belhaven University, Ridgeland, MS 39157, USA. E-mail: blay6307@bellsouth.net

Mark Hyman, MD

Chairman, Institute for Functional Medicine. Volunteer, Partners in Health Founder and Medical Director, The UltraWellness Center. 45 Walker Street, Lenox, MA 01240, USA. E-mail: mark@drhyman.com

Ivo Paclt, MUDr., CSc.

Associated Professor in Psychiatry. Psychiatric Department, 1st Faculty of Medicine, Charles University in Prague, Prague, Czech Republic, Ke Karlovu 11, Prague 2, 128 00 Czech Republic. E-mail: ivopaclt@seznam.cz and Subdepartment of Postgradual Training Child and Adolescent Psychiatry, Institute for Postgraduate Medical Education, Prague, Czech Republic.

Anna Strunecka, RNDr., DrSc.

Professor in Physiology. Department of Physiology, Faculty of Science, Charles University in Prague, Vinicna 7, 128 00 Prague 2, Czech Republic, and Institute of Medical Biochemistry, Laboratory of Neuropharmacology, 1st Faculty of Medicine, Charles University in Prague, Albertov 4, Prague 2, 128 00 Czech Republic. E-mail: strun@natur.cuni.cz

CELLULAR AND MOLECULAR BIOLOGY OF AUTISM SPECTRUM DISORDERS

CELLULAR AND MOLECULAR BIOLOGY OF
AUTISM SPECTRUM DISORDERS

Dedication

To scientists who search for the origin of autism.

To physicians who explore how to best help those suffering from autism.

To parents, relatives and friends who care for people with autism every day.

To all of us who desire to understand what autism means.

CHAPTER 1

Autism Spectrum Disorders: Clinical Aspects

Ivo Paclt[1,*] and Anna Strunecka[2]

[1]Psychiatric Department, 1st Faculty of Medicine, Charles University in Prague and Institute for Postgraduate Medical Education, Prague, Czech Republic and [2]Institute of Medical Biochemistry, Laboratory of Neuropharmacology, 1st Faculty of Medicine, Charles University in Prague, Prague, Czech Republic

Abstract: Autism spectrum disorders (ASD) are a group of related neurodevelopmental disorders, which includes autism (autistic disorder), Asperger syndrome, Rett syndrome, pervasive developmental disorder-not-otherwise specified (PDD-NOS), and childhood disintegrative disorder (CDD). This chapter provides the review of recent knowledge about clinical symptoms and criteria for diagnosis of heterogeneous symptoms of ASD. An alarming increase in the prevalence of ASD is of great concern to practicing pediatricians and psychiatrists. Some people attribute the increases over time in the frequency of ASD to factors such as new administrative classifications, changing diagnostic criteria, and heightened awareness. It is evident, that no single factor or a simple explanation can account for the increase. ASD are highly genetic and multifactorial, with many risk factors acting together. There is no therapy of the core symptoms of ASD at present. Several studies suggest that 50-75% of children with ASD are using complementary alternative medicine. Families and clinicians need access to theoretical and clinical evidence to assist them in the choice of therapies.

INTRODUCTION

ASD are a group of related neurodevelopmental disorders characterized by a collection of neurobehavioral and neurological dysfunctions, often occurring before age 36 months [1]. There are two major diagnostic classification systems in current use, the International Classification of Diseases, version 10 (ICD-10) and the Diagnostic and Statistical Manual of Mental Disorders 4th edition (DSM-IV). ICD-10 was endorsed by the Forty-third World Health Assembly in May 1990 and came into use in WHO Member States beginning in 1994 [2]. DSM-IV contains diagnostic criteria used in the United States and many other parts of the world [3]. These diagnostic classification systems have similar symptom criteria for diagnosis of ASD based on three general impairments: i) severe developmental deficits in socialization, ii) delayed or abnormal language and communication both verbal and non-verbal, iii) repetitive or unusual behaviors. Autistic individuals display behavioral symptoms that include ritualistic features, reliance on routines, and impairment of imaginative play. ASD include autism, childhood disintegrative disorder (CDD), Asperger syndrome, Rett syndrome, and pervasive developmental disorder-not-otherwise specified (PDD-NOS). The terms ASD and autism are often used interchangeably.

Asperger syndrome is characterized by higher intellectual and adequate language abilities; however the deficits in socialization, empathy and human relationships remain. Childhood disintegrative psychosis (CDD), also known as Heller's syndrome, shows normal development, generally up to an age of 2 years, comparable to other children of the same age. However, from around the age of 2 through the age of 10, language and play skills are lost almost completely. Rett syndrome also shows normal development up to an age of 2 years; then follows with autistic behavior and developmental delay, with abnormalities of muscle tone, coordination difficulties, loss of ability to walk, writhing limb movements, avoidance of eye contact, and hyperventilation. Rett syndrome is seen almost always in girls [4]. Rett syndrome is an X-linked ASD caused by mutations in *MECP2* gene, encoding methyl CpG-binding protein 2. The MeCP2 protein likely plays a role in forming connections (synapses) between nerve cells and seems to be critical for normal brain development [5]. Mutations in the CDKL5 gene cause an atypical form of Rett syndrome in females called the early-onset seizure variant [6]. PDD-NOS includes individuals who meet some of the criteria for autism or Asperger syndrome, but not all. People with PDD-NOS usually have fewer and milder symptoms than those with autism.

There has been an alarming 556% increase in the prevalence of ASD reported between 1991 and 1997, a prevalence higher than that of spina bifida, cancer, or Down syndrome [7]. The Centers for Disease Control and Prevention

*Address correspondence to: Ivo Paclt Ke Karlovu 11, 128 00 Prague 2, Czech Republic; Tel: +420 224965316; E-mail: ivopaclt@seznam.cz

(CDC) reported that between 4.2 to 12.1 per 1,000 children with an average of 9.0 per 1,000 children were identified with an ASD in the USA [8,9]. This translates to about 1 in 80 and 1 in 240 children with an average of about 1 in 110 identified with an ASD. There were ten communities who participated in an earlier 2002 surveillance year and the newer 2006 surveillance year. These communities observed an increase in identified ASD prevalence, ranging from 27% to 95% with an average increase of 57% (http://www.cdc.gov/ncbddd/autism/addm.html). Identified ASD prevalence increased across all sex, racial, ethnic and cognitive functioning subgroups. For all sites, the most consistent pattern was the increase among boys. ASD prevalence was four to five times higher for boys than for girls with 1 in 70 boys and 1 in 315 girls identified, on average, with one of the ASD syndromes.

The prevalence of parent-reported diagnosis of ASD among US children aged 3 to 17 years was estimated from the 2007 National Survey of Children's Health (sample size: 78037) [10]; a study that was nation-wide and included all US regions. It was based on parental reports during a phone survey with no corroboration of the diagnoses. The weighted current ASD point-prevalence was 110 per 10,000. This group of authors estimated that 673,000 US children have ASD. Odds of having ASD were 4 times as large for boys than girls. Non-Hispanic (NH) black and multiracial children had lower odds of ASD than NH white children. This observed point-prevalence, 1 in 91, is higher than previous US estimates. More inclusive survey questions, increased population awareness, and improved screening and identification by providers may partly explain this finding

A diagnosis survey was distributed to participating schools to be handed out to parents of all children aged 5-9 years in the UK [11]. The authors estimated the prevalence of ASD to be 157 per 10,000.

Increases over time in the frequency of ASD during the last two decades might be attributable to factors such as new administrative classifications, policy and practice changes, and heightened awareness. Some authors attribute this increase partially to changing diagnostic criteria. However, it seems that new environmental influences, namely the increase of excitotoxic factors and changes in vaccination program, play a substantial role [12]. It is evident, that no single factor or simple explanation can account for the increase. ASD are highly genetic and multifactorial, with many risk factors acting together. Some environmental and biological potential etiologic theories are questionable, exact interpretation of their results is very difficult and more accurate results must wait for future research and discussion. Others, such as Fombonne presented a prevalence of 0.7-72.6 per 10,000 as more accurate data [13]. Surveillance and screening strategies for early identification could enable early treatment and improved outcomes [14].

CLINICAL ASPECTS OF AUTISM

The basic symptoms of infantile autism are impairment in the use of facial expression and eye-to-eye gaze, lack of social or emotional reciprocity and impairment of social communication with both children and adults; bizarre behavior and marked disconnection from reality. Autism occurs in combination with other symptoms of developmental disorders mostly in combination with mental retardation or with other comorbid impairments such as epilepsy, attention deficit hyperactivity disorders (ADHD), disturbances of behavior, obsessive compulsive disorder in childhood, occasional schizophrenia, depressive and bipolar disorders in childhood [15]. Sleep problems, self-injurious behavior, and aggressiveness also occur frequently [16].

The basic models of psychiatric disturbances, namely autism, schizophrenia, and mental retardation, that are dependent on neurodevelopmental changes, can be derived from the Table **1**.

Table 1: Models of Neurodevelopmental Changes in Brain Maturation (modified according to Keshavan [17]).

	Early Development	**Late Development**	**Early and Late Development**
MRI Studies	decrease of cranial volume loss of gray and white matter	normal cranial volume, reduction of gray matter	decrease of cranial volume and loss of gray and white matter
Neuropathological Studies	reduction of all cells alteration of localization and cytoarchitecture	reduction of synaptic density, increase of cortical neuronal density, normal cell localization and cytoarchitecture	reduction of neuronal and synaptic density, increase of synaptic density in temporoparietal cortex, altered cell localization and cytoarchitecture
Clinical Studies	mental retardation autism	schizophrenia of adult age	schizophrenia in mental retarded patiens and in autistics

It is evident that the most serious and devastating disorders, such as mental retardation and infantile autism, have the most prominent organic changes of the central nervous system (CNS): loss of the both gray and white matter, reduction in the number of cells, and alterations in cytoarchitecture. The late neurodevelopmental model, clinically characterized as schizophrenia with an onset at puberty and adolescence, is characterized by a reduction of synaptic density and a rise in cortical neuronal density. However, atrophic changes appear during the disease course in most schizophrenics, which are seen in the area of extend brain ventricles.

Comorbidity of autism with mild and severe mental retardation occurs in 70% of patients (for review see [18]) and a combination with epilepsy also frequently occurs. Seizures are reported in 25-30% young adults with ASD and in 20% of children with autism. Approximately 41% of children with ASD with prominent language disturbances have seizures, with the petit mall, complex or parietal seizures, and myoclonic seizures being most often diagnosed. Girls suffer with seizures more frequently than boys.

Almost half of treated children with ASD have changes in their EEG [18]. These changes are in most cases severe, markedly heterogeneous, with diffused abnormalities, and a high occurrence of slow waves. Paroxysmal complexes of spikes and waves with unipolar or bilateral localization appear in many cases. There are prominent centrotemporale spikes and hypsarythmias in some children. Interestingly, these findings occur more frequently in the left hemisphere. Neurological and EEG abnormalities are more often seen in children with mental retardation.

Anthropometric measurements demonstrate increased head circumference and *post mortem* increased brain weight in autistic children. Abnormalities in various structures, namely amygdala, temporal lobe, and hippocampus, are reported using brain imaging methods. *Post mortem* studies found developmental abnormalities of some structures within the brainstem, particularly involving the inferior olives and facial motoric nucleus [19,20]. Purkinje cell number was reduced in all the adult cases, and this reduction was sometimes accompanied by gliosis [20]. Most studies describe changes in cerebellar structures [21] (see Chapter 2).

Neurodevelopmental changes are correlated with biological markers such as neurotransmitters, namely glutamate and γ-aminobutyric acid (GABA). These systems are important for the development of neural networks, with modulation of neuronal migration, proliferation and differentiation affecting the later disturbances of perception, attention, and emotion (see Chapter 3). There is marked change in the ratio of excitatory and inhibitory neurotransmitters in autism [22]. The most consistent finding has been that over 25% of autistic children and adolescents are hyperserotonemic [23]. Subjects with elevated whole blood serotonin levels have been shown to have elevated platelet serotonin transport into platelets and decreased serotonin 5-HT2 receptor binding. Serotonin was significantly increased in platelets from autistic children, while the amino acids aspartic acid, glutamine, and glutamic acid were significantly decreased in comparison with the controls [24]. It is suggested that the alterations of the amino acids in plasma and platelets from autistic children might represent a biochemical marker related to infantile autism (see Chapter 3 and 7).

However, serotonin plasma levels positively correlate between autistic children and their mothers, fathers, and siblings in a highly significant way [25,26]. Twenty-three of the 47 families studied had at least one hyperserotonemic member. Of these 23 families, 10 (43.5%) had two or more hyperserotonemic members; five families were identified, in which each family member studied had hyperserotonemia [25]. Moreover, the increased plasma serotonin levels correlated with psychopathological symptomatology within families. Most individuals with autism who are treated with potent serotonin transporter inhibitors have a reduction in ritualistic behavior and aggression. Reduction of central nervous system serotonin, induced by acute tryptophan depletion, causes a worsening of stereotyped behavior [23]. The dysregulation of noradrenalin and dopamine has also been reported in the brains of autistic individuals.

Several studies since the sixties tried to identify prenatal and postnatal factors with relevant connections to the etiology of ASD. Increased occurrences of complications during pregnancy or parturition, neonatal cyanosis, umbilical strangulation, and neonatal icterus were suspected. However, their contribution in the ASD etiology remains unclear.

ASD are highly genetic and multifactorial, with many risk factors acting together. Currently, genetic examination is highly recommended [7,27-29]. The concordance in monozygotic twins is 60%, in dizygotic 3% [7,30]. Many patients with genetic finding have significant autistic features (so called syndromologic autism). Many are mentally retarded. This can be seen in Rett syndrome, syndrome of fragile X-chromosome, neurofibromatosis, tuberosis sclerosis, Angleman syndrome, Cornelia de Lange syndrome, increased risk of sudden death, and untreated phenylketonuria.

DIAGNOSTIC PROCESS OF ASD

The clinical examination, not biological data, is important for autism diagnosis. The autism diagnosis is most often done between two and four years of age. The abnormal behavior can be seen in some children early, usually within a few months of life. However, the diagnostic process requires detailed and repeated clinical examination of anamnestic data, including exploration of the patient's and family's history. A comprehensive evaluation requires a thorough review that may include looking at the child's behavior and development and interviewing the parents. Psychotic disturbances, autism occurrence and mental retardation within the family, depressive and bipolar disorders, suicide and suicidal attempts are examined in familiar anamnesis.

In the personal history, we explore problems of social and communicative skills, receptive and expressive vocabulary, other language skills, and use of non-verbal communicative skills. We evaluate social communication, eye contact, structure of interests, imitative skills, interactive strategies, the pattern of behavior, stereotypes, resistance to change, unusual interests, idiosyncrasy to environment conditions, play and its peculiarities, material preferences, structure of eating behavior, symptoms of hyperactivity, impulsivity, aggressiveness, symptoms of mood disturbances, relationships with socially important relation, and adaptation to the social environment; namely to school.

Psychopathological symptoms include: altered activity, psychokinetic attenuation, hyperkinesis, existence of psychomotor agitation, movement stereotypia, verbigeration, echolalia, echopraxis, echomimic, automatic expressions of styling behavior, mannerisms, and flapping movements. We attempt to notice all unusual and bizarre behavior; parakinetic expressions in mimics, regressive parakinesis, things that represent ontogenetic and phylogenetically older patterns of behavior.

Pediatric examination, genetic testing, neurological testing with the use of selected imaging methods (CT and MRI) and EEG, hearing and vision screening, belongs in a detailed diagnosis. Psychological examination with complex intellect evaluation is important. Psychiatric and psychological examination can be completed with scale tests (CAR, ADIR). The final assessment and diagnosis thus requires the cooperation of psychiatrists, pediatricians, child neurologists, and child psychologists.

Autism is currently detected only at about three years of age. However, some deficits and disturbances connected with ASD are possible to observe at 12-18 months of age. The major impairment of toddlers is the lack of reciprocal emotional attention towards something or, particularly, towards somebody. Healthy toddlers develop dyadic interaction with caregivers (mostly with parents) at 3-4 months, even earlier. Baron-Cohen and co-workers [31] screened 41 18-month-old toddlers who were at high genetic risk for developing autism, and 50 randomly selected 18-month-olds, using a new instrument, the CHAT (Checklist for Autism in Toddlers [32]). More than 80% of the randomly selected 18-month-old toddlers passed on all items. Four children in the high-risk group failed on two or more of five key types of behavior: pretend play, protodeclarative pointing, joint-attention, social interest, and social play. These four toddlers received a diagnosis of autism by 30 months. Ostering and Dawson analyzed video records of play of one-year old infants. Eleven children, which were later diagnosed as autistics, were delayed in social behavior and shared attention. These infants had a lack of interest for objects, face of other persons, and reaction to its own name. IQ of these infants was not different from control healthy infants [33]. Recently, the Modified Checklist for Autism in Toddlers (M-CHAT) is widely used to screen toddlers with ASD risk [34].

Diagnostic Criteria for Autism

A comparison of patient psychopathological description, utilizing ICD-10 criteria, is most important for the diagnostic process. While the original Kanner criteria excluded from autism diagnosis patients with mental retardation, we recently began to differentiate patients with high functioning autism and low functioning autism (with mental retardation). The ICD-10 diagnosis of autism is based on similar criteria as DSM-IV. In addition, they have a fundamental continuity with the original description of autism made by Kanner and later modified by Rutter [35].

ICD-10 Criteria for Childhood Autism (F84.0) Postulate that:

A   Abnormal or impaired development is evident before the age of 3 years in at least one of the following areas:

    1. receptive or expressive language as used in social communication;

    2. the development of selective social attachments or of reciprocal social interaction;

    3. functional or symbolic play.

B   A total of at least six symptoms from (1), (2), and (3) must be present, with at least two from (1) and at least one from each of (2) and (3).

    1.   Qualitative impairment in social interaction:

        a) failure adequately to use eye-to-eye gaze, facial expression, body postures, and gestures to regulate social interaction;

        b) failure to develop (in a manner appropriate to mental age, and despite ample opportunities) peer relationships that involve a mutual sharing of interests, activities and emotions;

        c) lack of socio-emotional reciprocity as shown by an impaired or deviant response to emotions; or lack of modulation of behavior according to social context; or a weak integration of social, emotional, and communicative behaviors;

        d) lack of spontaneous seeking to share enjoyment, interests, or achievements with other people (e.g. a lack of showing, bringing, or pointing out to other people objects of interest to the individual).

    2.   Qualitative abnormalities in communication:

        a) delay in or total lack of, development of spoken language that is not accompanied or compensated through the use of gestures or mime as an alternative mode of communication (often preceded by a lack of communicative babbling);

        b) relative failure to initiate or sustain conversational interchange (at whatever level of language skill is present), in which there is reciprocal responsiveness to the communications of the other person;

        c) stereotyped and repetitive use of language or idiosyncratic use of words or phrases;

        d) lack of varied spontaneous make-believe play or (when young) social imitative play.

    3.   Restricted, repetitive, and stereotyped patterns of behavior, interests, and activities:

        a) an encompassing preoccupation with one or more stereotyped and restricted patterns of interest that are abnormal in content or focus; or one or more interests that are abnormal in their intensity and circumscribed nature though not in their content or focus;

        b) apparently compulsive adherence to specific, nonfunctional routines or rituals;

        c) stereotyped and repetitive motor mannerisms that involve either hand or finger flapping or twisting or complex whole body movements;

        d) preoccupations with part-objects of non-functional elements of play materials (such as their odor, the feel of their surface, or the noise or vibration they generate).

C   The clinical picture is not attributable to the other varieties of pervasive developmental disorders; specific development disorder of receptive language (F80.2) with secondary socio-emotional problems, reactive attachment disorder (F94.1) or disinhibited attachment disorder (F94.2); mental retardation (F70-F72) with some associated emotional or behavioral disorders; schizophrenia (F20.-) of unusually early onset; and Rett syndrome (F84.12).

DSM-IV Criteria for Autistic Disorder (299.0) postulate that:

A   A total of at least six items from (1), (2), and (3), with at least two from (1), and one each from (2) and (3):

    1. Qualitative impairments in social interaction:

        a) marked impairment in the use of multiple nonverbal behaviors such as eye-to-eye gaze, facial expression, body postures, and gestures to regulate social interaction;

        b) failure to develop peer relationships appropriate to developmental level;

        c) markedly impaired expression of pleasure in other people's happiness;

        d) lack of social or emotional reciprocity.

    2. Qualitative impairments in communication:

        a) delay in, or total lack of, the development of spoken language (not accompanied by an attempt to compensate through alternative modes of communication such as gestures or mime);

        b) in individuals with adequate speech, marked impairment in the ability to initiate or sustain a conversation with others;

        c) stereotyped and repetitive use of language or idiosyncratic language;

        d) lack of varied, spontaneous make-believe play or social imitative play appropriate to developmental level.

    3. Restricted repetitive and stereotyped patterns of behavior, interests, and activities:

        a) encompassing preoccupation with one or more stereotyped patterns of interest that is abnormal either in intensity or focus;

        b) apparently inflexible adherence to specific, nonfunctional routines or rituals;

        c) stereotyped and repetitive motor mannerisms (e.g., hand or finger flapping or twisting, or complex whole-body movements);

        d) persistent preoccupation with parts of object.

B   Delays or abnormal functioning in at least one of the following areas, with onset prior to age 3 years:

    1. social interaction;

    2. language as used in social communication;

    3. symbolic or imaginative play.

C   The disturbance is not better accounted for by Rett's Disorder or CDD.

More detail characteristics includes: verbal regression, new word learning problem, a lack of word meaning problems, monotonous speech diction, non-understanding and problem with perception emotional status of other people, executive deficits like planning abstracts thinking, cognitive flexibility by it these people don't answers external stimuli. They prefer detail from unit; perception reality may be fragmented [14].

The definition of DSM-IV is almost identical with ICD-10. The schematic and simplified comparison of ICD-10 and DSM IV is given in Fig. **1.** The comparison of various diagnostic guides for autism can be found in several papers, e.g. in Volkmar handbook of autism [36]. It is important to remember that both the ICD-10 and the DSM-IV are general diagnostic guides. Each case of autism is unique and symptoms will vary by individual.

ICD-10

A Abnormal or impaired development is evident before the age of 3 years in at least one of the following areas:

1 receptive or expressive language as used in social communication

2 the development of selective social attachments or of reciprocal social interaction

3 functional or symbolic play

B A total of at least six symptoms from 1), 2), and 3) must be present, with at least two from 1) and at least one from each of 2) and 3):

1 qualitative impairment in social interaction

a) failure to use eye-to-eye gaze, facial expression, body postures, and gestures to regulate social interaction; b) failure to develop peer relationships that involve a mutual sharing of interests, activities and emotions; c) lack of socio-emotional reciprocity; d) lack of spontaneous seeking to share enjoyment, (e.g. a lack of showing, bringing, or pointing out to other people objects of interest to the individual

2 qualitative abnormalities in communication:

a) delay in or lack of spoken language; b) relative failure to initiate or sustain conversation; c) stereotyped and repetitive use of language or idiosyncratic use of words; d) lack of varied spontaneous make-believe play or (when young) social imitative play

3 restricted, repetitive, and stereotyped patterns of behavior, interests, and activities:

a) an encompassing preoccupation with one or more stereotyped and restricted patterns of interest; b) apparently compulsive adherence to specific, nonfunctional routines or rituals; c) stereotyped and repetitive motor mannerisms; d) preoccupations with part-objects of non-functional elements of play materials

DSM-IV

A A total of at least six (or more) items from 1), 2), and 3), with at least two from 1) and one each from 2) and 3):

1 qualitative impairment in social interaction:

a) impairment in the use of multiple nonverbal behaviors (eye-to-eye gaze, facial expression body posture, and gestures to regulate social interaction); b) failure to develop peer relationships; c) a lack of showing, bringing, or pointing out objects of interest; d) lack of social or emotional reciprocity

2 qualitative impairments in communication

a) delay in spoken language; b) impairment in the ability to initiate or sustain a conversation; c) stereotyped and repetitive use of language or idiosyncratic language; d) lack of varied, spontaneous make-believe play or social imitative play

3 restricted, repetitive, and stereotyped patterns of behavior, interests, and activities

a) preoccupation with stereotyped and restricted patterns of interest; b) apparently inflexible adherence to nonfunctional routines or rituals; c) stereotyped and repetitive motor mannerisms; d) persistent preoccupation with parts of objects

B Delays or abnormal functioning in at least one of the following areas, with onset prior to age 3 years:

1 social interaction

2 language as used in social communication

3 symbolic or imaginative play

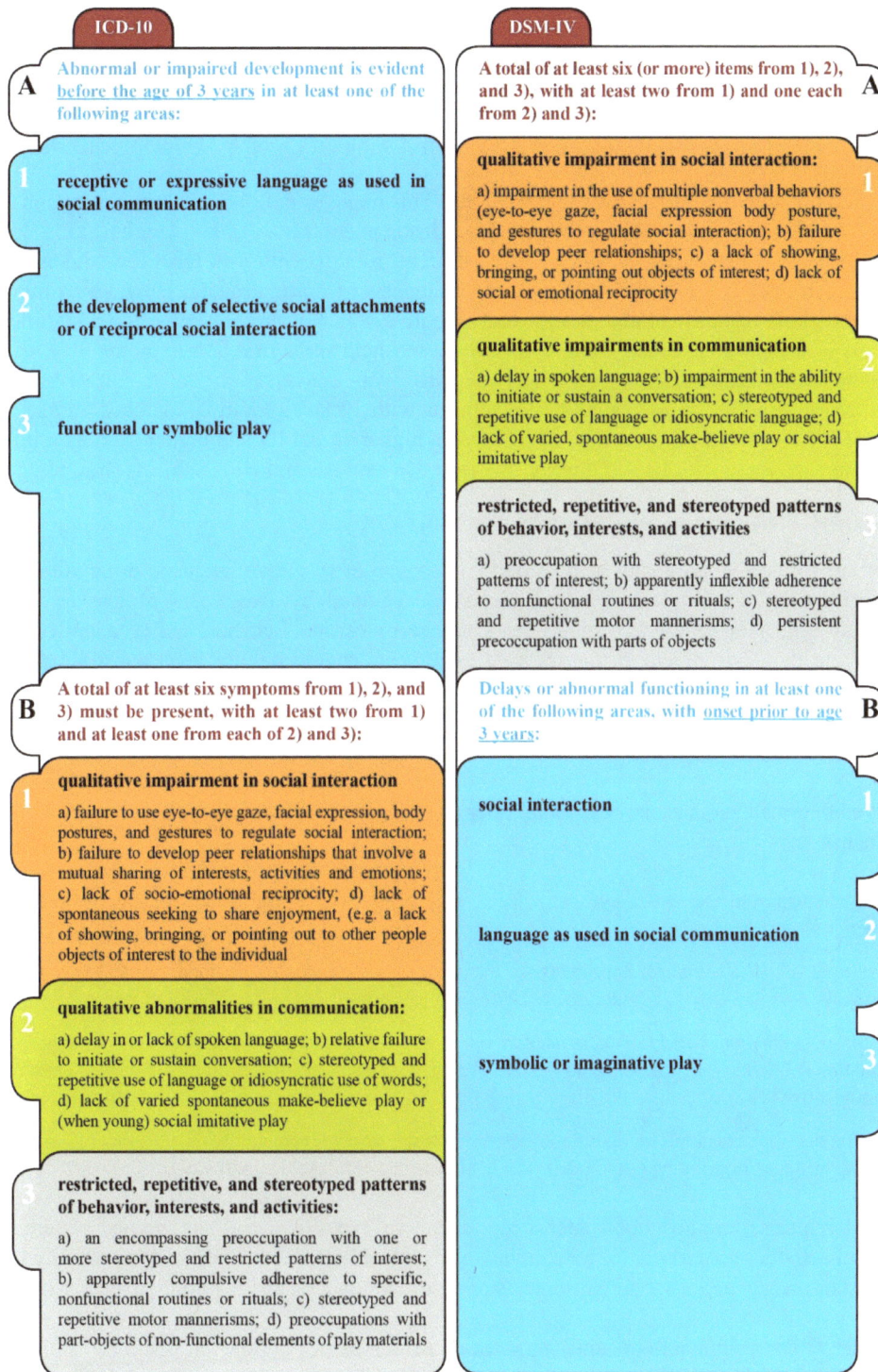

Figure 1: Comparison of diagnostic criteria for F84.0- childhood autism according to ICD-10 [2] and for 299.00 - autistic disorder according to DSM-IV [3].

Diagnostic Criteria for Asperger Syndrome

The differential diagnosis of Asperger syndrome (AS) and high functioning autism (HFA) is still controversial [37,38]. Some clinical workers use the term AS as synonymous with the HFA. Patients with AS have better language skills than that typical for HFA. Patients with AS more often have disturbances of learning, which are evident in psychological testing. Szatmari and co-workers compared both groups of patients and their results indicated that AS and HFA groups differed little, but large differences were observed on a battery of neuropsychological tests. When the AS and HFA with IQ above 85 were compared, outstanding differences in motor coordination, language comprehension, and facial recognition were observed. Finally, some evidence was found to suggest that the pattern of deficits seen in AS and HFA subjects varied by developmental level [38].

Rourke [39] coined the term non-verbal learning disability. This includes deficits in tactile perception, psychomotor coordination and problems in non-verbal communication. This is in contrast with good verbal skills and good verbal memory. Problems in communication of AS cases are different from patients with HFA [36]. AS is connected with higher intelligence, in comparison with Kanner autism. Diagnosis of AS is broader. However, a minimum of two symptoms of social skills impairment and at least one repetitive behavior are present without significant delays in developmental areas such as general language, cognitive or self-help skills in AS. Symptoms may include inability to develop peer relationships, inability to participate in a two-way conversation and repetitive body movements. Ghaziuddin and Mountain-Kimchi compared 22 AS subjects with 12 HFA controls, matched on age, sex and level of intelligence [40]. As a group, subjects with AS showed a higher verbal IQ and higher scores on information and vocabulary subtests than those with HFA. However, scores of several AS and HFA subjects showed a mixed pattern.

ICD-10 Diagnostic Criteria of Asperger Syndrome (F84.0)

A   A lack of any clinically significant general delay in spoken or receptive language or cognitive development. Diagnosis requires that single words should have developed by two years of age or earlier and that communicative phrases be used by three years of age or earlier. Self-help skills, adaptive behaviour and curiosity about the environment during the first three years should be at a level consistent with intellectual development. However, motor milestones may be somewhat delayed and motor clumsiness is usually seen (although not a necessary diagnostic feature). Isolated special skills, often related to abnormal preoccupations, are common, but are not required for diagnosis.

B   Qualitative abnormalities in reciprocal social interaction must be present in at least 2 from 4 aspects (criteria as for autism):

   a) failure adequately to use eye-to-eye gaze, facial expression, body postures, and gestures to regulate social interaction;

   b) failure to develop (in a manner appropriate to mental age, and despite ample opportunities) peer relationships that involve a mutual sharing of interests, activities and emotions;

   c) lack of socio-emotional reciprocity as shown by an impaired or deviant response to emotions; or lack of modulation of behavior according to social context; or a weak integration of social, emotional, and communicative behaviors;

   d) lack of spontaneous seeking to share enjoyment, interests, or achievements with other people (e.g. a lack of showing, bringing, or pointing out to other people objects of interest to the individual).

C   An unusually intense circumscribed interest or restrictive, repetitive, and stereotyped patterns of behavior, interests and activities (criteria as for autism; however, it would be less usual for these to include either motor mannerisms or preoccupations with part-objects or non-functional elements of play materials).

D   The disorder is not attributable to other varieties of pervasive developmental disorder; schizotypal disorder (F21); simple schizophrenia (F20.6); reactive and disinhibited attachment disorder of childhood (F94.1 and.2); obsessional personality disorder (F60.5); obsessive-compulsive disorder (F42).

DSM-IV Diagnostic Criteria For Asperger's Disorder (299.80)

A. Qualitative impairment in social interaction as manifested by at least two of the following:

   1. marked impairment in the use of multiple nonverbal behaviors such as eye-to-eye gaze, facial expression, body postures, and gestures to regulate social interaction;

    2. failure to develop peer relationships appropriate to developmental level;

    3. a lack of spontaneous seeking to share enjoyment, interests or achievements with other people (e.g. by a lack of showing, bringing, or pointing out objects of interest to other people);

    4. lack of social or emotional reciprocity.

B   Restricted repetitive and stereotyped patterns of behavior, interests, and activities, as manifested by at least one of the following:

    1. encompassing preoccupation with one or more stereotyped and restricted patterns of interest that is abnormal either in intensity or focus;

    2. apparently inflexible adherence to specific, nonfunctional routines or rituals;

    3. stereotyped and repetitive motor mannerisms (e.g.: hand or finger flapping or twisting, or complex whole-body movements);

    4. persistent preoccupation with parts of objects.

C   The disturbance causes clinically significant impairment in social, occupational, or other important areas of functioning.

D   There is no clinically significant general delay in language (e.g.: single words used by age 2 years, communicative phrases used by age 3 years).

E   There is no clinically significant delay in cognitive development or in the development of age- appropriate self-help skills, adaptive behavior (other than social interaction), and curiosity about the environment in childhood.

F   Criteria are not met for another specific PDD or Schizophrenia.

According to ICD-10, AS is a disorder of uncertain nosological validity, characterized by the same type of qualitative abnormalities of reciprocal social interaction that typify autism, together with a restricted, stereotyped, repetitive repertoire of interests and activities. It differs from autism primarily in the fact that there is no general delay or retardation in language or in cognitive development. This disorder is often associated with marked clumsiness. There is a strong tendency for the abnormalities to persist into adolescence and adult life. Psychotic episodes occasionally occur in early adult life.

According to DSM-IV, the basic diagnostic distinction between autism and Asperger's disorder is absence of clinically significant delays in language, cognitive development, and adaptive functioning in the Asperger's group. The rest of the diagnostic criteria (impairments in social interactions, restricted repetitive and stereotype patterns of behaviors) between autism and Asperger's disorder is identical. This makes it difficult to differentiate children with Asperger's disorder from those with HFA. AS is differentiated from HFA largely based on a history of "language delay" [41]. Recently, some researchers and clinicians argue that the current DSM-IV definition of AS is incorrect and a new updated definition is discussed [40,42,43]. Fig. **2** shows simplified comparison of diagnostic criteria of ICD-10 and DSM-IV for AS.

PROGNOSIS

Information on long-term prognosis in autism is limited. Outcome is known to be poor for those with an IQ below 50, but there have been few systematic studies of individuals with an IQ above this [37,38]. Billstedt *et al.* [44] followed individuals with autism from childhood to adulthood and found that only three from 83 individuals with autism were independent albeit leading fairly isolated lives. In patients with autism, it has been found that the best prognosis is found in children with language development before fifth year of age and with relatively good intellectual and social functions. Several studies followed the autistic children into adulthood [45-47]. Only a few individuals with ASD were independent, yet, as mentioned, were leading fairly isolated lives. One quarter of individuals were able to live with support, but most of them (76%) were described as dependent patients.

While most autistic patients live in specialized centers, patients with AS live with parents or use the services of specialized centers. Some studies demonstrate the relatively high proportion of independency is dependent on the patient intelligence [37]. Sixty-eight individuals meeting criteria for autism and with a performance IQ of 50 or above in childhood were followed up as adults [37]. Although a minority of adults had achieved relatively high levels of independence, most remained very dependent on their families or other support services. Stereotyped behaviors or interests frequently persisted into adulthood. Individuals with a childhood performance IQ of at least 70 had a significantly better outcome than those with an IQ below this. Some studies using different patient samples suggested a more optimistic outcome for a variable percentage of such patients, and appeared to be relatively independent of the patient′s IQ [37,45,46].

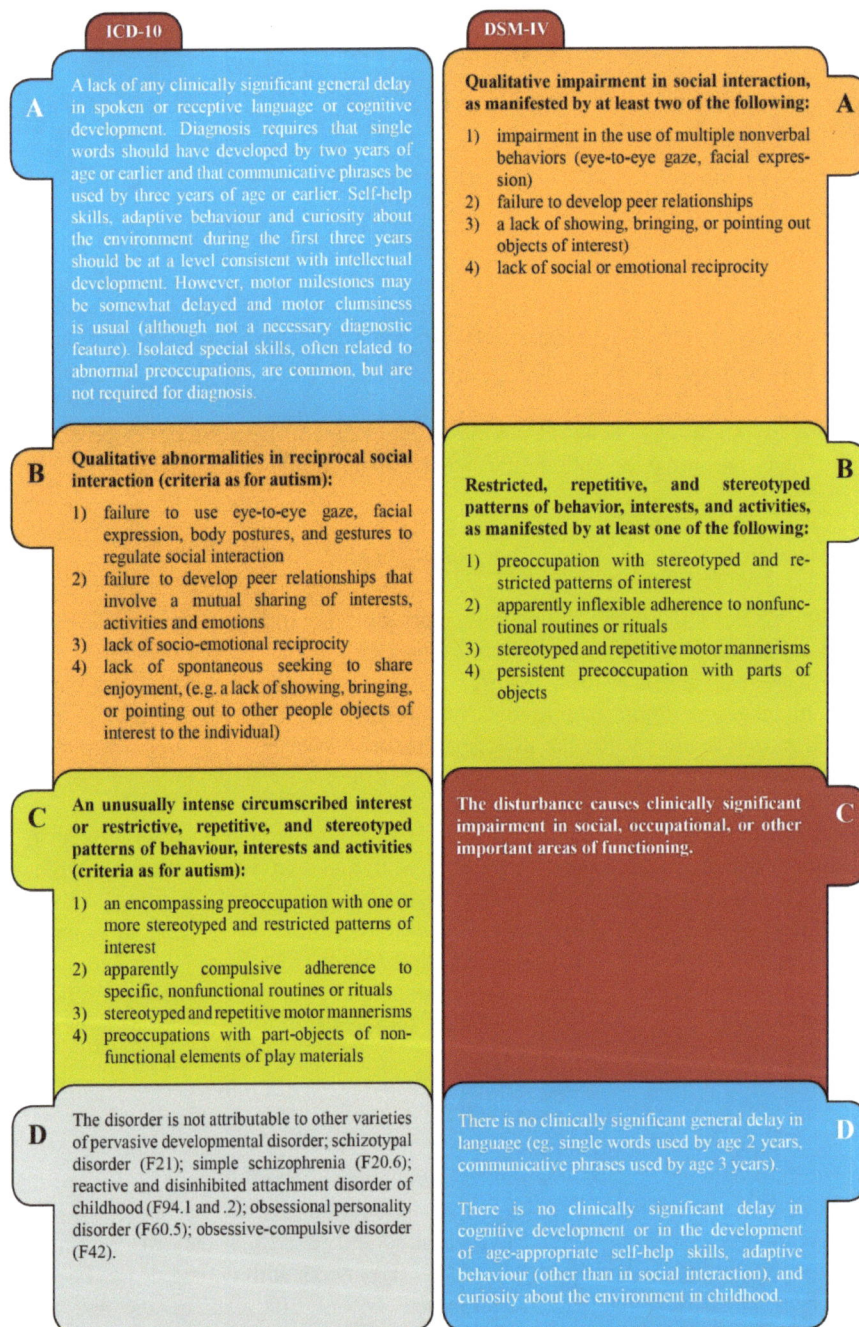

Figure 2: Diagnostic criteria for 84.5 Asperger syndrome from ICD-10 and diagnostic criteria for 299.90 Asperger′s disorder from DSM-IV.

Nevertheless, a good prognosis has been seen in only 20% of patients. Patients with AS probably have a better prognosis. Hutton [48] conducted a follow-up study extending to at least the age of 21 years of 135 individuals with an ASD diagnosed in childhood and an IQ of over 30. Of all participants, 16 percent developed a definite new psychiatric disorder. A further 6 percent developed a possible new disorder. Five individuals developed an obsessive-compulsive disorder and/or catatonia; eight an affective disorder with marked obsessive features; three complex affective disorders; four more straightforward affective disorders; one a bipolar disorder; and one an acute anxiety state complicated by alcohol excess [48]. Interestingly, there were no cases of schizophrenia. Schizophrenia is rare in the development of autistic individuals [49]. Volkmar and Cohen examined detailed case records of 163 adolescents and adults with well-documented histories of autism. These authors found that the frequency of schizophrenia among autistic patients (0.6%) is roughly comparable to the frequency of schizophrenia in the general population [50].

On the other hand, Lainhart and Folstein found that 35% of their autistic patients had the onset of affective disorder in childhood [51]. Of the cases mentioning family history, 50% had a family history of affective disorder or suicide. Changes in mood, self-attitude, and vital sense were rarely reported by the patients. On the other hand rare cases of schizophrenia in autistic people exists [52]. Psychotic symptoms, hallucinatory syndromes and other non-schizophrenic symptoms are frequent in a number of children and adults. Affective disorders are frequent in autistic children and adults [51]. Suicide attempts are frequent (in 3.4% person of study sample) in children with mental retardation or in autistic persons with high IQ abilities [53].

PHARMACOTHERAPY OF ASD

Pharmacotherapy of autism is often targeted to some dominant symptoms and/or symptoms of comorbid disorders, such as psychomotor instability and aggressiveness, hyperactivity and behavioral disorders, repetitive rituals and motor mannerisms, and sleeping disorders [54-58]. Haloperidol and risperidone were used in control studies of psychomotor instability and aggressiveness [59-61]. Forty autistic children (2-7 years) were treated with haloperidol (1.7 mg/day) for eight weeks. Some 69% of the children were improved; adverse effects appeared as acute dystonia and dyskinesia, which subsided with drug withdrawal [62]. Risperidone has been found efficacious for decreasing severe tantrums, aggression, and self-injurious behavior in children and adolescents with ASD [63]. Risperidone was used for treatment in two double-blind placebo control studies, altogether in 180 children; doses were approximately 1.8 mg/day and therapy was successful in 70– 87% [64,65]. McDougle and co-workers treated 101 autistic children at the age 5-17 years with risperidone (0.5-3.5 mg/day) in an 8-week double-blind, placebo-controlled trial and 63 autistic children in a 16-week open-label continuation study. Risperidone led to significant improvements in the restricted, repetitive, and stereotyped patterns of behavior, interests, and activities of autistic children but did not significantly change their deficit in social interaction and communication [60].

Research Units on the Pediatric Psychopharmacology Autism Network (RUPP) studied in 2005 72 children treated with methylfenidate (younger than 11 years of age) in double-blind control study (0.125-0.54 mg/kg three times a day). Methylfenidate was better than placebo in 50% of the children and side effects were prominent in 18% of patients (especially irritability) [66].

Hollander with co-workers examined the selective serotonin reuptake inhibitor liquid fluoxetine in the treatment of repetitive behaviors in children and adolescents with ASD [67,68]. In total, 45 child or adolescent patients with ASD were randomized into two acute, 8-week phases in a double-blind placebo-controlled crossover study of liquid fluoxetine (2.5-9.9 mg/day). Fluoxetine in low doses was more effective than placebo in the treatment of repetitive behaviors in childhood autism.

Buchsbaum *et al.* measured the effect of fluoxetine on regional cerebral metabolism in ASD [67]. Autism scores showed three of the patients had much improved and three were unchanged. Relative metabolic rates were significantly higher in the right frontal lobe following fluoxetine, especially in the anterior cingulate gyrus and the orbitofrontal cortex.

Sleep disorders of autistic patients are treated by melatonin (see Chapter 10). In a retrospective study, 107 children (2-18 years of age) with a confirmed diagnosis of ASD received melatonin (0.5-3 mg/day) [69]. Parents of 64 children (60%) reported improved sleep, although they continued to have concerns regarding sleep. Parents of 14 children (13%) continued to report sleep problems as a major concern, with only one child having worse sleep after starting melatonin (1%), and one child having undetermined response (1%). Only three children had mild side-effects after starting melatonin, which included morning sleepiness and increased enuresis. There was no reported increase in seizures after starting melatonin in children with pre-existing epilepsy and no new-onset seizures. The majority of children were taking psychotropic medications.

Some other targeted pharmacotherapeutic approaches have been investigated, such as the application of α-adrenergic agonists, atypical neuroleptics, such as aripiprazol and olanzapin. Aripiprazol was tested in double blind control-study in doses 5, 10, and 15mg/day for 8-week in patients with ASD. Aripiprazol was more effective than placebo and the diverse effects were sedation, tremor, and akathisia [58]. Interesting are preliminary studies with intranasal oxytocin application, but this treatment requires further studies [70,71] (see Chapter 11).

Promising treatments include melatonin, antioxidants, acetylcholinesterase inhibitors, and naltrexone. All of the reviewed treatments are currently considered off-label for ASD (i.e., not FDA-approved) and some have adverse effects. Further studies exploring these treatments are needed [82]. Research and practical experience has shown that the most effective treatment of ASD is a combination of specialized and supportive educational programming, communication training (such as speech/language therapy), social skills support and behavioral intervention [72-75]. Children with autism benefit from intensive, early intervention that focuses on increasing the frequency, form, and function of communicative acts.

COMPLEMENTARY AND ALTERNATIVE MEDICAL (CAM) TREATMENTS

Several studies suggest that 50-75% of children with ASD are using complementary alternative medicine (CAM) [76,77]. Many CAM treatments have not yet been tested in clinical trials and need further research. CAM used for autism can be divided by proposed mechanism: immune modulation, gastrointestinal support, supplements that affect neurotransmitter function, and nonbiological intervention. Levy *et al.* reported that almost one third of young children referred for evaluation of ASD were being treated with dietary therapies by their parents even before confirmation of diagnosis [78,79] Fig. **3**.

Figure 3: Children with diagnosis of ASD. (Courtesy of their parents; photos from documentation of Anna Strunecka).

Golnic and Ireland addressed national primary care physicians' attitudes and practices regarding biological CAM use in children with autism [80]. Physicians encouraged multi-vitamins (49%), essential fatty acids (25%), melatonin (25%) and probiotics (19%) and discouraged withholding immunizations (76%), chelation (61%), anti-infectives (57%), delaying immunizations (55%), and secretin (43%).

Levy and co-workers discussed the evidence supporting the most frequently used treatments, including categories of mind-body medicine, energy medicine, biologically-based therapy, manipulative and body-based practices [79]. Approximately half of families of children with ASD use a biologically based therapy. Levy *et al.* evaluated the biologically based CAM on principles of evidence-based medicine (EBM), integrating clinical expertise, patient (or family) values, and the best evidence for efficacy. These authors reviewed the existing literature and report the strength of the evidence as Grade A (randomized controlled trials, reviews and/or meta-analyses), Grade B (other evidence such as isolated well-designed controlled and uncontrolled studies), or Grade C (case reports or theories). It is necessary to consider the developmental aspects, since the improvement can also be the result of maturation of structures and functions.

Interestingly, grade A received secretin, a gastrointestinal hormone, which has the distinction of being one of the most extensively studied pharmacotherapeutic agent for autism. More than a dozen well designed, well-executed studies have been published, which failed to demonstrate efficacy of secretin for symptoms of autism. A recent Cochrane review reported 14 randomized controlled trials, with a total of 618 children. Nine studies used a crossover treatment design. The authors concluded that there is no evidence that single or multiple dose of intravenous secretin is effective for treatment of ASD.

Grade B received supplementation with vitamin B6 in combination with magnesium, and vitamin C despite that only one study, which reported positive results of decreased stereotyped behavior in a 30 week double-blind/ placebo controlled trial in 18 children with ASD has been conducted [81]. To date this study has not been replicated. Carnosine, which improved the expressive and receptive vocabulary, has documented usefulness in specific deficiency states and in the presence of certain medications, such as valproic acid, has never been evaluated as a treatment for motor or behavioral symptoms of autism. The supplementation with ω3-fatty acids, and gluten-free/ casein-free diet (GF/CF) were also evaluated by grade B studies.

Efficacy of folate and antioxidants, antibiotics (e.g., nonabsorbed rifaximin), antifungals and antimycotics, has been documented by case reports and received thus the grade C status. There are no reports of therapeutic use of transcranial magnetic stimulation or other magnet therapy for symptoms of ASD to date.

Scientists, clinicians, and families need access to well-designed clinical evidence to assist them in the choice of therapies.

REFERENCES

[1]   Greenspan S, Brazelton T, Cordero J, *et al.* Guidelines for early identification, screening, and clinical management of children with autism spectrum disorders. Pediatrics 2008; 121: 828-30.
[2]   WHO. International Classification of Diseases (ICD)-10. 10th Edition. Geneva, Switzerland: http://www.who.int/classifications/icd/en/; 1992.
[3]   Diagnostic and Statistical Manual of Mental Disorders DSM-IV. Fourth edition. Washington, D.C.: American Psychiatric Press; 1994.
[4]   Turk J, Graham P, Verhulst F. Child and adolescent psychiatry: A developmental approach. USA: Oxford University Press; 2007.
[5]   Lasalle JM, Yasui DH. Evolving role of MeCP2 in Rett syndrome and autism. Epigenomics 2009; 1: 119-30.
[6]   Carouge D, Host L, Aunis D, Zwiller J, Anglard P. CDKL5 is a brain MeCP2 target gene regulated by DNA methylation. Neurobiol Dis 2010; 38: 414-24.
[7]   Muhle R, Trentacoste SV, Rapin I. The genetics of autism. Pediatrics 2004; 113: e472-86.
[8]   Rice CE. Prevalence of autism spectrum disorders - autism and developmental disabilities monitoring network. Six sites, United States, 2000. MMWR Surveillance Summaries 2007; 56: 1-11.
[9]   Rice CE, Baio J, Van Naarden Braun K, *et al.* A public health collaboration for the surveillance of autism spectrum disorders. Paediatr Perinat Epidemiol 2007; 21: 179-90.

[10]   Kogan MD, Blumberg SJ, Schieve LA, *et al.* Prevalence of parent-reported diagnosis of autism spectrum disorder among children in the US, 2007. Pediatrics 2009; 124: 1395-403.

[11]   Baron-Cohen S, Scott FJ, Allison C, *et al.* Prevalence of autism-spectrum conditions: UK school-based population study. Br J Psychiatry 2009; 194: 500-9.

[12]   Blaylock RL, Strunecka A. Immune-glutamatergic dysfunction as a central mechanism of the autism spectrum disorders. Curr Med Chem 2009; 16: 157-70.

[13]   Fombonne E. Epidemiological surveys of pervasive developmental disorders. In: Volkmar F, editor. Autism and Pervasive Developmental Disorders. Cambridge: Cambridge University Press; 2007. p. 33-68.

[14]   Banaschewski T, Rohde L, editors. Biological Child Psychiatry: Recent Trends and Developments. Basel: Karger; 2008.

[15]   Reiren A. Overlap of autism with other neurodevelopmental syndromes. In: 56th Annual Meeting of the American Academy of Child and Adolescent Psychiatry; 2009; Hawai'i; 2009. p. 119.

[16]   Mayes S, Calhoun S. Variables related to sleep problems in children with autism. In: 56th Annual Meeting of the American Academy of Child and Adolescent Psychiatry; 2009; Hawai'i; 2009. p. 192.

[17]   Keshavan M, Murray R. Neurodevelopment and adult psychopathology: Cambridge University Press; 1997.

[18]   Poustka F. The neurobiology of autism. In: Volkmar F, editor. Autism and Pervasive Developmental Disorders. Cambridge: Cambridge University Press; 2007. p. 179-221.

[19]   Rodier PM, Ingram JL, Tisdale B, Nelson S, Romano J. Embryological origin for autism: developmental anomalies of the cranial nerve motor nuclei. J Comp Neurol 1996; 370: 247-61.

[20]   Bailey A, Luthert P, Dean A, *et al.* A clinicopathological study of autism. Brain 1998; 121 (Pt 5): 889-905.

[21]   Courchesne E, Yeung-Courchesne R, Press GA, Hesselink JR, Jernigan TL. Hypoplasia of cerebellar vermal lobules VI and VII in autism. N Engl J Med 1988; 318: 1349-54.

[22]   Cook EH. Autism: review of neurochemical investigation. Synapse 1990; 6: 292-308.

[23]   Cook EH, Leventhal BL. The serotonin system in autism. Curr Opin Pediatr 1996; 8: 348-54.

[24]   Rolf LH, Haarmann FY, Grotemeyer KH, Kehrer H. Serotonin and amino acid content in platelets of autistic children. Acta Psychiatr Scand 1993; 87: 312-6.

[25]   Cook EH, Jr., Leventhal BL, Freedman DX. Free serotonin in plasma: autistic children and their first-degree relatives. Biol Psychiatry 1988; 24: 488-91.

[26]   Leventhal BL, Cook EH, Jr., Morford M, Ravitz A, Freedman DX. Relationships of whole blood serotonin and plasma norepinephrine within families. J Autism Dev Disord 1990; 20: 499-511.

[27]   Bailey A, Le Couteur A, Gottesman I, *et al.* Autism as a strongly genetic disorder: evidence from a British twin study. Psychol Med 1995; 25: 63-77.

[28]   Folstein SE, Rosen-Sheidley B. Genetics of autism: complex aetiology for a heterogeneous disorder. Nat Rev Genet 2001; 2: 943-55.

[29]   Asher JE, Lamb JA, Brocklebank D, *et al.* A whole-genome scan and fine-mapping linkage study of auditory-visual synesthesia reveals evidence of linkage to chromosomes 2q24, 5q33, 6p12, and 12p12. Am J Hum Genet 2009; 84: 279-85.

[30]   Steffenburg S, Gillberg C, Hellgren L, *et al.* A twin study of autism in Denmark, Finland, Iceland, Norway and Sweden. J Child Psychol Psychiatry 1989; 30: 405-16.

[31]   Baron-Cohen S, Allen J, Gillberg C. Can autism be detected at 18 months? The needle, the haystack, and the CHAT. Br J Psychiatry 1992; 161: 839-43.

[32]   Robins DL, Fein D, Barton ML, Green JA. The Modified Checklist for Autism in Toddlers: an initial study investigating the early detection of autism and pervasive developmental disorders. J Autism Dev Disord 2001; 31: 131-44.

[33]   Ostering J, Dawson G. Early recogniton of children with autism: A study of first year birthday home video tapes. In: Meeting of the Society for Research in Child Development; 1993; New Orleans; 1993.

[34]   Pandey J, Verbalis A, Robins DL, *et al.* Screening for autism in older and younger toddlers with the Modified Checklist for Autism in Toddlers. Autism 2008; 12: 513-35.

[35]   Rutter M, Schopler E. Classification of pervasive developmental disorders: some concepts and practical considerations. J Autism Dev Disord 1992; 22: 459-82.

[36]   Volkmar F, editor. Autism and pervasive developmental disorders. Cambridge: Cambridge University Press; 2007.

[37]   Szatmari P, Bartolucci G, Bremner R, Bond S, Rich S. A follow-up study of high-functioning autistic children. J Autism Dev Disord 1989; 19: 213-25.

[38]   Szatmari P, Tuff L, Finlayson MA, Bartolucci G. Asperger's syndrome and autism: neurocognitive aspects. J Am Acad Child Adolesc Psychiatry 1990; 29: 130-6.

[39]   Rourke BP, Young GC, Leenaars AA. A childhood learning disability that predisposes those afflicted to adolescent and adult depression and suicide risk. J Learn Disabil 1989; 22: 169-75.

[40] Ghaziuddin M, Mountain-Kimchi K. Defining the intellectual profile of Asperger Syndrome: comparison with high-functioning autism. J Autism Dev Disord 2004; 34: 279-84.

[41] Bennett T, Szatmari P, Bryson S, *et al.* Differentiating autism and Asperger syndrome on the basis of language delay or impairment. J Autism Dev Disord 2008; 38: 616-25.

[42] Ghaziuddin M. Defining the behavioral phenotype of Asperger syndrome. J Autism Dev Disord 2008; 38: 138-42.

[43] Ghaziuddin M. Brief Report: Should the DSM V Drop Asperger Syndrome? J Autism Dev Disord 2010;

[44] Billstedt E, Gillberg IC, Gillberg C. Autism after adolescence: population-based 13- to 22-year follow-up study of 120 individuals with autism diagnosed in childhood. J Autism Dev Disord 2005; 35: 351-60.

[45] Rumsey JM, Rapoport JL, Sceery WR. Autistic children as adults: psychiatric, social, and behavioral outcomes. J Am Acad Child Psychiatry 1985; 24: 465-73.

[46] Larsen FW, Mouridsen SE. The outcome in children with childhood autism and Asperger syndrome originally diagnosed as psychotic. A 30-year follow-up study of subjects hospitalized as children. Eur Child Adolesc Psychiatry 1997; 6: 181-90.

[47] Howlin P, Goode S, Hutton J, Rutter M. Adult outcome for children with autism. J Child Psychol Psychiatry 2004; 45: 212-29.

[48] Hutton J, Goode S, Murphy M, Le Couteur A, Rutter M. New-onset psychiatric disorders in individuals with autism. Autism 2008; 12: 373-90.

[49] Mouridsen SE, Rich B, Isager T. Psychiatric morbidity in disintegrative psychosis and infantile autism: A long-term follow-up study. Psychopathology 1999; 32: 177-83.

[50] Volkmar FR, Cohen DJ. Comorbid association of autism and schizophrenia. Am J Psychiatry 1991; 148: 1705-7.

[51] Lainhart JE, Folstein SE. Affective disorders in people with autism: a review of published cases. J Autism Dev Disord 1994; 24: 587-601.

[52] Petty LK, Ornitz EM, Michelman JD, Zimmerman EG. Autistic children who become schizophrenic. Arch Gen Psychiatry 1984; 41: 129-35.

[53] Isager T, Mouridsen S, Rich B. Mortality and causes of death in pervasive developmental disorders. Autism 1999;3:7-16.

[54] Aman MG, Farmer CA, Hollway J, Arnold LE. Treatment of inattention, overactivity, and impulsiveness in autism spectrum disorders. Child Adolesc Psychiatr Clin N Am 2008; 17: 713-38.

[55] Johnson KP, Malow BA. Assessment and pharmacologic treatment of sleep disturbance in autism. Child Adolesc Psychiatr Clin N Am 2008; 17: 773-85.

[56] Soorya L, Kiarashi J, Hollander E. Psychopharmacologic interventions for repetitive behaviors in autism spectrum disorders. Child Adolesc Psychiatr Clin N Am 2008; 17: 753-71.

[57] Stigler KA, McDougle CJ. Pharmacotherapy of irritability in pervasive developmental disorders. Child Adolesc Psychiatr Clin N Am 2008; 17: 739-52.

[58] McDougle C, Carlson G. Advanced Psychopharmacology Update. Evidence-Based Treatment and Beyond. In: 56th Annual Meeting of the American Academy of Child and Adolescent Psychiatry; 2009; Hawai'i: AACAP; 2009. p. 186.

[59] McCracken JT, McGough J, Shah B, *et al.* Risperidone in children with autism and serious behavioral problems. N Engl J Med 2002; 347: 314-21.

[60] McDougle CJ, Scahill L, Aman MG, *et al.* Risperidone for the core symptom domains of autism: results from the study by the autism network of the research units on pediatric psychopharmacology. Am J Psychiatry 2005; 162: 1142-8.

[61] Scahill L, Koenig K, Carroll DH, Pachler M. Risperidone approved for the treatment of serious behavioral problems in children with autism. J Child Adolesc Psychiatr Nurs 2007; 20: 188-90.

[62] Campbell M, Geller B, Cohen I. Current status of drug research and treatment with autistic children. J Pediatr Psychol 1977; 2: 156-61.

[63] Munarriz R, Bennett L, Goldstein I. Risperidone in children with autism and serious behavioral problems. N Engl J Med 2002; 347: 1890-1; author reply -1.

[64] RUPP (Research Units on Pediatric Psychopharmacology Autism Network). Risperidone in children with autism for serious behavioral problems. N Engl J Med 2002; 347: 314-21.

[65] Shea S, Turgay A, Carroll A, *et al.* Risperidone in the treatment of disruptive behavioral symptoms in children with autistic and other pervasive developmental disorders. Pediatrics 2004; 114: e634-41.

[66] RUPP (Research Units on Pediatric Psychopharmacology Autism Network). A randomized, double blind, placebo-controlled, crossover trial of methylphenidate in children with hyperactivity associated with pervasive developmental disorders. Arch Gen Psychiatry 2005; 62: 1266-74.

[67] Buchsbaum MS, Hollander E, Haznedar MM, *et al.* Effect of fluoxetine on regional cerebral metabolism in autistic spectrum disorders: a pilot study. Int J Neuropsychopharmacol 2001; 4: 119-25.

[68] Hollander E, Phillips A, Chaplin W, *et al.* A placebo controlled crossover trial of liquid fluoxetine on repetitive behaviors in childhood and adolescent autism. Neuropsychopharmacology 2005; 30: 582-9.

[69]   Andersen IM, Kaczmarska J, McGrew SG, Malow BA. Melatonin for insomnia in children with autism spectrum disorders. J Child Neurol 2008; 23: 482-5.

[70]   Hollander E, Novotny S, Hanratty M, *et al.* Oxytocin infusion reduces repetitive behaviors in adults with autistic and Asperger's disorders. Neuropsychopharmacology 2003; 28: 193-8.

[71]   Hollander E, Bartz J, Chaplin W, *et al.* Oxytocin increases retention of social cognition in autism. Biol Psychiatry 2007; 61: 498-503.

[72]   Bellini S, Peters JK. Social skills training for youth with autism spectrum disorders. Child Adolesc Psychiatr Clin N Am 2008; 17: 857-73.

[73]   Foxx RM. Applied behavior analysis treatment of autism: the state of the art. Child Adolesc Psychiatr Clin N Am 2008; 17: 821-34.

[74]   Minshawi NF. Behavioral assessment and treatment of self-injurious behavior in autism. Child Adolesc Psychiatr Clin N Am 2008; 17: 875-86.

[75]   Paul R. Interventions to improve communication in autism. Child Adolesc Psychiatr Clin N Am 2008; 17: 835-56.

[76]   Hanson E, Kalish LA, Bunce E, *et al.* Use of complementary and alternative medicine among children diagnosed with autism spectrum disorder. J Autism Dev Disord 2007; 37: 628-36.

[77]   Hyman M. Autism: Is it all in the head? Altern Ther Health Med 2008; 14: 12-5.

[78]   Levy SE, Hyman SL. Novel treatments for autistic spectrum disorders. Ment Retard Dev Disabil Res Rev 2005; 11: 131-42.

[79]   Levy SE, Hyman SL. Complementary and alternative medicine treatments for children with autism spectrum disorders. Child Adolesc Psychiatr Clin N Am 2008; 17: 803-20.

[80]   Golnik AE, Ireland M. Complementary alternative medicine for children with autism: a physician survey. J Autism Dev Disord 2009; 39: 996-1005.

[81]   Dolske MC, Spollen J, McKay S, Lancashire E, Tolbert L. A preliminary trial of ascorbic acid as supplemental therapy for autism. Prog Neuropsychopharmacol Biol Psychiatry 1993; 17: 765-74.

[82]   Rossignol DA. Novel and emerging treatments for autism spectrum disorders: a systematic review. Ann Clin Psychiatry 2009; 21: 213-36.

CHAPTER 2

The Cerebellum in Autism Spectrum Disorders

Russell L. Blaylock, MD

Institute for Theoretical Neuroscience, LLC, and Visiting Professor of Biology, Belhaven University, Ridgeland, MS 39157, USA

Abstract: The cerebellum is the most commonly affected part of the brain in autistic spectrum disorders (ASDs). The histopathological changes strongly indicate selected damage to particular cell groups and lobules of the cerebellum rather than diffuse injury. A number of studies have shown injury and abnormal development of the vermis of the cerebellum, with a predominance of neuronal loss among Purkinje cells and granule cells. In addition, one see abnormal pathway development indicating intrauterine damage or damage occurring during the early postnatal period. Several studies have shown abnormalities of glutamate receptors (GluRs) of various kinds, including metabotropic GluRs (mGluRs). In this chapter, I review the histopathologic findings within the ASD cerebellum and demonstrate evidence for immunoexcitotoxicity affecting cerebellar neurodevelopment as well as evidence for early neurodegeneration. Newer studies have shown that the cerebellum may have significant cognitive and higher cortical functions, either by way of its connections to prefrontal-limbic areas or more indirect pathways.

INTRODUCTION

Taken together 90 to 95% of the studies examining the pathological changes in ASD have disclosed damage to the cerebellum [1-3]. Molecular abnormalities have been found in all studies. Most studies reporting no pathological changes were either magnetic resonance imaging (MRI) scan studies or earlier pathological studies that did not examine the entire neuropil. The most commonly reported pathological change was a loss of Purkinje cells, even though some studies did not report such a loss [4,5].

Other studies have reported small cerebellar lobes, the most common being the inferior vermis [6-9] and neocerebellar cortex, posterior lobe [8]. The majority of such studies were MRI scan studies, with the older reports using scanning techniques and technology that may impose questions of anatomical accuracy. Clearly, pathological and anatomical changes in the cerebellum constitute the most common finding in the autistic brain.

PATHOLOGICAL CHANGES IN CEREBELLUM AND BRAINSTEM IN ASD

The most commonly affected area of the nervous system in autistic individuals is the cerebellum, with over 90% of studies showing abnormalities. These include not only a loss of Purkinje cells (Fig. **1**), but also a loss of granule cells, alterations in synaptogenesis, low levels of reelin, dramatic lowering of glutamate dehydrogenase (GDH), low γ-amino butyric acid (GABA) levels, microgliosis, small cerebellar vermis and/or hemispheres and alterations in the cellular structure of cerebellar nuclei and inferior olive.

In 1980 Williams and co-workers did the first extensive neuropathological analysis of the cerebellum in 4 male autistic individuals ages 12 to 33 years and one female autistic child age 3 years [10]. All were mentally retarded and two had seizures. Only one of the cases demonstrated a significant loss of Purkinje cells, but the person was severely mentally retarded and suffered from recurrent seizures, which itself can reduce Purkinje cell counts [11].

Ritvo and co-workers six years later examined the brains of 4 autistics individuals, all three mentally retarded and no seizures, and found a decrease in the number of Purkinje cells in the cerebellar hemispheres and vermis [12]. Two other studies reported no cerebellar abnormalities [13,14].

Fatemi and co-workers were the first to examine Purkinje cell size [15]. They examined 5 adult male autistics and 5

Address correspondence to: Institute for Theoretical Neuroscience, LLC, and Visiting Professor of Biology, Belhaven University, Ridgeland, MS 39157, USA; E-mail: Blay6307@bellsouth.net

age-matched controls and reported a 24% reduction in the size of the Purkinje cells in the autistic patients. As with the observation of several studies reporting small, packed neurons in the limbic area and other cortical zones in autistics, this could indicate a loss of dendrites and synapses.

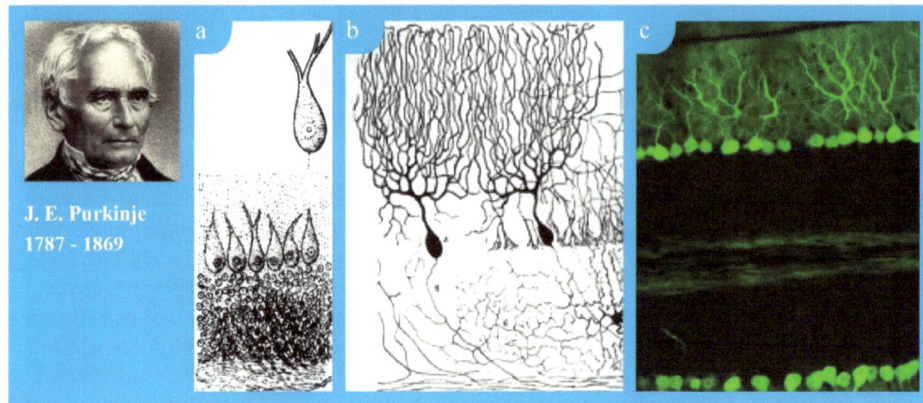

Figure 1: Purkinje and Purkinje cells. Czech scientist Jan Evangelista Purkinje observed and described the largest neurons in the cerebellum in 1837 (a). Later, Ramon y Cajal in Spain used the Golgi staining technique and was able to describe the rich branching of Purkinje cells (b). Purkinje cells under fluorescence microscope (c) (Masako Suzuki).

Of 29 cases of autistic brains examined pathologically in 8 studies, 21 demonstrated Purkinje cell loss [10,12-18] and several studies demonstrated pathological changes in the inferior olive or other cerebellar nuclei and brainstem [16,20-22]. White matter abnormalities have also been found in the autistic cerebellar projection tracts, which includes a significant reduction in the size of the left superior cerebellar peduncle and middle cerebellar peduncles bilaterally [23].

In 1994, Courchesne and co-workers described two subgroups of autistic patients among 50 autistics from ages 2 to 40 years scanned using MRI technology [24]. One subgroup demonstrated hypoplasia of the vermian lobules VI and VII and the others hyperplasia of the same lobules. They reviewed previous studies that did not find vermian abnormalities and discovered several shortcomings in technique and technology. When they reexamined these cases using more advanced techniques they found the same two subgroupings. The hyperplastic form was seen in 11% of those examined and hypoplastic form in 89%. They also found that verbal IQ was correlated with the degree of the involvement of the vermian lobules VI and VII. Those with the greatest reduction in size of these lobules more often had IQs below 70. All patients with the hyperplastic form had IQs below 70. Histological examination of the inferior vermis demonstrated a significant loss of Purkinje cells (50 to 60% reduction) and reductions in the numbers of granule cells as well.

Cerebellar Circuitry in Autism

In the past the cerebellum was considered to have only motor functions and play a major role in regulating motor actions. Newer studies are showing that it also plays a major role in cognitive, behavioral, and higher order non-motor functions [25-30]. The cerebellum has a single output via efferents of the Purkinje cells, which is in contact with a complex array of dendritic connections from basket and stellate cells.

The cerebellar cortex contains a sheet of Purkinje cells which are influence by two major inputs—direct action from climbing fibers arising from cells in the inferior olive and indirect action from mossy fibers arising from granule cell parallel fibers and cortical interneurons. The efferent axons of the Purkinje cells terminate on cerebellar nuclei. In general, the cerebellum is divided into the vermis, paravermis and lateral cerebellum and is further divided into a number of lobules, which have a complex molecular interplay. Lateral cerebellar hemisphere efferents terminate in the dentate nucleus, paramedian in the globose/embiliform nuclei and the vermis in the fastigial nucleus.

A number of studies have shown substantial connections from the association and limbic corticies and hypothalamus to the cerebellum [27,31-34]. Kelley and Strick have shown that the afferents entering the cerebellum from these association cortices terminate in the higher-order cerebellum, primarily lobules VI and VII [35]. A recent meta-analysis of neuroimaging studies of the neocerebellum confirms its role in emotion, language, working memory and executive functions [28,36].

Confirmation of higher, non-motor functions of the cerebellum also comes from studies of lesions to the cerebellum [29, 37]. In one such study 27 children undergoing surgical removal of cerebellar tumors (medulloblastomas, astrocytomas and ependymoma) researchers found abnormalities in expressive language, deficits in visual-spatial functions, word-finding difficulties and other cognitive dysfunctions [37]. Some developed preservation problems, a neurological problems often seen in autistics. Interestingly, many of the expressive language defects are also seen in autistic children, such as giving only brief responses to questions, a general lack of elaboration when engaged in conversation, word-finding difficulties and a general reluctance to engage in conversations.

It is obvious from existing studies that cerebrocerebellar circuits that link complex association cortices and paralimbic regions with the cerebellum strongly suggest a major role in cerebellar regulation of neural circuits subserving higher behavioral function. These studies also suggest that the cerebellum regulates thought process, especially as regards their appropriateness, speed of cognition and temporal ordering of thoughts so as to complete a mental picture appropriate to the situation.

It is thought that the cerebellum, primarily the anterior lobes, contains internal models of motor function, which can be temporally altered by learning [38,39]. This system of alterable network models is also thought to exist for non-motor functions [40]. Anatomical studies of monkeys demonstrate significant interconnections between the cerebellar hemispheres and prefrontal (BA 9) and parietal (BA5) association cortices [41]. Thinking is the ability to manipulate something in our minds, which requires a great deal of coordination, integration of higher network function from association cortices and focusing of attention.

According to this model, the prefrontal cortex (PFC) is the controller. That is, it performs executive functions that are critical for controlling conscious thought and uses set internal goals to accomplish these functions [42]. Within the PFC, three major regions are utilized to carry out these mental models—the dorsolateral, medial, and orbitofrontal cortex. The dorsolateral PFC controls abstract reasoning and problem solving and the orbitofrontal and medial cortices control affect and motivational functions [43]. Lesions to the PFC are known to cause impulsive, inappropriate and disorganized behavior. According to Miller and co-workers, the PFC generates patterns of activity representing goals and methods one can utilize to attain these goals [44]. These network neurons within the cerebellum influence activity throughout widespread areas of the brain and utilize memory retrieval, response execution, integration of sensation/perception as well as emotional evaluation.

Also involved in cerebellar control of emotion and cognition is the mental model within the temporo-parietal cortex. According to the theory, the temporo-parietal cortices maintain a small-scale model of reality that is used to reason and explain current events and to anticipate future events. This can involve abstract visualization or even non-visualized representations. Most of these processes are occurring in the temporo-parietal association areas, especially the inferotemporal cortex [45,46].

As for cerebellar mental models, the neural networks involved are microcomplexes, which are acting as learning machines composed of intricate circuitry. It is the climbing fibers from the inferior olive that actively modify the input-output signals for error correction. They can induce long-term depression (LTD) in the co-activated Purkinje cell synapses. This allows an updating of the internal cerebellar models.

Just as with error correction with motor models, these alterable cognitive and behavioral cerebellar models are essential for ongoing cognitive function. It has been shown that area BA46 of PFC and crus I and II of the cerebellar hemispheres, via the pontine nuclei, receive inputs from the anterior cingulate gyrus and project to the temporo-parietal cortex [34,47]. Thought process from the cerebellum are considered to be more implicit than explicit. Imaging studies demonstrate co-activation of cerebellar hemispheres with PFC and temporo-parietal cortex during mental task. It has also been observed that there is significantly more cerebellar activation with tasks that require greater attention.

Cerebellar Activation in ASD

Allen and Courchesne found that autistics demonstrated significantly less cerebellar activation with attention tasks than did normals [49]. In their study they found that autistics demonstrated a reduction in functional activation in the parietal lobes, which could involve cerebro-ponto-cerebellar projections and cerebello-thalamo-cortical reciprocal connections. In their control subjects, matched for age and gender, they found bilateral superior posterior cerebellar hemisphere activation during task requiring attention. For activation magnitude, the group difference was statistically significant for lobule VI ipsilateral for the moving hand.

Using parametric test to remove outliers, they observed a significant reduction in performance-matched activation extent and magnitude during attention task that primarily involved the contralateral lobule VI and approached significance for lobule VII and right superior cerebellar hemisphere. When matched for the task version, activation was significantly reduced in the left superior hemisphere lobule VIIa and approached significance for contralateral lobule VI and right superior hemisphere lobule VIIa. These differences became more obvious when they subtracted interference from motor activation.

It is known that attention impairments are the most common forms of cognitive deficits in autism [50]. Autistics have difficulty shifting attention when faced with new tasks and new spatial locations. Attentional difficulties are more common in those with the greatest hypoplasia and seen with lesions to the neocerebellum. Interestingly, nicotinic type receptors play a role in attention and a recent study found abnormalities in these receptors in the autistic cerebellum [50].

There is evidence that the cerebellum plays a significant role in language function that goes beyond motor speech. For example, during silent word generation tests using fMRI scanning, one sees co-activation of the left PFC, left dorsolateral PFC and right cerebellar hemisphere [51]. Imaging studies have demonstrated a strong compensatory system when faced with focal network dysfunction. This disconnection between clinical function and functional anatomical networks indicates that extensive crowding out of neural systems occurs in autism in order to compensate for anatomical abnormalities. Allen and Courchesne noted that rather than the normal activation within the ipsilateral paleocerebellum, in autistics one sees a spreading into the contralateral posterior regions that are normally used for attention [48].

Previous studies have suggested that the cerebellum acts as a major modulating system for behavior, language and cognitive function [52-54]. More recently, mainly using functional imaging, researchers have concluded the cerebellum, when disrupted, can produce a condition called cerebellar cognitive affective syndrome, which consist of a collection of cognitive disorders [29]. These include, deficient planning, problems with set-shifting, impaired abstract reasoning, impaired working memory, decreased verbal fluency, visuo-spatial disorganization, personality changes (flattened affect and disinhibited or inappropriate behavior), linguistic difficulties (dysprodia, agrammatism or mild anomia) and an overall lowering of intellectual function, none of which can be explained by abnormalities in motor function.

The behavioral presentation is more pronounced when the damage is more widespread and involves the posterior cerebellar hemispheres. Damage to the vermis is characterized by pronounced defects in affect, such as irritability, impulsivity, disinhibition and poor behavioral modulation. The linguistic defects appear to be more common with right hemisphere damage [55]. Riva and Giorgi in their study of children undergoing tumor resection of the cerebellum noted autistic-like features in some with vermal lesions [56]. This included autistic-like stereotypical performance, obsessive rituals, difficulty understanding social cues, aversion to being touched and a number of frontal lobe executive impairments, such as preservation, disinhibition, impaired working memory and poor abstract reasoning.

It is interesting to note that studies have shown that lower cerebellar volume in schizophrenic patients was strongly associated with psychotic symptoms and duration of negative symptoms [57]. The vermis appears to be a site of major involvement in schizophrenia [58,59]. Previously, autism was referred to as juvenile schizophrenia. Decreased activity has been seen in the PFC, inferior temporal and parietal corticies and increased activity in the cerebellum, thalamus and retrosplenal corticies in medication free schizophrenia patients undergoing PET scanning [60].

Alterations in cerebellar activation seen when schizophrenic patients perform word list recall indicate significant compensatory mechanism at play as shown by PET scanning [61]. There is a considerable overlap in the findings in schizophrenia and autism [2,3,10,16,18,22,26].

Schmahmann has proposed the term "dysmetria of thought" to describe the behavioral and cognitive disturbances seen with the cerebellar cognitive affective syndrome [28]. He describes this as abnormality in the normal cerebellar modulation of behavior, which utilizes a dampening mechanism to maintain smooth functioning of higher-order cognitive processes. Based on these extensive studies, the cerebellum can be divided into functional lobes, with varying degrees of overlap. In this schema, the anterior lobes are concerned with motor control, the posterior lobes with higher order behaviors and the lateral posterior lobes are involved in cognitive operations. The vermis is the equivalent of the limbic cerebellum.

The neurobehavioral and cognitive effect of cerebellar lesions depends on the age during which the lesion occurred. Older children appeared to recover better than younger children and infants with localized cerebellar lesions. One case of resection of a well-circumscribed tumor involving the midline cerebellum in a 22 year old, right handed college student demonstrated significant "dysmetria of thought" type deficits [29]. Post-operatively she underwent a dramatic change in her personality, with minimal ataxia findings. The patient perseverated both motor and verbal responses, prosody was poor and she mispronounced her words. She was notably impulsive, emotionally labile and almost childlike. A number of cognitive dysfunctions were compromised including visual memory, visuo-spatial integration, pattern-based reasoning and reasoning about social situations. Arousal and alertness was not affected, remote and semantic memory was preserved and new learning was only mildly affected.

SPECT scanning demonstrated hypoperfusion in most of the rostral two-thirds of the left temporal lobe, left prefrontal region and bilaterally in the parietal lobes. Hyperfusion was seen in the left thalamus. With feed-forward projections from the corticoponto-cerebellar loop and the feedback projections from the cerebellothalamic and thalamocortical system, one can appreciate the effects on brain activation with localized cerebellar lesions. Schmahmann and Pandyo demonstrated strong and highly organized projections to the pons arising from the association areas in the dorsolateral and dorsomedial prefrontal cortex [47]. Glickstein demonstrated similar projections from the posterior parietal area [62].

Cerebellar diaschisis leading to hypoperfusion in the parietal, temporal and prefrontal cortices has been documented using PET/SPECT scanning [29,63]. This would also explain some of the autistic studies demonstrating focal areas of reduced brain perfusion, even though it may also be explained by delayed brain maturation [64,65].

Newer studies have clarified the circuitry of the non-motor functions of the cerebellum. For example, Habas and co-workers have shown that crus I and II make significant contributions to parallel cortico-cerebellar loops that are involved in executive control, salience detection, episodic memory and self-reflection and that the largest portion of the neocerebellum plays a major role in the executive control network [66].

In this paper they discuss four major neural networks connected to the cerebellum-the sensorimotor network, the default mode network, the executive network and the salience network. The study involved functional anatomic parcellation of the neocerebellum across these intrinsic connectivity networks and functional connectivity was based on parallel activation of these networks. While this is not absolute proof of functional connectivity, when considered in light of anatomical studies of connectivity between the cerebellum, association cortices, limbic system and hypothalamus, as well as lesion studies, one has strong circumstantial evidence for connectivity.

Of particular interest when considering ASDs is the relationship of the cerebellum to default mode networks and the prefrontal cortex. The default network consists of the posterior cingulate/precuneus, medial PFC/progenual cingulate cortices, temporo-parietal region, and medial temporal lobes with intimate links to lobule IX within the cerebellum.

The default network becomes operative when one is no longer engaged with the external world or in task operations — it is considered an internal, self-aware system of mentation. Characteristically it involves episodic memory retrieval, self-reflection, mental imagery and stream-of consciousness processing [67]. For the details concerning connectivity between the cortical association areas and the cerebellar lobules I would suggest the reader refer to reference 66 for greater detail.

Based on the known circuitry of the cerebellum and the results of a number of functional studies, including MEG and electrophysiological studies, there is considerable evidence that the cerebellum is playing a major role in higher brain functions including social behavior and that lesions to the cerebellum can account for most of the clinical findings of ASDs. Some have proposed a brainstem theory to explain the development of autism, for which there is some evidence, but the strongest link would be to injury to the pontine nuclei, and inferior olive caused by immunoexcitotoxicity [68]. Yet compelling evidence indicates that there exist major disruptions of neuronal function and development in the association areas of the brain and abnormalities of cerebral and cerebellar connectivity as well.

EVIDENCE FOR IMMUNOEXCITOTOXICITY IN THE AUTISTIC CEREBELLUM

The first extensive examination of the autistic brain for inflammatory pathology was by Vargas and co-workers [69]. In their study of autistic persons from age 5 years to 44 years, they reported widespread microglial and astrocytic activation throughout the cortex and cerebellum with the most intense activation within the cerebellum.

Six of the 15 autistic patients in the study had epilepsy and 12 of 15 were mentally retarded (one unknown status). Epilepsy is often associated with hypoxia/ischemia and thus microglial activation and Purkinje cell loss [70]. Phenytoin can also result in a loss of Purkinje neurons [71]. Nine of the fifteen died of conditions associated with hypoxia/ischemia and this could be the source of microglial activation, since such activation is known to occur rapidly after systemic or primary brain insults. Against this explanation is the fact that 8 of the 12 control patients of roughly matched age, also died of hypoxic/ischemic conditions and none demonstrated significant microglial activation.

Histologically they found a patchy loss of neurons in the Purkinje cells layer and granule cell layer in 9 of 10 of the cerebella, with one showing almost complete loss of Purkinje cells and marked loss of granule cells. The most extensive microglial activation occurred in the cerebellum of the autistics. Only one case demonstrated no loss of Purkinje cells. Occasional microglial nodules were seen in the granule cell layer and white matter.

Importantly, confocal microscopy demonstrated that the activated microglia and astrocytes were closely associated with degenerating Purkinje cells, granule cells and axons. They found no association between the degree of gliosis and age, developmental regression or mental retardation. Those with epilepsy had a statistically significant higher degree of microglial activation in the cerebellum as compared with those not having epilepsy. It is also significant to appreciate that the autistic cerebella had a marked accumulation of perivascular macrophages and monocytes. It is known that microglia derived from migrating macrophages are more neurodestructive than are intrinsic brain microglia. Also, the magnitude of the astroglial activation was no different in those with or without epilepsy, again ruling out significant overall seizure associated astroglial activation.

In the cerebrum, activated microglia and astrocytes were found throughout the cortex, especially at the grey-white junction. They noted marked microglial reaction in the medial frontal gyrus and anterior cingulate region as well as throughout the white matter tracts. In four of nine cases there was a panlaminar distribution of the activated microglia.

Taken together, this study suggest that the most intense immune reaction occurs in the cerebellum, with significant reaction occurring throughout several association cortical layers, grey/white matter junctions and white matter tracts. No mention was made of the brainstem in the Vargas et al study, so we do not know if microglial reaction had occurred or if it was clustered around the circumventricular organs, in particular the area postrema.

Previous studies suggested that at least a subgroup of ASD patients had anti-brain antibodies that were either not seen or were rare in non-ASD children—either normally developing or other non-ASD neurodevelopmental disorders. Dalton found that maternal IgG antibodies reacting to rodent Purkinje cells of a mother having multiple autistic children. Injection of the maternal serum into mouse pups produced behavioral effects [72]. Other studies have also identified anti-brain antibodies in autistic children, but most such studies are small and not all children possessed the identified antibody [73-75].

Recently, Martin and co-workers repeated the studies using four rhesus monkeys and found that monkeys injected with IgG antibodies from mothers with autistic children developed characteristic autistic-like stereotypies and hyperactivity [76]. Serum injected from mothers with normally developing children produced none of these effects. Based on these studies, several researchers have proposed an autoimmune etiology for autism.

Another study looked at two special immune bands, a 37KDa band and a 73 KDa band and found that 26.2% of autistic children reacted to the 37 KDa band, whereas only 8.1% of normally developing children and 2.5% of non-autistic developmentally delayed children reacted [77]. The greatest sensitivity was in autistic children reacting to both the 37KDa band and 73KDa band and this showed a strong correlation to the regressive phenotype (87% positive). They found no link to a history of autoimmunity in the mothers.

IgG is normally detectible in the fetal circulation, from the mother, as early as 18 weeks gestation and in levels equal to that of the mother. It is also known that the IgG elevations produced by vaccinations can last for many years. Willis and co-workers examined 63 autistics and 23 siblings for reaction to a 52KDa band against human cerebellar protein. They used 63 typically developed children and 21 delayed developmental children as controls and found that 21% of the autistic children that were positive for the 52KDa protein reacted against human cerebellar protein versus 2% (1/63) in the TD and none of the sibling controls. Immunohistochemical staining identified the Golgi cells (an interneuron) as the site of the immune reaction. There was no relation between behavioral outcome and the immune reaction.

A number of neurological conditions have been linked to autoimmune reactions, including schizophrenia, obsessive-compulsive disorder, systemic lupus erythematosus and PANDAS [77-82]. It is important to note that in these disorders, proinflammatory cytokines are also elevated and are more likely to be the cause of neurological dysfunction that immunoglobulins. Growing evidence indicates that neurological dysfunction is more the result of bystander injury than autoimmune reactions against specific neurologic structures [83]. This entails damage to surrounding neurons, dendrites and synapses caused by a massive release of reactive oxygen species, reactive nitrogen species, lipid peroxidation products (LPP) and excitotoxins from immune activated microglia and astrocytes.

It should also be appreciated that most of the children examined in these studies did not have evidence of an autoimmune reaction, indicating that a smaller subset of ASD children have a genetic propensity for autoimmune related CNS injury. A considerable amount of evidence indicates that systemic immune activation can rapidly activate the brain's primary immune cell, the microglia, and that when excessive or prolonged activation occurs in a neurodestructive mode of activation, neural dysfunction and neurodestruction can occur [84-87].

It has been shown that elevation in systemic interleukin-1ß (IL-1ß) can increase brain IL-1ß, with the source being primarily microglial in origin [88]. IL-1ß is a major stimulus for activation of microglia [89,90]. Systemic elevations in tumor necrosis factor-α (TNF-α) have also been shown to elevate brain mRNA for TNF-α and IL-1ß in the hypothalamus, hippocampus and somatosensory cortex, without increasing IL-10, a major anti-inflammatory cytokine [88]. In studies of priming of microglia, no concomitant elevations in *tumor growth factor*-ß (TGF-ß) were seen, again indicating a preponderance of pro-inflammatory cytokine activation.

A number of conditions are known can activate microglia, including psychological stress, oxidative stress, inflammatory prostaglandins and environmental toxins, such as lead, mercury, aluminum, cadmium, glutamate, and pesticides/herbicides. Likewise a number of conditions relevant to the autistic brain increase the generation of reactive oxygen species/reactive nitrogen species (ROS/RNS), including seizures, hypoxia/ischemia, mitochondrial dysfunction and immunoexcitotoxicity [91-93]. Glutamate re-uptake proteins are redox sensitive and high levels of ROS/RNS and LPP, especially 4-hydroxynonenal and acrolein are known to trigger excitotoxicity by causing an accumulation of excitotoxic amino acids, such as glutamate, aspartate and metabolic products of homocysteine [93,94]. Elevated levels of ROS/RNS and LPP have been found in autistic brains [95].

It has also been shown that both excitatory amino acids and immune cytokines play a role in brain development, neuronal migration, differentiation, dendritic arborization and synaptogenesis [96-100]. Likewise, the glutamate transporter proteins also play a major role in neurodevelopment [101]. In both cases there is a timed rise and fall in cytokine levels, glutamate levels and glutamate transport proteins during various developmental stages that persist postnatally. Both proinflammatory cytokines and glutamate have trophic effects during neurodevelopment and also play a central role in pruning of malformed and excessive dendritic, axonal and synaptic connections [97,99,102]. Marret and co-workers demonstrated dramatic developmental disarray in cortical architectonics cause by ibotenate, a N-methyl-D-aspartic acid receptor (NMDAR) agonist, dosing during critical period of brain embryogenesis [103].

Within the developing cerebellum one finds an array of glutamate receptors (GluRs), both ionotropic and metabotropic among all the neuronal cell types and white matter pathways. These receptors undergo a developmental profile that varies with subunit insertion, such as NMDAR2A, which later during brain development switches to NMDAR2C subunit [104]. The alteration of GluRs subunits plays a significant role in brain neurophysiology and development. It has also been shown that various cytokines, particularly TNF-α, IL-1 and IL-6, play a major role in GluRs sensitivity by altering glutamate release, upregulating glutaminase and altering NMDA and α-amino-3-hydroxy-5-methyl-4-isoxazole propionic acid (AMPA) receptor trafficking, all of which markedly increase the risk of excitotoxicity (see Immunoexcitotoxicity chapter) [105-110].

Under conditions of chronic inflammation, one sees hypersensitivity of GluRs function and hence a greater likelihood of immunoexcitotoxicity. Priming of microglia and infiltration of macrophages significantly increases immunoexcitotoxicity as well [111]. The autistic child is exposed to a number of priming and microglial stimulating events, such as recurrent exposure to infections (either prenatal or postnatal), exposure to a great number of sequential vaccinations, seizures (clinical and subclinical) and exposure to environmental agents known to activate microglia—such as, lead, mercury, aluminum, fluoride, industrial chemicals, certain pharmaceutical agents, pesticides and herbicides.

It is my view that there are two forms of ASDs and possibly more. The classical infantile autism occurs most likely during the second trimester and results in significant abnormalities in development of widespread areas of the brain, including the cerebellum, brainstem, and cerebrum. This form of autism presents a more severe clinical picture and is less likely to respond to treatment. In addition, the incidence of this form of autism has changed very little since first being described by Leo Kanner in 1943 [112]. A second form of ASD appears to be the result of post-natal events and has less severe pathology, especially involving the brainstem. In many ways this parallels the developmental pathophysiology of schizophrenia, a closely related disorder. It may be that in the post-natal form of autism, immunoexcitotoxicity plays a major role in producing physiological dysfunction rather than neuronal loss or severe malformation. That is, it may more involve synaptic loss or dysfunction and dendritic alterations that are reversible. It is obvious there is a graded spectrum of severity of brain involvement. It is also obvious that there is a progressive immunoexcitotoxic process in cases of regressive autism.

It is known that both schizophrenia and autism are dramatically increased in the offspring of children born to mothers who developed viral infections around mid-term to late pregnancy [113,114]. Experimental studies have shown that maternal infection is not necessary, only immune stimulation. For example, Bell and co-workers found that exposing Fischer 344 and Lewis rats to 0.1 mg/kg lipopolysaccharide (LPS) at E15 produced a 20-fold increase in placental TNF-α and a 5-fold increase in fetal brain levels [115]. Interferon-γ was elevated only in the fetal brain (20-fold) and IL-6 was elevated only in the placenta (10-fold). Interestingly, IL-10 in this study was mildly elevated in the placenta and slightly decreased in the fetal brain. Such a profile significantly raises the proinflammatory status of the child's brain.

Studies, both experimental and clinical, have shown that the virus does not pass through the placenta and is not found in the fetal brain. Interleukin-6 has been shown to be essential for the behavioral effects of maternal infection or immune stimulation using poly(I:C) or LPS [116]. IL-6 does not pass through the placenta except during mid-term.

Another study specifically looked at the effect of maternal immune stimulation with either the influenza virus or poly(I:C) and found that both produced a localized loss of Purkinje cells in lobule VII in the cerebellum in both neonates and as adults [117]. They also demonstrated delayed migration of cerebellar granule cells in lobules VII and VIII. The changes in the cerebellum were strikingly similar to that seen in autistic brains.

Forkhead box P2 (*FOXP2*) gene has been implicated in cases of autism and schizophrenia [118,119]. Fatemi and co-workers demonstrated that infecting dams at E9 upregulated *FOXP2* in the cerebellum at P35 and in the hippocampi of the offspring [120]. FOXP2 is a putative transcription factor which binds to the DNA and overexpression of the gene has been associated with speech and language deficits [121].

Immune stimulation by viruses in dams has been shown to induce neurotransmitter abnormalities in the offspring. In virally exposed mice, serotonin levels in the cerebella at postnatal day 14 and 35 demonstrate significant reductions [123]. Taurine was also found to be significantly reduced in the cerebella. Serotonin is thought to play a significant role in brain development [123].

It is also interesting to note that the timing of the infection is critical to the final neurological outcome. For example, maternal viral infections and immune stimulation at E9 were related to gene alterations affecting reelin, glutamate/GABAergic system and astroglial activation, which were not seen if infected at E18 [124-126]. Reelin has been shown to be lowered in both schizophrenia and autism [127]. Reelin is a glycoprotein secreted by granule cells that plays a major role in neuronal migration during brain development and maintains synaptic plasticity throughout postnatal life. Several studies have shown abnormally low levels of reelin in autistic sera and brains [128]. Fatemi and co-workers examined the levels of reelin and anti-apoptotic **Bcl-2 protein** (Bcl-2), the survival protein, in autistic cerebella and demonstrated a reduction in reelin bands of 43%, 44% and 44% as well as a reduction in **Bcl-2** of 34% to 51% [129].

Studies have shown that NR2B subunits-containing NMDARs are abundant in synapses during early development and are gradually replaced by NR2A-containing NMDARs [130]. Composition of NMDAR subunits are essential for refinement of excitatory synapses as the brain develops and as it becomes more mature. A recent study demonstrated that the essential molecular control for subunit changes was reelin and involved integrin as well [131]. The lateral mobility of the NMDARs, essential for NMDAR function, was regulated by reelin and integrin. Because reelin in the cerebellum, unlike the frontal lobe, is synthesized by a group of glutamatergic granule cells, disruptions in glutamate levels could also affect reelin levels in the developing cerebellum [132].

Studies have shown that GABA receptors are also dysfunctional in examined autistic brains [133,134]. GABA receptors are divided into three classes, GABAA $_{A-C,}$ each linked via G-proteins to K^+ and Ca^{2+} channels. The GABA receptors are composed of two subunits—GABRA1-3 and GABBR1-2. They found that GABR1-3 were decreased in cerebella of autistics versus controls. GABRA1 was decreased by 63% and GABRB3 by 57%. In BA 40 (parietal area) all GABA subunits were significantly reduced and in BA9 (frontal) only GABRA1 was significantly reduced. In addition to low levels of GABA$_A$ receptors, they also found low levels of GABA$_B$ receptor subunits. In these studies researchers demonstrated a 67% reduction of GABBR1 and 46% reduction in GABBR2 in autistic cerebella as compared to healthy controls. GABBR1 was reduced in BA 40 (parietal area) by 71% and BA9 (frontal area) by 70%. GABBR2 was not reduced in either cortical area.

These studies support previous studies showing that glutamic acid decarboxylase (GAD) 65 mRNA levels were decreased in the cerebellum of autistic brains [135]. GAD is used to synthesize GABA from glutamate. It has also been shown that monosodium glutamate (MSG) can alter GAD function, such that the balance between GluR activation and GABA receptor modulation is toward an excitotoxic background [136]. Normally, excitotoxic damage to the cerebellum stimulates GABA receptor activity to reduce the risk of excitotoxicity, but with lowered GABA receptor function in autism, excitotoxicity becomes much more likely.

A new study found significant reductions in cerebellar nicotinic acetylcholine receptors (nAChRs) in the cerebellum of 8 adults with autism, all mentally retarded and 5 of 8 having epilepsy [137]. No changes were seen in muscarinic acetylcholine receptors M1 or M2 in the cerebellum, but have been reported in the parietal lobe. They found extensive loss of Purkinje cells in the cerebellar hemispheres and thinning of the molecular layer in the crown and depth of the folia. Cases with epilepsy demonstrated severe loss of Purkinje cells in the culmen and declive (lobules II-VI) of the vermis. The study suggests that most of the deficit was with post-synaptic high-affinity nAChRs on

Purkinje cells and granules cells. Importantly, nAChRs have been shown to govern the release of other neurotransmitters, such as GABA and glutamate [138]. For example, the presence of presynaptic nAChRs can modulate glutamate release from mossy fibers.

CONCLUSIONS

Taken together, theses studies indicate that during brain inflammation one sees significant alterations in specific neurotransmitters (both excitatory and inhibitory), reelin and survival proteins, which result in abnormal neurogenesis, Purkinje and granule cell loss, abnormal connectivity and synaptogenesis in the autistic brain, in particular the cerebellum.

The interaction of pro-inflammatory cytokines with GluRs appears to be playing a central role. Microglial activation and alterations in astrocytes reactivity are major regulators of both immune mediators (cytokines, chemokines, interferons) and a number of excitatory neurotransmitters and molecules, such as glutamate, aspartate, quinolinic acid, and homocysteine metabolites.

A majority of studies have shown evidence of excitotoxicity, including high levels of ROS/RNS, LPP and elevated levels of inflammatory prostaglandins. Many of the findings in the autistic brain can be explained by chronic inflammation and resulting immunoexcitotoxicity, including alterations in multiple neurotransmitters, abnormal ratios of excitatory amino acid receptor activity to inhibitory GABA receptor activity, microglial activation, astrocytes activation, white matter overgrowth, abnormal cortical architectonics, and elevations in pro-inflammatory cytokines and chemokines.

The cerebellum appears to be a major site of damage and abnormal neurodevelopment in ASDs. Newer evidence indicates that the cerebellum plays a major role in non-motor functions, such as behavior, social interaction, repetitive behaviors, and cognition. With extensive reciprocal connections to major association cortices and the prefrontal cortex, it appears that the cerebellum plays a significant role in coordinating the major association cortices, balancing attention and developing ongoing mental maps as the brain undergoes maturation and development. Disruption of these maps and dysmetria of thought can produce most of the behaviors seen in autism.

REFERENCES

[1]   Bauman M, Kemper T. Developmental cerebellar abnormalities: a consistent finding in early infantile autism. Neurology 1986; 36 S1: 190.
[2]   Kemper TL, Bauman M. Neuropathology of infantile autism. J Neuropathol Exp Neurol 1998; 57: 645-52.
[3]   Allen G, Courchesne E. Differential effects of developmental cerebellar abnormality on cognitive and motor functions in the cerebellum: an fMRI study of autism. Am J Psychiatry 2003; 160: 262-73.
[4]   Bailey A, Luthert P, Bolton P, *et al.* Autism and megalencephaly. Lancet 1993; 341: 1225-6.
[5]   Guerin P, Lyon G, Barthelemy C, *et al.* Neuropathological study of a case of autistic syndrome with severe mental retardation. Dev Med Child Neurol 1996; 38: 203-11.
[6]   Murakami JW, Courchesne E, Press GA, Yeung-Courchesne R, Hesselink JR. Reduced cerebellar hemisphere size and its relationship to vermal hypoplasia in autism. Arch Neurol 1989; 46: 689-94.
[7]   Courchesne E, Saitoh O, Yeung-Courchesne R, *et al.* Abnormality of cerebellar vermian lobules VI and VII in patients with infantile autism: identification of hypoplastic and hyperplastic subgroups with MR imaging. AJR Am J Roentgenol 1994; 162: 123-30.
[8]   Webb SJ, Sparks BF, Friedman SD, *et al.* Cerebellar vermal volumes and behavioral correlates in children with autism spectrum disorder. Psychiatry Res 2009; 172: 61-7.
[9]   Hodge SM, Makris N, Kennedy DN, *et al.* Cerebellum, language, and cognition in autism and specific language impairment. J Autism Dev Disord 2010; 40: 300-16.
[10]   Williams RS, Hauser SL, Purpura DP, DeLong GR, Swisher CN. Autism and mental retardation: neuropathologic studies performed in four retarded persons with autistic behavior. Arch Neurol 1980; 37: 749-53.
[11]   Fujikawa DG, Itabashi HH, Wu A, Shinmei SS. Status epilepticus-induced neuronal loss in humans without systemic complications or epilepsy. Epilepsia 2000; 41: 981-91.
[12]   Ritvo ER, Freeman BJ, Scheibel AB, *et al.* Lower Purkinje cell counts in the cerebella of four autistic subjects: initial findings of the UCLA-NSAC Autopsy Research Report. Am J Psychiatry 1986; 143: 862-6.

[13]   Bailey A, Luthert P, Dean A, *et al.* A clinicopathological study of autism. Brain 1998; 121 (Pt 5): 889-905.

[14]   Kemper TL, Bauman ML. Neuropathology of infantile autism. Mol Psychiatry 2002; 7 Suppl 2: S12-3.

[15]   Fatemi SH, Halt AR, Realmuto G, *et al.* Purkinje cell size is reduced in cerebellum of patients with autism. Cell Mol Neurobiol 2002; 22: 171-5.

[16]   Bauman ML. Microscopic neuroanatomic abnormalities in autism. Pediatrics 1991; 87: 791-6.

[17]   Kemper TL, Bauman ML. The contribution of neuropathologic studies to the understanding of autism. Neurol Clin 1993; 11: 175-87.

[18]   Courchesne E, Townsend J, Saitoh O. The brain in infantile autism: posterior fossa structures are abnormal. Neurology 1994; 44: 214-23.

[19]   Whitney ER, Kemper TL, Bauman ML, Rosene DL, Blatt GJ. Cerebellar Purkinje cells are reduced in a subpopulation of autistic brains: a stereological experiment using calbindin-D28k. Cerebellum 2008; 7: 406-16.

[20]   Rodier PM, Ingram JL, Tisdale B, Nelson S, Romano J. Embryological origin for autism: developmental anomalies of the cranial nerve motor nuclei. J Comp Neurol 1996; 370: 247-61.

[21]   Courchesne E. Brainstem, cerebellar and limbic neuroanatomical abnormalities in autism. Curr Opin Neurobiol 1997; 7: 269-78.

[22]   Jou RJ, Minshew NJ, Melhem NM, Keshavan MS, Hardan AY. Brainstem volumetric alterations in children with autism. Psychol Med 2009; 39: 1347-54.

[23]   Brito AR, Vasconcelos MM, Domingues RC, *et al.* Diffusion tensor imaging findings in school-aged autistic children. J Neuroimaging 2009; 19: 337-43.

[24]   Courchesne E, Saitoh O, Townsend JP, *et al.* Cerebellar hypoplasia and hyperplasia in infantile autism. Lancet 1994; 343: 63-4.

[25]   Schmahmann JD. From movement to thought: Anatomic substrates of the cerebellar contribution to cognitive processing. Hum Brain Mapp 1996; 4: 174-98.

[26]   Desmond JE, Gabrieli JD, Wagner AD, Ginier BL, Glover GH. Lobular patterns of cerebellar activation in verbal working-memory and finger-tapping tasks as revealed by functional MRI. J Neurosci 1997; 17: 9675-85.

[27]   Schmahmann JD, Sherman JC. The cerebellar cognitive affective syndrome. Brain 1998; 121 (Pt 4): 561-79.

[28]   Schmahmann JD. Disorders of the cerebellum: ataxia, dysmetria of thought, and the cerebellar cognitive affective syndrome. J Neuropsychiatry Clin Neurosci 2004; 16: 367-78.

[29]   Bellebaum C, Daum I. Cerebellar involvement in executive control. Cerebellum 2007; 6: 184-92.

[30]   Turner BM, Paradiso S, Marvel CL, *et al.* The cerebellum and emotional experience. Neuropsychologia 2007; 45: 1331-41.

[31]   Haines DE, Dietrichs E, Mihailoff GA, McDonald EF. The cerebellar-hypothalamic axis: basic circuits and clinical observations. Int Rev Neurobiol 1997; 41: 83-107.

[32]   Middleton FA, Strick PL. Dentate output channels: motor and cognitive components. Prog Brain Res 1997; 114: 553-66.

[33]   Middleton FA, Strick PL. Cerebellar projections to the prefrontal cortex of the primate. J Neurosci 2001; 21: 700-12.

[34]   Dum RP, Strick PL. An unfolded map of the cerebellar dentate nucleus and its projections to the cerebral cortex. J Neurophysiol 2003; 89: 634-9.

[35]   Kelly RM, Strick PL. Cerebellar loops with motor cortex and prefrontal cortex of a nonhuman primate. J Neurosci 2003; 23: 8432-44.

[36]   Stoodley CJ, Schmahmann JD. Functional topography in the human cerebellum: a meta-analysis of neuroimaging studies. Neuroimage 2009; 44: 489-501.

[37]   Levisohn L, Cronin-Golomb A, Schmahmann JD. Neuropsychological consequences of cerebellar tumour resection in children: cerebellar cognitive affective syndrome in a paediatric population. Brain 2000; 123 (Pt 5): 1041-50.

[38]   Kawato M, Furukawa K, Suzuki R. A hierarchical neural-network model for control and learning of voluntary movement. Biol Cybern 1987; 57: 169-85.

[39]   Wolpert D, Miall R, Kawato M. Internal models in the cerebellum. Trends Cogn Sci 1998; 2: 338-47.

[40]   Schmahmann JD. An emerging concept. The cerebellar contribution to higher function. Arch Neurol 1991; 48: 1178-87.

[41]   Sasaki K, Kawaguchi S, Oka H, Sakai M, Mizuno N. Electrophysiological studies on the cerebellocerebral projections in monkeys. Exp Brain Res 1976; 24: 495-507.

[42]   Ito M. Control of mental activities by internal models in the cerebellum. Nat Rev Neurosci 2008; 9: 304-13.

[43]   Happaney K, Zelazo PD, Stuss DT. Development of orbitofrontal function: current themes and future directions. Brain Cogn 2004; 55: 1-10.

[44]   Miller EK, Cohen JD. An integrative theory of prefrontal cortex function. Annu Rev Neurosci 2001; 24: 167-202.

[45]   Tanaka K. Inferotemporal cortex and object vision. Annu Rev Neurosci 1996; 19: 109-39.

[46] Sheinberg DL, Logothetis NK. The role of temporal cortical areas in perceptual organization. Proc Natl Acad Sci U S A 1997; 94: 3408-13.

[47] Schmahmann JD, Pandya DN. Anatomical investigation of projections to the basis pontis from posterior parietal association cortices in rhesus monkey. J Comp Neurol 1989; 289: 53-73.

[48] Allen G, Muller RA, Courchesne E. Cerebellar function in autism: functional magnetic resonance image activation during a simple motor task. Biol Psychiatry 2004; 56: 269-78.

[49] Allen G, Courchesne E. Attention function and dysfunction in autism. Front Biosci 2001; 6: D105-19.

[50] Schlosser R, Hutchinson M, Joseffer S, *et al.* Functional magnetic resonance imaging of human brain activity in a verbal fluency task. J Neurol Neurosurg Psychiatry 1998; 64: 492-8.

[51] Lee M, Martin-Ruiz C, Graham A, *et al.* Nicotinic receptor abnormalities in the cerebellar cortex in autism. Brain 2002; 125: 1483-95.

[52] Dow RS. Some novel concepts of cerebellar physiology. Mt Sinai J Med 1974; 41: 103-19.

[53] Snider RS, Maiti A. Cerebellar contributions to the Papez circuit. J Neurosci Res 1976; 2: 133-46.

[54] Watson PJ. Nonmotor functions of the cerebellum. Psychol Bull 1978; 85: 944-67.

[55] Silveri MC, Leggio MG, Molinari M. The cerebellum contributes to linguistic production: a case of agrammatic speech following a right cerebellar lesion. Neurology 1994; 44: 2047-50.

[56] Riva D, Giorgi C. The cerebellum contributes to higher functions during development: evidence from a series of children surgically treated for posterior fossa tumours. Brain 2000; 123 (Pt 5): 1051-61.

[57] Wassink TH, Andreasen NC, Nopoulos P, Flaum M. Cerebellar morphology as a predictor of symptom and psychosocial outcome in schizophrenia. Biol Psychiatry 1999; 45: 41-8.

[58] Loeber RT, Cintron CM, Yurgelun-Todd DA. Morphometry of individual cerebellar lobules in schizophrenia. Am J Psychiatry 2001; 158: 952-4.

[59] Ichimiya T, Okubo Y, Suhara T, Sudo Y. Reduced volume of the cerebellar vermis in neuroleptic-naive schizophrenia. Biol Psychiatry 2001; 49: 20-7.

[60] Andreasen NC, O'Leary DS, Flaum M, *et al.* Hypofrontality in schizophrenia: distributed dysfunctional circuits in neuroleptic-naive patients. Lancet 1997; 349: 1730-4.

[61] Crespo-Facorro B, Paradiso S, Andreasen NC, *et al.* Recalling word lists reveals "cognitive dysmetria" in schizophrenia: a positron emission tomography study. Am J Psychiatry 1999; 156: 386-92.

[62] Glickstein M, May JG, 3rd, Mercier BE. Corticopontine projection in the macaque: the distribution of labelled cortical cells after large injections of horseradish peroxidase in the pontine nuclei. J Comp Neurol 1985; 235: 343-59.

[63] Botez MI, Leveille J, Lambert R, Botez T. Single photon emission computed tomography (SPECT) in cerebellar disease: cerebello-cerebral diaschisis. Eur Neurol 1991; 31: 405-12.

[64] Zilbovicius M, Garreau B, Samson Y, *et al.* Delayed maturation of the frontal cortex in childhood autism. Am J Psychiatry 1995; 152: 248-52.

[65] Burroni L, Orsi A, Monti L, *et al.* Regional cerebral blood flow in childhood autism: a SPET study with SPM evaluation. Nucl Med Commun 2008; 29: 150-6.

[66] Habas C, Kamdar N, Nguyen D, *et al.* Distinct cerebellar contributions to intrinsic connectivity networks. J Neurosci 2009; 29: 8586-94.

[67] Raichle ME, Snyder AZ. A default mode of brain function: a brief history of an evolving idea. Neuroimage 2007; 37: 1083-90; discussion 97-9.

[68] McGinnis WR. Could oxidative stress from psychosocial stress affect neurodevelopment in autism? J Autism Dev Disord 2007; 37: 993-4.

[69] Vargas DL, Nascimbene C, Krishnan C, Zimmerman AW, Pardo CA. Neuroglial activation and neuroinflammation in the brain of patients with autism. Ann Neurol 2005; 57: 67-81.

[70] Savic I, Seitz RJ, Pauli S. Brain distortions in patients with primarily generalized tonic-clonic seizures. Epilepsia 1998; 39: 364-70.

[71] Rapport RL, 2nd, Shaw CM. Phenytoin-related cerebellar degeneration without seizures. Ann Neurol 1977; 2: 437-9.

[72] Dalton P, Deacon R, Blamire A, *et al.* Maternal neuronal antibodies associated with autism and a language disorder. Ann Neurol 2003; 53: 533-7.

[73] Singer HS, Morris CM, Williams PN, *et al.* Antibrain antibodies in children with autism and their unaffected siblings. J Neuroimmunol 2006; 178: 149-55.

[74] Zimmerman AW, Connors SL, Matteson KJ, *et al.* Maternal antibrain antibodies in autism. Brain Behav Immun 2007; 21: 351-7.

[75] Singer HS, Morris CM, Gause CD, *et al.* Antibodies against fetal brain in sera of mothers with autistic children. J Neuroimmunol 2008; 194: 165-72.

[76]   Martin LA, Ashwood P, Braunschweig D, *et al.* Stereotypies and hyperactivity in rhesus monkeys exposed to IgG from mothers of children with autism. Brain Behav Immun 2008; 22: 806-16.

[77]   Huerta PT, Kowal C, DeGiorgio LA, Volpe BT, Diamond B. Immunity and behavior: antibodies alter emotion. Proc Natl Acad Sci U S A 2006; 103: 678-83.

[78]   Pandey RS, Gupta AK, Chaturvedi UC. Autoimmune model of schizophrenia with special reference to antibrain antibodies. Biol Psychiatry 1981; 16: 1123-36.

[79]   Kiessling LS, Marcotte AC, Culpepper L. Antineuronal antibodies: tics and obsessive-compulsive symptoms. J Dev Behav Pediatr 1994; 15: 421-5.

[80]   Leonard HL, Swedo SE. Paediatric autoimmune neuropsychiatric disorders associated with streptococcal infection (PANDAS). Int J Neuropsychopharmacol 2001; 4: 191-8.

[81]   Snider LA, Swedo SE. PANDAS: current status and directions for research. Mol Psychiatry 2004; 9: 900-7.

[82]   Jones AL, Mowry BJ, Pender MP, Greer JM. Immune dysregulation and self-reactivity in schizophrenia: do some cases of schizophrenia have an autoimmune basis? Immunol Cell Biol 2005; 83: 9-17.

[83]   McGeer PL, McGeer EG. Local neuroinflammation and the progression of Alzheimer's disease. J Neurovirol 2002; 8: 529-38.

[84]   Rockwood K, Cosway S, Carver D, *et al.* The risk of dementia and death after delirium. Age Ageing 1999; 28: 551-6.

[85]   Holmes C, El-Okl M, Williams AL, *et al.* Systemic infection, interleukin 1beta, and cognitive decline in Alzheimer's disease. J Neurol Neurosurg Psychiatry 2003; 74: 788-9.

[86]   Perry VH, Newman TA, Cunningham C. The impact of systemic infection on the progression of neurodegenerative disease. Nat Rev Neurosci 2003; 4: 103-12.

[87]   Nguyen MD, D'Aigle T, Gowing G, Julien JP, Rivest S. Exacerbation of motor neuron disease by chronic stimulation of innate immunity in a mouse model of amyotrophic lateral sclerosis. J Neurosci 2004; 24: 1340-9.

[88]   Churchill L, Taishi P, Wang M, *et al.* Brain distribution of cytokine mRNA induced by systemic administration of interleukin-1beta or tumor necrosis factor alpha. Brain Res 2006; 1120: 64-73.

[89]   Basu A, Krady JK, O'Malley M, *et al.* The type 1 interleukin-1 receptor is essential for the efficient activation of microglia and the induction of multiple proinflammatory mediators in response to brain injury. J Neurosci 2002; 22: 6071-82.

[90]   Basu A, Krady JK, Levison SW. Interleukin-1: a master regulator of neuroinflammation. J Neurosci Res 2004; 78: 151-6.

[91]   Blanc EM, Keller JN, Fernandez S, Mattson MP. 4-hydroxynonenal, a lipid peroxidation product, impairs glutamate transport in cortical astrocytes. Glia 1998; 22: 149-60.

[92]   Blaylock R. Interaction of cytokines, excitotoxins, and reactive nitrogen and oxygen species in autism spectrum disorders. J Amer Nutr Assoc 2003; 6: 21-35.

[93]   Blaylock R. Chronic microglial activation and excitotoxicity secondary to excessive immune stimulation: possible factors in Gulf War Syndrome and autism. J Am Phys Surg 2004; 9: 46-51.

[94]   McCracken E, Valeriani V, Simpson C, *et al.* The lipid peroxidation by-product 4-hydroxynonenal is toxic to axons and oligodendrocytes. J Cereb Blood Flow Metab 2000; 20: 1529-36.

[95]   McGinnis WR. Oxidative stress in autism. Altern Ther Health Med 2004; 10: 22-36; quiz 7, 92.

[96]   Komuro H, Rakic P. Modulation of neuronal migration by NMDA receptors. Science 1993; 260: 95-7.

[97]   Dziegielewska KM, Moller JE, Potter AM, *et al.* Acute-phase cytokines IL-1beta and TNF-alpha in brain development. Cell Tissue Res 2000; 299: 335-45.

[98]   Nacher J, McEwen BS. The role of N-methyl-D-asparate receptors in neurogenesis. Hippocampus 2006; 16: 267-70.

[99]   Schlett K. Glutamate as a modulator of embryonic and adult neurogenesis. Curr Top Med Chem 2006; 6: 949-60.

[100]   Ghiani CA, Beltran-Parrazal L, Sforza DM, *et al.* Genetic program of neuronal differentiation and growth induced by specific activation of NMDA receptors. Neurochem Res 2007; 32: 363-76.

[101]   Matsugami TR, Tanemura K, Mieda M, *et al.* From the Cover: Indispensability of the glutamate transporters GLAST and GLT1 to brain development. Proc Natl Acad Sci U S A 2006; 103: 12161-6.

[102]   Suzuki M, Nelson AD, Eickstaedt JB, *et al.* Glutamate enhances proliferation and neurogenesis in human neural progenitor cell cultures derived from the fetal cortex. Eur J Neurosci 2006; 24: 645-53.

[103]   Marret S, Gressens P, Evrard P. Arrest of neuronal migration by excitatory amino acids in hamster developing brain. Proc Natl Acad Sci U S A 1996; 93: 15463-8.

[104]   Monyer H, Burnashev N, Laurie DJ, Sakmann B, Seeburg PH. Developmental and regional expression in the rat brain and functional properties of four NMDA receptors. Neuron 1994; 12: 529-40.

[105]   Qiu Z, Sweeney DD, Netzeband JG, Gruol DL. Chronic interleukin-6 alters NMDA receptor-mediated membrane responses and enhances neurotoxicity in developing CNS neurons. J Neurosci 1998; 18: 10445-56.

[106] Pickering M, Cumiskey D, O'Connor JJ. Actions of TNF-alpha on glutamatergic synaptic transmission in the central nervous system. Exp Physiol 2005; 90: 663-70.

[107] Lai AY, Swayze RD, El-Husseini A, Song C. Interleukin-1 beta modulates AMPA receptor expression and phosphorylation in hippocampal neurons. J Neuroimmunol 2006; 175: 97-106.

[108] Takeuchi H, Jin S, Wang J, *et al.* Tumor necrosis factor-alpha induces neurotoxicity via glutamate release from hemichannels of activated microglia in an autocrine manner. J Biol Chem 2006; 281: 21362-8.

[109] Pais TF, Figueiredo C, Peixoto R, Braz MH, Chatterjee S. Necrotic neurons enhance microglial neurotoxicity through induction of glutaminase by a MyD88-dependent pathway. J Neuroinflammation 2008; 5: 43.

[110] Takeuchi H, Jin S, Suzuki H, *et al.* Blockade of microglial glutamate release protects against ischemic brain injury. Exp Neurol 2008; 214: 144-6.

[111] Blaylock RL, Strunecka A. Immune-glutamatergic dysfunction as a central mechanism of the autism spectrum disorders. Curr Med Chem 2009; 16: 157-70.

[112] Dancey TE. Early infantile autism, 1943-1955; discussion of paper presented by Leo Kanner, M.D. Psychiatr Res Rep Am Psychiatr Assoc 1957; 66-88.

[113] Shi L, Tu N, Patterson PH. Maternal influenza infection is likely to alter fetal brain development indirectly: the virus is not detected in the fetus. Int J Dev Neurosci 2005; 23: 299-305.

[114] Brown AS. Prenatal infection as a risk factor for schizophrenia. Schizophr Bull 2006; 32: 200-2.

[115] Bell MJ, Hallenbeck JM, Gallo V. Determining the fetal inflammatory response in an experimental model of intrauterine inflammation in rats. Pediatr Res 2004; 56: 541-6.

[116] Smith SE, Li J, Garbett K, Mirnics K, Patterson PH. Maternal immune activation alters fetal brain development through interleukin-6. J Neurosci 2007; 27: 10695-702.

[117] Shi L, Smith SE, Malkova N, *et al.* Activation of the maternal immune system alters cerebellar development in the offspring. Brain Behav Immun 2009; 23: 116-23.

[118] Gong X, Jia M, Ruan Y, *et al.* Association between the FOXP2 gene and autistic disorder in Chinese population. Am J Med Genet B Neuropsychiatr Genet 2004; 127B: 113-6.

[119] Sanjuan J, Tolosa A, Gonzalez JC, *et al.* Association between FOXP2 polymorphisms and schizophrenia with auditory hallucinations. Psychiatr Genet 2006; 16: 67-72.

[120] Fatemi SH, Reutiman TJ, Folsom TD, Sidwell RW. The role of cerebellar genes in pathology of autism and schizophrenia. Cerebellum 2008; 7: 279-94.

[121] Hurst JA, Baraitser M, Auger E, Graham F, Norell S. An extended family with a dominantly inherited speech disorder. Dev Med Child Neurol 1990; 32: 352-5.

[122] Fatemi SH, Reutiman TJ, Folsom TD, *et al.* Maternal infection leads to abnormal gene regulation and brain atrophy in mouse offspring: implications for genesis of neurodevelopmental disorders. Schizophr Res 2008; 99: 56-70.

[123] Sodhi MS, Sanders-Bush E. Serotonin and brain development. Int Rev Neurobiol 2004; 59: 111-74.

[124] Fatemi SH, Emamian ES, Kist D, *et al.* Defective corticogenesis and reduction in Reelin immunoreactivity in cortex and hippocampus of prenatally infected neonatal mice. Mol Psychiatry 1999; 4: 145-54.

[125] Fatemi SH, Emamian ES, Sidwell RW, *et al.* Human influenza viral infection in utero alters glial fibrillary acidic protein immunoreactivity in the developing brains of neonatal mice. Mol Psychiatry 2002; 7: 633-40.

[126] Fatemi SH, Araghi-Niknam M, Laurence JA, *et al.* Glial fibrillary acidic protein and glutamic acid decarboxylase 65 and 67 kDa proteins are increased in brains of neonatal BALB/c mice following viral infection in utero. Schizophr Res 2004; 69: 121-3.

[127] Maloku E, Covelo IR, Hanbauer I, *et al.* Lower number of cerebellar Purkinje neurons in psychosis is associated with reduced reelin expression. Proc Natl Acad Sci U S A 2010; 107: 4407-11.

[128] Fatemi SH, Snow AV, Stary JM, *et al.* Reelin signaling is impaired in autism. Biol Psychiatry 2005; 57: 777-87.

[129] Fatemi SH, Stary JM, Halt AR, Realmuto GR. Dysregulation of Reelin and Bcl-2 proteins in autistic cerebellum. J Autism Dev Disord 2001; 31: 529-35.

[130] Seeburg PH, Burnashev N, Kohr G, *et al.* The NMDA receptor channel: molecular design of a coincidence detector. Recent Prog Horm Res 1995; 50: 19-34.

[131] Groc L, Choquet D, Stephenson FA, *et al.* NMDA receptor surface trafficking and synaptic subunit composition are developmentally regulated by the extracellular matrix protein Reelin. J Neurosci 2007; 27: 10165-75.

[132] Sinagra M, Gonzalez Campo C, Verrier D, *et al.* Glutamatergic cerebellar granule neurons synthesize and secrete reelin in vitro. Neuron Glia Biol 2008; 4: 189-96.

[133] Fatemi SH, Reutiman TJ, Folsom TD, Thuras PD. GABA(A) receptor downregulation in brains of subjects with autism. J Autism Dev Disord 2009; 39: 223-30.

[134] Fatemi SH, Folsom TD, Reutiman TJ, Thuras PD. Expression of GABA(B) receptors is altered in brains of subjects with autism. Cerebellum 2009; 8: 64-9.

[135] Yip J, Soghomonian JJ, Blatt GJ. Decreased GAD67 mRNA levels in cerebellar Purkinje cells in autism: pathophysiological implications. Acta Neuropathol 2007; 113: 559-68.

[136] Urena-Guerrero ME, Orozco-Suarez S, Lopez-Perez SJ, Flores-Soto ME, Beas-Zarate C. Excitotoxic neonatal damage induced by monosodium glutamate reduces several GABAergic markers in the cerebral cortex and hippocampus in adulthood. Int J Dev Neurosci 2009; 27: 845-55.

[137] Ray MA, Graham AJ, Lee M, *et al.* Neuronal nicotinic acetylcholine receptor subunits in autism: an immunohistochemical investigation in the thalamus. Neurobiol Dis 2005; 19: 366-77.

[138] Didier M, Berman SA, Lindstrom J, Bursztajn S. Characterization of nicotinic acetylcholine receptors expressed in primary cultures of cerebellar granule cells. Brain Res Mol Brain Res 1995; 30: 17-28.

CHAPTER 3

Dysregulation of Glutamatergic Neurotransmission in Autism Spectrum Disorders

Anna Strunecka

Department of Physiology, Faculty of Science, Charles University in Prague, Prague, Czech Republic

Abstract: Despite the great number of observations being made concerning cellular and molecular dysfunctions associated with autism spectrum disorders (ASD), an integrative and unifying mechanism to explain the heterogeneous symptoms and etiology of ASD has not been proposed in the major scientific literature. We offer the explanation of potential etiology of ASD as dysregulation of glutamatergic neurotransmission with underlying interactions between chronic microglial activation and the excitotoxic cascade playing the central role. This chapter summarizes current knowledge of the structural and functional diversity of glutamate receptors (GluRs) and excitatory amino acid transporters. Recent research of the autism genome also supports the view that abnormalities in genes connected with glutamate neurotransmission and disturbed regulation of glutamate pathways may be directly involved in ASD. We further suggest that the increasing prevalence of ASD during the last decades might reflect the synergistic action of an increased burden of new excitotoxic factors. In this chapter we discuss the effects of dietary excitatory amino acids, mainly glutamate and aspartate, which could exacerbate the pathological and clinical symptoms of ASD. The mechanism of excitotoxicity is the topic of the next chapter.

INTRODUCTION

It is generally accepted that glutamate is the principal excitatory neurotransmitter in the brain, acting at more than a half of its synapses. Several diverse pieces of evidence spurred an interest in acidic amino acids glutamate and aspartate, which were found in high concentrations throughout the brain [1-3]. Advances in glutamate subunit typing have greatly advanced our understanding of the normal physiology and pathophysiology of glutamate neurotransmission. The ability to express and manipulate cloned GluRs subunits is leading to huge advances in our understanding of these receptors (for a review see [4]). The development of selective antagonists, crucial to the subtype classification, allowed the fundamental importance of GluRs in synaptic activity throughout the CNS to be realized.

GluRs activity is required for fast synaptic transmission as well as for synaptic plasticity, learning and memory, motor coordination, pain transmission, and brain development [5-7]. GluRs are not only pivotal for normal brain function but they are also important drug targets. Several authors have hypothesized that alterations of glutamatergic neurotransmission play a key role in the brain pathophysiology.

Blaylock and Strunecka brought evidence that most heterogeneous symptoms of ASD have a common set of events closely connected with dysregulation of glutamatergic neurotransmission in the brain with enhancement of excitatory receptor function by pro-inflammatory immune cytokines as the underlying mechanism [8].

GLUTAMATERGIC NEUROTRANSMISSION

The GluRs system consists of three ionotropic receptors (NMDA, AMPA, and kainate) and three metabotropic receptors (mGluRs) types, with a number of cloned subtypes. The various GluRs differ in structure and physiology. Heroic studies have shown that the GluRs system is composed of a very complex set of subunits and receptor types. It is the patterns of subunit assembly within the various GluRs that produce the regional differences in the brain's response to glutamate stimulation. While mGluRs belong to the G protein coupled receptors (GPCRs) and mediate changes in intracellular signaling pathways, activation of ionotropic receptors AMPA, kainate, and NMDA opens ion channels for sodium ions and Ca^{2+} [9-11]. GluRs are heterogeneously distributed throughout the brain. They

---

*Address correspondence to: Anna Strunecka, Faculty of Sciences, Charles University in Prague, Vinicna 7, 128 00 Prague 2, Czech Republic; E-mail: strun@natur.cuni.cz

have been found on neurons, astrocytes, microglia, oligodendrocytes, and neuronal pre- and post synaptic sites as well as axons.

Ionotropic GluRs

Following from their initial studies, Curtis and Watkins [3] tested a large series of acidic amino acids. Among these earliest compounds was N-methyl-aspartate, the D-isomer (NMDA), of which has played a prominent role in defining receptor nomenclature. NMDA proved to be greater than 10 times more potent than L-glutamate itself whereas the L-isomer was similar in potency to L- and D-glutamate. By the early 1970s, excitatory amino acids receptors were tentatively divided into 'glutamate-preferring' and 'aspartate-preferring' categories, with kainate and NMDA being regarded as the key exogenous ligands [4]. It was generally considered that L-aspartate was the likely transmitter at NMDA receptors, largely because of its greater sensitivity to NMDA antagonists than L-glutamate.

Kainate is a glutamate analogue, isolated from the seaweed *Digenea simplex*. Another potent excitant eventually playing a similar nomenclature role, quisqualate, was discovered. A nomenclature consisting of three ionotropic GluRs subtypes, namely NMDA, quisqualate, and kainate receptors thus arose. Since quisqualate was found later to act also at metabotropic GluRs [12], the 'quisqualate' nomenclature was changed to AMPA (α-amino-3-hydroxy-5-methylisoxazolepropionic acid).

During 1980s, the concept of the NMDA, AMPA, and kainate subclasses of ionotropic GluRs became well entrenched and this terminology has been firmly established [4]. Subsequent molecular studies demonstrated the potential diversity of ionotropic GluRs. The cloning and expression of 16 genes coding for ionotropic GluRs subunits was realized. The use of site-directed mutagenesis, receptor hybridization techniques, molecular modeling of binding sites, X-ray crystal structure, high throughput screens, patch clamp electrophysiology, transgenic animals, etc. has revealed the enormous potential for combinations and co-assembly of subunits of these receptors [4,13]. Various nomenclatures were introduced by the laboratories that cloned the subunits. For these reasons, the journal Neuropharmacology and The International Union of Basic and Clinical Pharmacology (IUPHAR) have joined forces to address the nomenclature of ligand-gated ion channels (LGICs) that are activated by neurotransmitters [13]. They confirmed that tetrameric ionotropic GluRs are subdivided into NMDA, AMPA, and kainate receptor subfamilies. The new nomenclature for ionotropic GluRs subunits was recommended. It was agreed to adopt Glu, the three letter amino acid code for glutamate, rather than GLU, to identify the neurotransmitter.

Activation of ionotropic AMPA, kainate, and NMDA receptors opens ion channels for sodium ions and Ca^{2+}. Homeostatic control of the Ca^{2+} entering through tetrameric NMDA receptors is of paramount importance. NMDA receptor channels are highly permeable to Ca^{2+}; thus over-activation leads to excitotoxic neuronal cell death. NMDA receptors have been implicated in schizophrenia, epilepsy, ischemic brain damage, and neurodegenerative disorders. Thus, NMDA receptors are recently a major target for drug design [5,6,14,15].

NMDA receptors are unique in that they require the simultaneous binding of two neurotransmitters, the co-agonists L-glutamate and glycine, together with the alleviation of a voltage-dependent blockade by magnesium ions (Mg^{2+}). It has been suggested that, unless the receptors were protected by a very high affinity transport process, the glycine site would be fully occupied in physiological conditions. The glycine binding sites became the target of considerable pharmacological activity since antagonists of glycine binding sites possess anticonvulsant and neuroprotective properties [16].

A great deal of interest is now also directed towards the AMPA receptors, which control fast transmission. Newer studies are showing that during development AMPA receptors lack GluR2 subunits and this makes them Ca^{2+} permeable and therefore more excitable by glutamate [4].

Metabotropic GluRs

Metabotropic GluRs are a group of seven-transmembrane-domain proteins that couple to G proteins. They activate heterotrimeric G proteins in response to ligand stimulation. Heterotrimeric G proteins are constructed of three types of subunits, an α-subunit uniquely capable of binding and degrading GTP and a tightly knit complex of β- and γ-subunits. The nomenclature now popularly known as Gαq, Gαs, and Gαi classes determines the interaction of the α-

subunit with various effector molecules [16,17]. Gαs means that the α-subunit of heterotrimeric G protein interacts with adenylyl cyclase (AC) and stimulates the production of cyclic adenosine monophosphate (cAMP), while Gαi inhibits the production of cAMP. Gαq has been used for a class of G proteins, which activate phospholipase C (PLC), the effector of phosphoinositide signaling system [18,19]. It is generally accepted that phosphatidylinositol 4, 5-bisphosphate (PIP$_2$) from the plasma membrane is hydrolyzed by PLC and yields inositol 1, 4, 5-trisphosphate (1, 4, 5-IP$_3$) and diacylglycerol (DAG) after receptor stimulation. Both products of this hydrolysis catalyzed by PLC have a second-messenger role. 1, 4, 5-IP$_3$ binds to receptors in membranes of endoplasmic reticulum, which results in a release of Ca^{2+} into the cytosol. DAG activates protein kinase C (PKC), which affects the activity of many enzymes via protein phosphorylation (Fig 1). Calcium ions (Ca^{2+}) enhance the enzymatic activity of PLC. GPCRs are the largest family of cell surface receptors

Metabotropic GluRs comprise a whole family with currently eight members grouped into three classes according to their amino acid sequence identity and pharmacological profile. They are G-protein coupled, either positively linked to PLC (class I) or negatively linked to AC (class II and III) [20]. Metabotropic GluRs thus control the levels of second messengers such as 1, 4, 5-IP3, DAG, Ca^{2+}, and cAMP. They elicit the release of arachidonic acid via intracellular Ca^{2+} ([Ca^{2+}]$_i$) mobilization from intracellular stores such as mitochondria and endoplasmic reticulum [21]. This facilitates the release of glutamate. There is an intimate interaction between the mGluRs and NMDA receptors, allowing rapid modulation of excitatory synaptic transmission. Among other effects, mGluRs are known to induce phosphorylation of ionotropic GluRs. Moreover, it has been found that, for example, mGluR1 can cause various cell responses via coupling with different types of G protein [22, 23]. The interactions between various types and subtypes of GluRs, second messenger molecules, eicosanoid metabolites, reactive oxygen species (ROS), reactive nitrogen species (RNS), lipid peroxidation products (LPP), and phosphorylating enzymes make control of these systems extremely complex. Studies of Martin-Ruiz *et al.* [24] have disclosed that GluRs also interact with serotonergic, adrenonergic, and cholinergic neural nets as well.

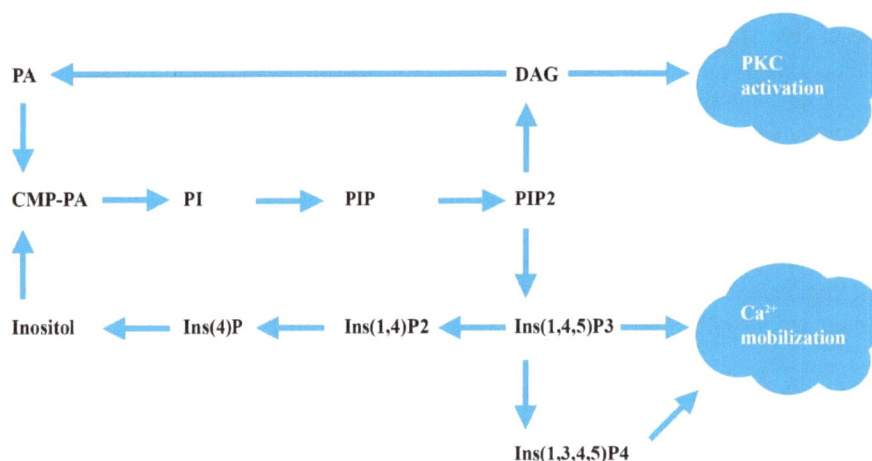

Figure 1: A simplified scheme of the phosphoinositide signaling system. A great amount of experimental evidence supports the general concept that phosphatidylinositol 4,5-bisphosphate (PIP$_2$) from the plasma membrane is hydrolyzed and yields inositol 1,4,5-trisphosphate (Ins(1,4,5)P$_3$) and diacylglycerol (DAG) upon receptor stimulation. Both products of this hydrolysis catalyzed by phospholipase C (PLC) have second messenger role. Ins(1,4,5)P$_3$ binds to a receptor in membranes of endoplasmic reticulum, which results in a release of Ca^{2+} into the cytosol. The Ins(1,4,5)P$_3$ receptor composes an Ins(1,4,5)P$_3$-gated Ca^{2+}-channel. DAG activates protein kinase C (PKC). The coupling between the receptor and PLC is mediated by G-proteins. In this "dual" second messenger hypothesis, Ins(1,4,5)P$_3$ is the link between PIP$_2$ and Ca^{2+}.

Metabotropic GluRs are found on many cell types including neurons, microglia, astrocytes, oligodendrocytes, T- and B-cell lymphocytes, osteoblasts, hepatocytes, and endothelial cells, among others. These receptors have a number of effects on cells that can influence outcome after trauma, regulate inflammation, alter endothelial permeability, and influence neuronal and oligodendroglial cell death. Hence, mGluRs also represent a pharmacological path to a relatively subtle amelioration of neurotoxicity because they serve a modulatory rather than a direct role in excitatory glutamatergic transmission.

Glutamate Homeostasis in the Brain: Brain Handling of Glutamate

Glutamate concentrations in plasma are 30-100 µmol/L; in whole brain, they are 10,000-12,000 µmol/L but only 0.5-2 µmol/L in extracellular fluids [25]. Because of the extreme toxicity of extracellular glutamate, even in very small concentrations, there exists a very complex system to prevent extracellular glutamate accumulation. This includes the glutamate reuptake system, which utilizes five **excitatory amino acid transporters** (EAAT1-5), the first two of which are referred to as glutamate transporter-1 (GLT-1) and glutamate aspartate transporter (GLAST) [26-28]. These sodium and energy-dependent glutamate transporters can move glutamate in either direction, that is, toward astrocyte uptake or into the extracellular space, depending on physiological and pathological conditions. The glutamate transporters also provide glutamate for synthesis of glutathione and protein, GABA, and for energy production. Glutamate transporters are thus involved in both brain physiological activity and many pathological states.

Astrocytes are the major regulators of glutamate homeostasis in the brain [26]. Astroglial cells clear extracellular glutamate through the glutamate transporters, GLT-1 and GLAST, and subsequently convert the incorporated glutamate into glutamine by the enzyme **glutamine synthetase** (GS), an exclusively glial enzyme, and then export glutamine to neurons (Fig. 2). Recently, Zou *et al.* [29] demonstrated that inhibition of GS in astrocytes significantly impaired glutamate uptake and glutamine release. Conversely, induction of GS expression in astrocytes by gene transfer significantly enhanced the glutamate uptake and glutamine release.

Astrocytes contain a relatively high activity, of **phosphate activated glutaminase** although it is significantly lower than that of synaptosomes, cultures of cerebellar granule cells or of cortical neurons [30,31]. One of the products, glutamate, inhibits the enzyme strongly, whereas the other product, ammonia, has only a slight inhibitory action on the enzyme.

Glutamate can also be cleared from the extracellular space by metabolic conversion of glutamate to α-ketoglutarate. **Glutamate dehydrogenase** (GDH) is located in the mitochondria and is an important branch-point enzyme between carbon and nitrogen metabolism.

Glutamate can be converted to GABA, the major inhibitory neurotransmitter in the brain. This reaction is catalyzed by the pyridoxal phosphate-dependent **glutamic acid decarboxylase** (glutamate decarboxylase) (GAD).

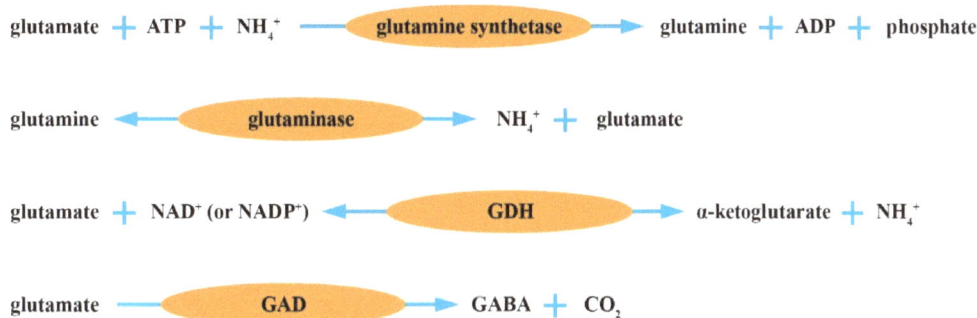

glutamate $+$ ATP $+$ NH_4^+ ⟶ glutamine synthetase ⟶ glutamine $+$ ADP $+$ phosphate

glutamine ⟵ glutaminase ⟶ NH_4^+ $+$ glutamate

glutamate $+$ NAD^+ (or $NADP^+$) ⟵ GDH ⟶ α-ketoglutarate $+$ NH_4^+

glutamate ⟶ GAD ⟶ GABA $+$ CO_2

Figure 2: Metabolic transformations of glutamate. GDH - glutamate dehydrogenase; GAD - glutamate decarboxylase.

In mammals, GAD exists in two isoforms encoded by two different genes - *GAD1* and *GAD2*. These isoforms are GAD67 and GAD65 with molecular weights of 67 and 65 kDa. GAD has been immunocytochemically localized in the somata and dendrites of certain neurons in rat cerebellum and Ammons horn following colchicine injections into these two brain regions. In the cerebellum, the GAD-positive reaction product was observed in the somata and proximal dendrites of Purkinje, Golgi II, basket and stellate neurons [32].

Liang *et al.* [33] demonstrated that the astrocytic ***glutamate-glutamine cycle*** serves as a major contributor to GABA in active inhibitory synapses in hippocampal area CA1. Interestingly, in the hippocampus, the astrocytic glutamate transporter is responsible for at least 80% of glutamate clearance and the majority of uptake-dependent synaptic inactivation.

Akbarian *et al.* [34] quantified levels of mRNA for the GAD67 isoform and the number and laminar distribution of GAD mRNA-expressing neurons in the dorsolateral prefrontal cortex of schizophrenics and matched controls, using in situ hybridization-histochemistry, densitometry, and cell-counting methods. Schizophrenics showed a pronounced decrease in GAD mRNA levels in neurons of layer I (40%), layer II (48%) and an overall 30% decrease in layers III to VI. There were also strong overall reductions in GAD mRNA levels. Substantial dysregulation of GAD mRNA expression, coupled with downregulation of reelin is observed in schizophrenia and bipolar disorders [35, 36].

Dysregulation of the glutamate reuptake system can result in excitotoxicity and abnormal development of the CNS [37]. Activation of astrocytes contributes to and reinforces an inflammatory cascade. Astrocytes comprise approximately half of the volume of the adult mammalian brain and are the primary trophic supportive elements. By transforming to a reactive state, important astrocytic functions, which include the uptake of glutamate and release of glutamine; the uptake of glutathione precursors, and release of glutathione, may be impaired. Moreover, Lehman *et al.* [38] provided evidence that several forms of acute brain injury are associated with the increased expression of GS and decreased expression of GLT-1 and/or GLAST, eventually leading to the accumulation of excitotoxic extracellular glutamate concentrations (Fig. **3**.).

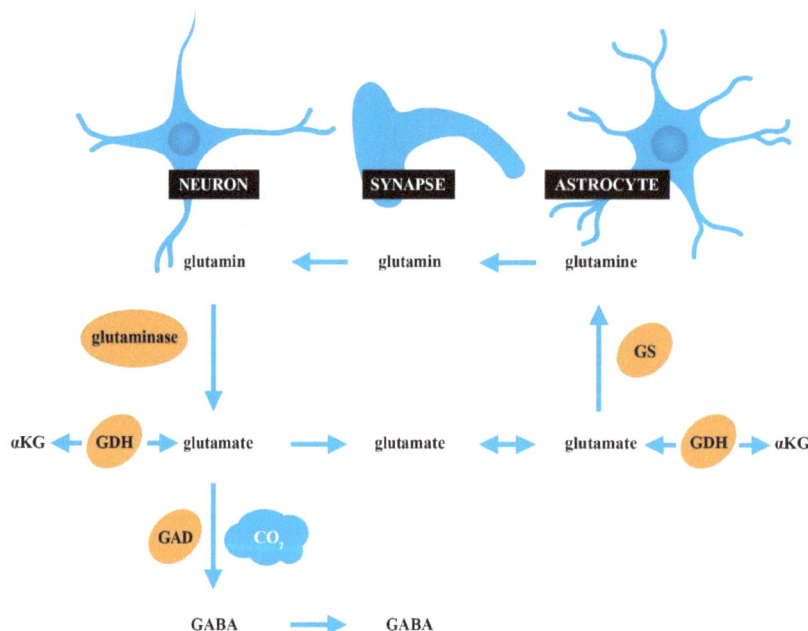

Figure 3: Schematic representation of glutamate/glutamine neurotransmitter cycling between astrocytes and neurons. The neurotransmitter glutamate is recycled through an astrocytic-neuronal glutamate-glutamine cycle in which synaptic glutamate is taken up by astrocytes, metabolized to glutamine, and transferred to neurons for conversion back to glutamate and subsequent release. GS - glutamine synthetase; GDH - glutamate dehydrogenase; GAD - glutamate decarboxylase.

Hominid Features of Human Astrocytes

Oberheim with co-workers [39] described interesting comparison of human and rat astrocytes. These authors reported that protoplasmic astrocytes in human neocortex are 2.6-fold larger in diameter and extend 10-fold more glial fibrillary acidic protein (GFAP) -positive primary processes than their rodent counterparts. In cortical slices prepared from acutely resected surgical tissue, protoplasmic astrocytes propagate Ca^{2+} waves approximately fourfold faster than rodent. Human astrocytes also transiently increase $[Ca^{2+}]_i$ in response to glutamatergic and purinergic receptor agonists. The human neocortex also harbors several anatomically defined subclasses of astrocytes not represented in rodents. These include a population of astrocytes that reside in layers 5-6 and extend long fibers characterized by regularly spaced varicosities. Another specialized type of astrocyte, the interlaminar astrocyte, abundantly populates the superficial cortical layers and extends long processes without varicosities to cortical layers 3 and 4. Human fibrous astrocytes resemble their rodent counterpart but are larger in diameter. Thus, human cortical astrocytes are both larger, and structurally more complex and more diverse, than those of rodents. Oberheim *et al.* [39] posit that this astrocytic complexity has permitted the increased functional competence of the adult human brain.

The additional discovery that glutamate release from astrocytes is controlled by molecules linked to inflammatory reactions, such as the cytokine TNF-α and prostaglandins, suggests that glia-to-neuron signaling may be sensitive to changes in production of these mediators in pathological conditions. Local inflammatory reactions, characterized by astrocytic and microglial activation, has been reported in several neurodegenerative disorders [40]. When TNF-α was applied to cultures of astrocytes, it significantly reduced GS expression and inhibited glutamate-induced GS activation resulting in increased excitotoxicity to neurons [29].

Considerable amounts of evidence obtained by several groups during the past years demonstrated the existence of a bidirectional communication between astrocytes and neurons. Importantly, astrocyte $[Ca^{2+}]_i$ elevations can trigger the release of gliotransmitters, which modulate neuronal activity as well as synaptic transmission and plasticity [41,42]. Astrocytes are now widely regarded as cells that propagate Ca^{2+} over long distances in response to stimulation and release gliotransmitters (for example glutamate) in a Ca^{2+}-dependent manner to modulate a host of important brain functions [43,44]. It is speculated that Ca^{2+} signaling may play a role in astrocytic functions related to the blood brain barrier (BBB), including blood flow regulation, metabolic trafficking, and water homeostasis [45, 46]. The concept that astrocytes release gliotransmitters to affect synaptic transmission has been a paradigm shift in neuroscience research over the past decade.

Importantly, mGluRs are expressed in glial cells including astrocytes, oligodendrocytes, and microglia [47]. The recognition that astrocytes undergo elevations in $[Ca^{2+}]_i$ following activation of GPCRs by synaptically released neurotransmitters indicate that astrocytes may be active participants in brain information processing. Studies of astrocyte-neuron interactions have shown that Ca^{2+} signaling is a potent modulator of the strength of both excitatory and inhibitory synapses. The concept that astrocytes possess a mechanism for rapid cell communication has not been incorporated, however, into the supportive functions of astrocytes.

Lavenex *et al.* [48] demonstrated that the expression of genes associated with glycolysis and glutamate metabolism in astrocytes and the coverage of excitatory synapses by astrocytic processes undergo significant suppression in the CA1 field of the monkey hippocampus during postnatal development. Interestingly, the hippocampus is also the brain structure that is most sensitive to hypoxic-ischemic episodes. These authors suggested that a developmental decrease in astrocytic processes could underlie the selective vulnerability of CA1 during hypoxic-ischemic episodes in adulthood, its decreased susceptibility to febrile seizures with age, as well as contribute to the emergence of selective, adult-like memory function.

THE ROLE OF DYSREGULATION OF GLUTAMATERGIC NEUROTRANSMISSION IN ASD ETIOLOGY

We suggested the explanation of potential etiology of ASD as dysregulation of glutamatergic neurotransmission, with underlying interactions between chronic microglial activation and the excitotoxic cascade playing the central role [8,49,50-52] (see Chapter 4). Given the major role of glutamate in brain development, Carlsson [53] proposed a hypoglutamatergic hypothesis for autism, based on an interaction between overactivity of 5-HT2A (serotonin) receptors and NMDA receptors. His hypothesis was based on behavioral effects associated with glutamate antagonist and the evidence that affected anatomical areas of the brain in autism contain abundant GluRs. However, the hypoglutamatergic hypothesis does not explain the high incidence of seizures in autistic patients and/or the findings of elevated glutamate levels in the blood and cerebrospinal fluid (CSF) of autistic children.

Contrawise, some other authors suggest a hyperglutamatergic hypothesis based on findings of increased serum level of glutamate in children and adults with ASD [54-56], the reduction of the levels of rate-limiting enzymes GAD65 and GAD67, and the increased gliosis in the brains of autistic subjects [57,58].

During the developmental period of synaptogenesis, also known as the brain growth spurt period, neurons are very sensitive to specific disturbances in their synaptic environment. The brain growth spurt occurs in different species at different times relative to birth. In rats and mice it is a postnatal event, but in humans it extends from the 6th month of gestation to several years after birth. Thus, there is a period in fetal and neonatal human development, lasting for several years, during which immature CNS neurons are exquisitely sensitive to environmental agents that can trigger widespread neurodegeneration by inducing specific abnormal changes in the synaptic environment. During this period, abnormal increase in NMDA receptor activity triggers excitotoxic neurodegeneration. Only a transient disturbance, lasting for a few hours, is sufficient to trigger either excitotoxic or apoptotic neurodegeneration during this developmental period [59]. The contribution of immunoexcitotoxicity as the central mechanism in ASD etiology is explicitly explained in the following chapter.

ASD are highly genetic and multifactorial, with many risk factors acting together. Recent research of the autism genome support further the view that abnormalities in genes connected with GluRs and dysregulation of glutamate pathways may be directly involved in ASD etiopathology. We further suggest that the increasing prevalence of ASD during the last decades might reflect the synergistic action of increased burden of new ecotoxicological factors, which include excitotoxic amino acids, mainly glutamate and aspartate, fluoride in combination with aluminum ions (Al^{3+}), mercury, and the increasing number of vaccines in the period of rapid postnatal brain development. These items will be discussed in several other chapters. Here we will review and analyze the evidence for disturbance of glutamate homeostasis in the brain.

Abnormalities in Genes Connected with Disturbance of Glutamate Homeostasis in ASD

Autism appears to be the most highly genetic of the psychiatric disorders, as evidenced by the high risk of autism in additional children in families with an autistic child and the concordance rate for monozygotic twins being much higher than that of dizygotic twins. However, it is evident, that ASD do not follow a simple Mendelian mode of transmission but are clearly a polygenic. A commonly accepted genetic model involves several genes (between five and ten) that interact to produce the disorder. Candidate genes for studies of autism range from genes that are thought to play a role in neurodevelopment, neurotransmission, and behavior [60,61]. Copy number variants (CNVs) associated with autism seem to show variable expressivity, also leading to other phenotypes, such as schizophrenia, mental retardation/developmental delay, and epilepsy. Initial genome-wide single nucleotide polymorphism (SNP) association studies have each identified a single novel associated locus with modest effect.

Guilmatre with a team of co-workers [62] investigated 28 candidate loci previously identified by comparative genomic hybridization studies for gene dosage alteration in 247 cases with mental retardation, 260 cases with ASD, 236 cases with schizophrenia or schizoaffective disorder, and in 236 controls. Recurrent or overlapping CNVs were found in cases at 39.3% of the selected loci. The collective frequency of CNVs at these loci is significantly increased in cases with autism, schizophrenia, and mental retardation compared with controls. Most of these CNVs, which contain genes involved in neurotransmission or in synapse formation and maintenance, are present in the three pathological conditions (schizophrenia, autism, and mental retardation), supporting the existence of shared biological pathways in these neurodevelopmental disorders.

GAD Deficiency in ASD

Selective loss of Purkinje cells and the cerebellar atrophies are the neurological abnormalities most consistently found in persons diagnosed with ASD. Fatemi with co-workers first reported that GAD65 was reduced by 48% and 50% in parietal and cerebellar areas of autistic brains versus controls, respectively [63, 64]. The similar extent of reduction was found for GAD67 (by 61% and 51% in parietal and cerebellar areas). Fatemi suggested that GAD deficiency may be due to or associated with abnormalities in levels of glutamate/GABA, or transporter/receptor density in autistic brain. Because Purkinje cells are involved in motor coordination, working memory and learning, loss of these cells are likely to cause symptoms defining behavioral parameters of ASD. Yip *et al.* [65] quantified GAD67 mRNA, the most abundant isoform in Purkinje cells, using in situ hybridization in adult autistic and control cases. Their results indicate that GAD67 mRNA level was reduced by 40% in the autistic group. These authors thus suggest that reduced Purkinje cell GABA input to the cerebellar nuclei potentially disrupts cerebellar output to higher association cortices affecting motor and/or cognitive function and supports the previous reports of alterations in the GABAergic system in limbic and cerebro-cortical areas in autistic brains.

The involvement of alterations in the gene encoding GAD has been supported by recent studies [58, 66]. Rout and Dhossche [67] postulated a hypothesis for the development of ASD according to which the extent of Purkinje cell loss is triggered by GAD-antibody. These authors suggest that identification and characterization of GAD-antibodies from pregnant mothers with a family history of autism, from children with autistic siblings, and individuals diagnosed with autism may allow us to find preventive and new therapeutic avenues.

Reelin

Reelin is a protein with a crucial role in the regulation of neuronal migration and synaptic plasticity. Reduced reelin levels were found in blood and frontal and cerebellar cortices of autistic patients by Fatemi's group. Reelin genes (RELN) were shown to be associated with ASD. Disruption of the reelin and GABAergic signaling systems have been observed in psychiatric disorders including autism, schizophrenia, bipolar disorder, and major depression [68]. The recent observations show that commonly used psychotropic medications (clozapine, fluoxetine, haloperidol, lithium, olanzapine, and valproic acid) alter levels of reelin and GAD65/67.

Metabotropic GluRs Genes

A significant association between GluR6 gene, located on chromosome 6q21, and autism was found [69]. Data sets of 107 parent-offspring trios indicated significant maternal transmission disequilibrium. In contrast to maternal transmission, paternal transmission of GluR6 alleles was as expected in the absence of linkage, suggesting a maternal effect such as imprinting.

Serajee *et al.* [70] demonstrated the partial duplication of the mGluR8 gene at 7q31 and its possible association with autism. A high incidence of mutation of the *GRM8* gene controlling the mGluR8 receptor subunit negatively modulates glutamate neurotransmission. Mutation of this gene increases glutamate hyperactivity and thus excitotoxicity. This receptor subunit is located on a number of anatomical areas of the brain affected in autism, including the lateral reticular thalamic nucleus, pyriform cortex and to a lesser degree the cerebellum, caudate and hippocampus. The in situ hybridization results indicate a predominantly glial cell expression of mGluR8 in human brain [71]. Moreover, in the subsequent study, Serajee and co-workers [72] genotyped 196 trios (581 individuals), which include at least two affected members with a diagnosis of autism, Asperger's syndrome, or PDD-NOS. They found the nominally positive association of three positional candidate genes that all act in the phosphoinositide signaling pathway. It has been mentioned that the phosphoinositide signaling system has key role for transduction of signals for mGluRs. Moreover, phosphoinositides are involved in numerous cellular processes, such as neuronal differentiation, neuroprotection, transduction of signals from growth factors, numerous neurotransmitters and hormones [73]. Disturbance of this signaling cascade and metabolism of inositol-phosphate is connected with schizophrenia and AD [74-76]. The exact implication of the observation that SNPs in three phosphoinositide signaling pathway genes, *INPP1, PIK3CG,* and *TSC2* are in linkage disequilibrium with autism is not clear at present. However, the associations of these three candidate genes with autism have neurobiological plausibility.

Ionotropics GluRs Genes

Purcell *et al.* [77] collected brain samples from a total of ten individuals with autism and 23 matched controls, mainly from the cerebellum. They found that mRNA levels of genes *EAAT1* and *AMPA* receptor were significantly increased in autism. However, AMPA receptor density was decreased in the cerebellum of individuals with autism. Ramanathan *et al.* [78] detected abnormalities in genes controlling AMPA receptors as well as glycine receptors (*GLRA3 and GLRB*), which play a critical role in ionotropic GluR control, in a single case of autism.

Gene Encoding the Mitochondrial Aspartate/Glutamate Carrier

A strong association between autism and *SLC25A12*, a gene encoding the mitochondrial aspartate/glutamate carrier (also called aralar or citrin) in neurons has been reported [79,80]. Two SNPs that showed evidence for divergent distribution between autistic and nonautistic subjects were identified, both within *SLC25A12*, localized on chromosome 7q21.3. In the second stage, the two SNPs in *SLC25A12* were further genotyped in 411 autistic families, and linkage and association tests were carried out in the 197 informative families. A strong association of autism with SNPs within the *SLC25A12* gene was demonstrated. These findings were confirmed by study of Segurado *et al.* [81] investigating 158 Irish affected child-parent trios (442 individuals). Nevertheless, the functional relevance of such variants remains obscure.

Glutamate Transport Proteins

The Autism Genome Project compared the genomes of 1, 181 families, each of which had at least two autistic offsprings [82]. Linkage and CNV analyzes implicate chromosome 11p12-p13 among other candidate loci, which has been linked to glutamate transport proteins [82].

Neurexin

This largest genome scan ever conducted implicated the deletion of *neurexin 1* to be involved in ASD pathology. Neurexin 1 is believed to be involved in building glutamate synapses. Neurexins team with previously implicated neuroligins (NLGN) for glutamatergic synaptogenesis, highlighting glutamate-related genes as promising candidates for contributing to ASD [82]. NLGN is a postsynaptic cell adhesive molecule that upon binding to neurexins on the presynaptic cell surface provides the physical stability of the synapses. Linkage and copy number variation analyses also indicate the *NLGN* gene as a strong candidate for autism [69, 82].

It is thus obvious from all these genetic studies that genetic influence on glutamate function is playing a role in the etiology of ASD.

EFFECTS OF EXOGENOUS EXCITATORY AMINO ACIDS: THE IMPACT FOR UNDERSTANDING ASD ETIOLOGY

Despite significant neuropathological findings and anatomical alterations in classic autism, an increasing number of new cases appearing since the early 1980s include a large number who do not show dramatic changes in brain gross anatomy as seen in classic autism. The difference in severity appears to vary with the stage at which the immunoexcitotoxic insult arises and its intensity. Postnatal injury is more likely to produce one of the lesser ASD syndromes.

Glutamate and aspartate are components of the daily diet of pregnant mothers, newborns and infants, as well as children. Also of concern is the observation that autistic children tend to prefer junk type foods, most of which contain significant amounts of excitotoxic additives. Behavioral studies have shown that exposure to excess glutamate during critical periods of brain development can produce prolonged alterations in behavior [83]. Treatment with glutamate during the early postnatal period in rats resulted in defects in adjusting to a new environment 21 and 60 days later, something commonly found in ASD [84]. The behavioral effects are more common in male animals, with few effects being found in the females. Affected males showed little social interest in their littermates, demonstrated defects in novelty and perceptual mechanisms and an inability to focus attention, again all characteristics of the child with one of the ASD.

Increased Plasma Level of Glutamate and Aspartate in ASD

Moreno-Fuenmayor *et al.* [85] first reported increased serum level of glutamate and aspartate in fourteen autistic children, all below 10 years of age. All affected children had lower plasma levels of glutamine. These authors suggested that increased glutamatemia may be dietary in origin or may arise endogenously for several reasons, among others, metabolic derangements in glutamate metabolism perhaps involving vitamin B6, defects or blockage of GluRs at the neuronal compartment, or alterations in the function of the neurotransmitters transporters. Fatemi [63] proposed that the observed reductions in GAD67 and GAD65 levels may account for reported increases of glutamate in blood and platelets of autistic subjects. Patients with autism or Asperger syndrome, their siblings and parents all had higher plasma glutamic acid level than controls, with reduced plasma glutamine [54]. Shinohe *et al.*

[55] measured serum levels of amino acids in 18 male adult patients with autism and age-matched 19 male healthy subjects and confirmed that serum levels of glutamate was significantly higher than that of controls (89.2 μM in autistic vs. 61.1 μM in controls). Increased arterial glutamate levels (5- to 10-fold higher in comparison with controls) were reported in connection with pathophysiological conditions such as acquired immunodeficiency syndrome (AIDS) and malignancies [86,87].

BBB Permeability to Glutamate

It has been generally accepted that the BBB is impermeable to glutamate, even at high concentrations, except in a few small areas that have fenestrated capillaries (CVO) [25]. Contrary to these findings, the early studies of Olney and colleagues [88-93] demonstrated that the most frequently encountered food excitotoxin glutamate, which is commercially added to many foods, can freely penetrate certain brain regions and rapidly destroy neurons by hyperactivating the NMDA receptors. Of special concern as well is the discovery that glutamate, by activating the NMDA receptors on the BBB, can disrupt the barrier, leading to free access of blood-borne toxins into the CNS.

Olney and co-workers described brain damage in infant mice following oral intake of glutamate or aspartate and demonstrated brain-damaging potential of protein hydrolyzates and neurotoxic effects of glutamate [37,88,91]. Their experiments show that glutamate induced neuronal necrosis in the infant mouse hypothalamus; brain damage in infant primates, and several histological changes. In newborn mice subcutaneous injections of monosodium glutamate (MSG) induced acute neuronal necrosis in several regions of the developing brain.

Orally ingested MSG has been shown to raise blood glutamate levels as much as 20 to 45-fold higher than baseline values [94,95]. It has also been shown, using radiolabeled [^{3}H]-glutamate, that MSG can pass through the placenta and preferentially accumulate in the fetal brain [96]. Likewise, Olney *et al.* [90] has shown that the immature brain is approximately four-times more sensitive to glutamate excitotoxicity as the adult brain. An explanation for hypersensitivity of the immature brain lies in the observation that during brain development the NMDA receptor is more sensitive to glutamate and less responsive to magnesium protection. Frieder and Grimm [97,98] have demonstrated that feeding MSG to pregnant rats can lead to severe alterations in learning that only affected the male rats, again demonstrating a male preponderance of toxic effects. It has also been shown that sensitivity to glutamate increases after birth. A number of studies have demonstrated that MSG feeding early in life can lead to prolonged free radical generation in the brain, which can last until adult stages of life [99]. In addition, some studies [98,100] have demonstrated alteration in hippocampal architecture and synaptic development following neonatal exposure to MSG in animal models.

Importantly, Olney reported, that adult animals treated with exogenous glutamate showed marked obesity and female sterility [101]. Studies of food consumption failed to demonstrate hyperphagia to explain the obesity. These important findings are almost neglected at present.

Exogenous Aspartate Produces Hyperactivity

The early studies brought interesting observations in connection with the effects of exogenous aspartate. Davies and Johnston observed in 1976 that both D- and L-aspartate produce gross hyperactivity when injected intraperitoneally into immature rats [102] All animals given D-aspartate became very active after 35-70min. Bouts of hyperactivity was evident in all rats until they were killed about 4 h later. This finding indicates that both isoforms of aspartate enter brain after intraperitoneal administration and induce hyperactivity. Another study observed that exposure to D-aspartate potentiated, by approximately 15-fold, the effects of L-glutamate *in vitro* [103].

CONCLUSIONS

Our review brings evidence that numerous animal experiments, genetic and metabolic studies of ASD patients support the hypothesis that dysregulation of glutamate function is playing a role in the etiology of ASD. The summary is given in Table **1**.

Table 1: Observed Alterations in ASD, Which May Be Connected with Dysfunctions of Glutamatergic Neurotransmission and Impairments of Glutamate Homeostasis.

Level	Alterations
genetic	reduction of GAD65 and GAD67 mRNA in Purkinje cells GluR6 is abundantly expressed in the hippocampus, basal ganglion and cerebellum, mutation of the *GRM8* gene abnormalities in genes for AMPA receptor and glycine receptors gene mutation of reelin deletion of the *neurexin 1* gene aspartate/glutamate carrier *SLC25A12* gene
biochemical	decreased protein levels of GAD higher concentration of glutamate/glutamine in the amygdala-hippocampal region lower levels of gray matter NAA and Glx increased serum levels of glutamate increased serum levels of aspartate
pathophysiological	seizure activity high levels of androgens early onset of puberty insomnia – sleep problems GI disturbances
neuropathological	abnormalities in the architecture of the brain affecting cortical, subcortical, limbic and cerebellar structures, asymmetrical enlargement of the amygdala; hypoplasia of the inferior vermis of the cerebellum with loss of Purkinje cells, increased gliosis (GFAP)
behavioral and neurological	loss of eye contact, deficiencies in socialization, abnormal mind function, language dysfunction, repetitive behaviors, difficulties with executive prefrontal lobe functions

REFERENCES

[1] Berl S, Waelsch H. Determination of glutamic acid, glutamine, glutathione and gamma-aminobutyric acid and their distribution in brain tissue. J Neurochem 1958; 3: 161-9.

[2] Curtis DR, Phillis JW, Watkins JC. The chemical excitation of spinal neurones by certain acidic amino acids. J Physiol 1960; 150: 656-82.

[3] Curtis DR, Watkins JC. Acidic amino acids with strong excitatory actions on mammalian neurones. J Physiol 1963; 166: 1-14.

[4] Lodge D. The history of the pharmacology and cloning of ionotropic glutamate receptors and the development of idiosyncratic nomenclature. Neuropharmacology 2009; 56: 6-21.

[5] Holden C. Psychiatric drugs. Excited by glutamate. Science 2003; 300: 1866-8.

[6] Konradi C, Heckers S. Molecular aspects of glutamate dysregulation: implications for schizophrenia and its treatment. Pharmacol Ther 2003; 97: 153-79.

[7] Simonyi A, Schachtman TR, Christoffersen GR. The role of metabotropic glutamate receptor 5 in learning and memory processes. Drug News Perspect 2005; 18: 353-61.

[8] Blaylock RL, Strunecka A. Immune-glutamatergic dysfunction as a central mechanism of the autism spectrum disorders. Curr Med Chem 2009; 16: 157-70.

[9] Beattie EC, Carroll RC, Yu X, *et al.* Regulation of AMPA receptor endocytosis by a signaling mechanism shared with LTD. Nat Neurosci 2000; 3: 1291-300.

[10] De Blasi A, Conn PJ, Pin J, Nicoletti F. Molecular determinants of metabotropic glutamate receptor signaling. Trends Pharmacol Sci 2001; 22: 114-20.

[11] Lavreysen H, Dautzenberg FM. Therapeutic potential of group III metabotropic glutamate receptors. Curr Med Chem 2008; 15: 671-84.

[12] Sladeczek F, Pin JP, Recasens M, Bockaert J, Weiss S. Glutamate stimulates inositol phosphate formation in striatal neurones. Nature 1985; 317: 717-9.

[13] Collingridge GL, Olsen RW, Peters J, Spedding M. A nomenclature for ligand-gated ion channels. Neuropharmacology 2009; 56: 2-5.

[14]   Dingledine R, Borges K, Bowie D, Traynelis SF. The glutamate receptor ion channels. Pharmacol Rev 1999; 51: 7-61.

[15]   Leveque JC, Macias W, Rajadhyaksha A, *et al.* Intracellular modulation of NMDA receptor function by antipsychotic drugs. J Neurosci 2000; 20: 4011-20.

[16]   Kemp JA, Leeson PD. The glycine site of the NMDA receptor--five years on. Trends Pharmacol Sci 1993; 14: 20-5.

[17]   Morris AJ, Malbon CC. Physiological regulation of G protein-linked signaling. Physiol Rev 1999; 79: 1373-430.

[18]   Berridge MJ, Irvine RF. Inositol trisphosphate, a novel second messenger in cellular signal transduction. Nature 1984; 312: 315-21.

[19]   Irvine RF. Inositide evolution - towards turtle domination? J Physiol 2005; 566: 295-300.

[20]   Riedel G, Wetzel W, Reymann KG. Comparing the role of metabotropic glutamate receptors in long-term potentiation and in learning and memory. Prog Neuropsychopharmacol Biol Psychiatry 1996; 20: 761-89.

[21]   Tsai VW, Scott HL, Lewis RJ, Dodd PR. The role of group I metabotropic glutamate receptors in neuronal excitotoxicity in Alzheimer's disease. Neurotox Res 2005; 7: 125-41.

[22]   Kubo Y, Tateyama M. Towards a view of functioning dimeric metabotropic receptors. Curr Opin Neurobiol 2005; 15: 289-95.

[23]   Tateyama M, Kubo Y. Coupling profile of the metabotropic glutamate receptor 1alpha is regulated by the C-terminal domain. Mol Cell Neurosci 2007; 34: 445-52.

[24]   Martin-Ruiz R, Ugedo L, Honrubia MA, Mengod G, Artigas F. Control of serotonergic neurons in rat brain by dopaminergic receptors outside the dorsal raphe nucleus. J Neurochem 2001; 77: 762-75.

[25]   Hawkins RA. The blood-brain barrier and glutamate. Am J Clin Nutr 2009; 90: 867S-74S.

[26]   Seal RP, Amara SG. Excitatory amino acid transporters: a family in flux. Annu Rev Pharmacol Toxicol 1999; 39: 431-56.

[27]   Leighton BH, Seal RP, Shimamoto K, Amara SG. A hydrophobic domain in glutamate transporters forms an extracellular helix associated with the permeation pathway for substrates. J Biol Chem 2002; 277: 29847-55.

[28]   Susarla BT, Seal RP, Zelenaia O, *et al.* Differential regulation of GLAST immunoreactivity and activity by protein kinase C: evidence for modification of amino and carboxyl termini. J Neurochem 2004; 91: 1151-63.

[29]   Zou J, Wang YX, Dou FF, *et al.* Glutamine synthetase down-regulation reduces astrocyte protection against glutamate excitotoxicity to neurons. Neurochem Int 2010; 56: 577-84.

[30]   Kvamme E, Svenneby G, Hertz L, Schousboe A. Properties of phosphate activated glutaminase in astrocytes cultured from mouse brain. Neurochem Res 1982; 7: 761-70.

[31]   Hogstad S, Svenneby G, Torgner IA, *et al.* Glutaminase in neurons and astrocytes cultured from mouse brain: kinetic properties and effects of phosphate, glutamate, and ammonia. Neurochem Res 1988; 13: 383-8.

[32]   Ribak CE, Vaughn JE, Saito K. Immunocytochemical localization of glutamic acid decarboxylase in neuronal somata following colchicine inhibition of axonal transport. Brain Res 1978; 140: 315-32.

[33]   Liang SL, Carlson GC, Coulter DA. Dynamic regulation of synaptic GABA release by the glutamate-glutamine cycle in hippocampal area CA1. J Neurosci 2006; 26: 8537-48.

[34]   Akbarian S, Kim JJ, Potkin SG, *et al.* Gene expression for glutamic acid decarboxylase is reduced without loss of neurons in prefrontal cortex of schizophrenics. Arch Gen Psychiatry 1995; 52: 258-66.

[35]   Woo TU, Walsh JP, Benes FM. Density of glutamic acid decarboxylase 67 messenger RNA-containing neurons that express the N-methyl-D-aspartate receptor subunit NR2A in the anterior cingulate cortex in schizophrenia and bipolar disorder. Arch Gen Psychiatry 2004; 61: 649-57.

[36]   Benes FM, Lim B, Matzilevich D, *et al.* Regulation of the GABA cell phenotype in hippocampus of schizophrenics and bipolars. Proc Natl Acad Sci U S A 2007; 104: 10164-9.

[37]   Olney JW, Ho OL, Rhee V, DeGubareff T. Letter: Neurotoxic effects of glutamate. N Engl J Med 1973; 289: 1374-5.

[38]   Lehmann C, Bette S, Engele J. High extracellular glutamate modulates expression of glutamate transporters and glutamine synthetase in cultured astrocytes. Brain Res 2009; 1297: 1-8.

[39]   Oberheim NA, Takano T, Han X, *et al.* Uniquely hominid features of adult human astrocytes. J Neurosci 2009; 29: 3276-87.

[40]   Vesce S, Rossi D, Brambilla L, Volterra A. Glutamate release from astrocytes in physiological conditions and in neurodegenerative disorders characterized by neuroinflammation. Int Rev Neurobiol 2007; 82: 57-71.

[41]   Araque A. Astrocytes process synaptic information. Neuron Glia Biol 2008; 4: 3-10.

[42]   Perea G, Navarrete M, Araque A. Tripartite synapses: astrocytes process and control synaptic information. Trends Neurosci 2009; 32: 421-31.

[43]   Agulhon C, Petravicz J, McMullen AB, *et al.* What is the role of astrocyte calcium in neurophysiology? Neuron 2008; 59: 932-46.

[44]   Fiacco TA, Agulhon C, McCarthy KD. Sorting out astrocyte physiology from pharmacology. Annu Rev Pharmacol Toxicol 2009; 49: 151-74.

[45] Simard M, Arcuino G, Takano T, Liu QS, Nedergaard M. Signaling at the gliovascular interface. J Neurosci 2003; 23: 9254-62.

[46] Wang X, Takano T, Nedergaard M. Astrocytic calcium signaling: mechanism and implications for functional brain imaging. Methods Mol Biol 2009; 489: 93-109.

[47] D'Antoni S, Berretta A, Bonaccorso CM, *et al.* Metabotropic glutamate receptors in glial cells. Neurochem Res 2008; 33: 2436-43.

[48] Lavenex P, Sugden SG, Davis RR, Gregg JP, Lavenex PB. Developmental regulation of gene expression and astrocytic processes may explain selective hippocampal vulnerability. Hippocampus 2009;

[49] Blaylock R. Chronic microglial activation and excitotoxicity secondary to excessive immune stimulation: possible factors in Gulf War Syndrome and autism. J Am Phys Surg 2004; 9: 46-51.

[50] Blaylock RL. A possible central mechanism in autism spectrum disorders, part 1. Altern Ther Health Med 2008; 14: 46-53.

[51] Blaylock RL. A possible central mechanism in autism spectrum disorders, part 2: immunoexcitotoxicity. Altern Ther Health Med 2009; 15: 60-7.

[52] Blaylock RL. A possible central mechanism in autism spectrum disorders, part 3: the role of excitotoxin food additives and the synergistic effects of other environmental toxins. Altern Ther Health Med 2009; 15: 56-60.

[53] Carlsson ML. Hypothesis: is infantile autism a hypoglutamatergic disorder? Relevance of glutamate - serotonin interactions for pharmacotherapy. J Neural Transm 1998; 105: 525-35.

[54] Aldred S, Moore KM, Fitzgerald M, Waring RH. Plasma amino acid levels in children with autism and their families. J Autism Dev Disord 2003; 33: 93-7.

[55] Shinohe A, Hashimoto K, Nakamura K, *et al.* Increased serum levels of glutamate in adult patients with autism. Prog Neuropsychopharmacol Biol Psychiatry 2006; 30: 1472-7.

[56] Hashimoto K, Shinohe, A., Mori, N. Reply to: The hyperglutamatergic hypothesis of autism. Progress in Neuropsychopharmacology and Biological Psychiatry 2007; Nov 13: Epub.

[57] Laurence JA, Fatemi SH. Glial fibrillary acidic protein is elevated in superior frontal, parietal and cerebellar cortices of autistic subjects. Cerebellum 2005; 4: 206-10.

[58] Fatemi SH. The hyperglutamatergic hypothesis of autism. Prog Neuropsychopharmacol Biol Psychiatry 2008; 32: 911, author reply 2-3.

[59] Olney JW. New insights and new issues in developmental neurotoxicology. Neurotoxicology 2002; 23: 659-68.

[60] Folstein SE, Rosen-Sheidley B. Genetics of autism: complex aetiology for a heterogeneous disorder. Nat Rev Genet 2001; 2: 943-55.

[61] Chakrabarti B, Dudbridge F, Kent L, *et al.* Genes related to sex steroids, neural growth, and social-emotional behavior are associated with autistic traits, empathy, and Asperger syndrome. Autism Res 2009; 2: 157-77.

[62] Guilmatre A, Dubourg C, Mosca AL, *et al.* Recurrent rearrangements in synaptic and neurodevelopmental genes and shared biologic pathways in schizophrenia, autism, and mental retardation. Arch Gen Psychiatry 2009; 66: 947-56.

[63] Fatemi SH, Halt AR, Stary JM, *et al.* Glutamic acid decarboxylase 65 and 67 kDa proteins are reduced in autistic parietal and cerebellar cortices. Biol Psychiatry 2002; 52: 805-10.

[64] Fatemi SH, Araghi-Niknam M, Laurence JA, *et al.* Glial fibrillary acidic protein and glutamic acid decarboxylase 65 and 67 kDa proteins are increased in brains of neonatal BALB/c mice following viral infection in utero. Schizophr Res 2004; 69: 121-3.

[65] Yip J, Soghomonian JJ, Blatt GJ. Decreased GAD67 mRNA levels in cerebellar Purkinje cells in autism: pathophysiological implications. Acta Neuropathol 2007; 113: 559-68.

[66] Peedicayil J, Thangavelu P. Purkinje cell loss in autism may involve epigenetic changes in the gene encoding GAD. Med Hypotheses 2008; 71: 978.

[67] Rout UK, Dhossche DM. A pathogenetic model of autism involving Purkinje cell loss through anti-GAD antibodies. Med Hypotheses 2008; 71: 218-21.

[68] Fatemi SH, Reutiman TJ, Folsom TD. Chronic psychotropic drug treatment causes differential expression of Reelin signaling system in frontal cortex of rats. Schizophr Res 2009; 111: 138-52.

[69] Jamain S, Betancur C, Quach H, *et al.* Linkage and association of the glutamate receptor 6 gene with autism. Mol Psychiatry 2002; 7: 302-10.

[70] Serajee FJ, Zhong H, Nabi R, Huq AH. The metabotropic glutamate receptor 8 gene at 7q31: partial duplication and possible association with autism. J Med Genet 2003; 40: e42.

[71] Malherbe P, Kratzeisen C, Lundstrom K, *et al.* Cloning and functional expression of alternative spliced variants of the human metabotropic glutamate receptor 8. Brain Res Mol Brain Res 1999; 67: 201-10.

[72]   Serajee FJ, Nabi R, Zhong H, Mahbubul Huq AH. Association of INPP1, PIK3CG, and TSC2 gene variants with autistic disorder: implications for phosphatidylinositol signalling in autism. J Med Genet 2003; 40: e119.

[73]   Berridge MJ. Unlocking the secrets of cell signaling. Annu Rev Physiol 2005; 67: 1-21.

[74]   Strunecka A, Ripova D. What can the investigation of phosphoinositide signaling system in platelets of schizophrenic patients tell us? Prostaglandins Leukot Essent Fatty Acids 1999; 61: 1-5.

[75]   Strunecká A, Patočka J. Aluminofluoride Complexes in the Etiology of Alzheimer´s disease. In: Atwood D, Roesky C, editors. Structure and Bonding. New Developments in Biological Aluminum Chemistry. Germany: Springer-Verlag; 2003. p. 139-81.

[76]   Ripova D, Platilova V, Strunecka A, Jirak R, Hoschl C. Alterations in calcium homeostasis as biological marker for mild Alzheimer's disease? Physiol Res 2004; 53: 449-52.

[77]   Purcell AE, Jeon OH, Zimmerman AW, Blue ME, Pevsner J. Postmortem brain abnormalities of the glutamate neurotransmitter system in autism. Neurology 2001; 57: 1618-28.

[78]   Ramanathan S, Woodroffe A, Flodman PL, *et al.* A case of autism with an interstitial deletion on 4q leading to hemizygosity for genes encoding for glutamine and glycine neurotransmitter receptor sub-units (AMPA 2, GLRA3, GLRB) and neuropeptide receptors NPY1R, NPY5R. BMC Med Genet 2004; 5: 10.

[79]   Ramoz N, Reichert JG, Smith CJ, *et al.* Linkage and association of the mitochondrial aspartate/glutamate carrier SLC25A12 gene with autism. Am J Psychiatry 2004; 161: 662-9.

[80]   Ramoz N, Cai G, Reichert JG, Silverman JM, Buxbaum JD. An analysis of candidate autism loci on chromosome 2q24-q33: evidence for association to the STK39 gene. Am J Med Genet B Neuropsychiatr Genet 2008; 147B: 1152-8.

[81]   Segurado R, Conroy J, Meally E, *et al.* Confirmation of association between autism and the mitochondrial aspartate/glutamate carrier SLC25A12 gene on chromosome 2q31. Am J Psychiatry 2005; 162: 2182-4.

[82]   Szatmari P, Paterson AD, Zwaigenbaum L, *et al.* Mapping autism risk loci using genetic linkage and chromosomal rearrangements. Nat Genet 2007; 39: 319-28.

[83]   Kubo T, Kohira R, Okano T, Ishikawa K. Neonatal glutamate can destroy the hippocampal CA1 structure and impair discrimination learning in rats. Brain Res 1993; 616: 311-4.

[84]   Dubovicky M, Tokarev D, Skultetyova I, Jezova D. Changes of exploratory behaviour and its habituation in rats neonatally treated with monosodium glutamate. Pharmacol Biochem Behav 1997; 56: 565-9.

[85]   Moreno-Fuenmayor H, Borjas L, Arrieta A, Valera V, Socorro-Candanoza L. Plasma excitatory amino acids in autism. Invest Clin 1996; 37: 113-28.

[86]   Droge W, Eck HP, Naher H, Pekar U, Daniel V. Abnormal amino-acid concentrations in the blood of patients with acquired immunodeficiency syndrome (AIDS) may contribute to the immunological defect. Biol Chem Hoppe Seyler 1988; 369: 143-8.

[87]   Hack V, Stutz O, Kinscherf R, *et al.* Elevated venous glutamate levels in (pre)catabolic conditions result at least partly from a decreased glutamate transport activity. J Mol Med 1996; 74: 337-43.

[88]   Olney JW, Ho OL. Brain damage in infant mice following oral intake of glutamate, aspartate or cysteine. Nature 1970; 227: 609-11.

[89]   Olney JW, Ho OL, Rhee V. Cytotoxic effects of acidic and sulphur containing amino acids on the infant mouse central nervous system. Exp Brain Res 1971; 14: 61-76.

[90]   Olney JW, Sharpe LG, Feigin RD. Glutamate-induced brain damage in infant primates. J Neuropathol Exp Neurol 1972; 31: 464-88.

[91]   Olney JW, Ho OL, Rhee V. Brain-damaging potential of protein hydrolysates. N Engl J Med 1973; 289: 391-5.

[92]   Olney JW. Excitatory transmitter neurotoxicity. Neurobiol Aging 1994; 15: 259-60.

[93]   Olney JW. Excitotoxins in foods. Neurotoxicology 1994; 15: 535-44.

[94]   Olney JW. Glutamate, a neurotoxic transmitter. J Child Neurol 1989; 4: 218-26.

[95]   Plaitakis A, Flessas P, Natsiou AB, Shashidharan P. Glutamate dehydrogenase deficiency in cerebellar degenerations: clinical, biochemical and molecular genetic aspects. Can J Neurol Sci 1993; 20 Suppl 3: S109-16.

[96]   Yu T, Zhao Y, Shi W, Ma R, Yu L. Effects of maternal oral administration of monosodium glutamate at a late stage of pregnancy on developing mouse fetal brain. Brain Res 1997; 747: 195-206.

[97]   Frieder B, Grimm VE. Prenatal monosodium glutamate (MSG) treatment given through the mother's diet causes behavioral deficits in rat offspring. Int J Neurosci 1984; 23: 117-26.

[98]   Frieder B, Grimm VE. Prenatal monosodium glutamate causes long-lasting cholinergic and adrenergic changes in various brain regions. J Neurochem 1987; 48: 1359-65.

[99]   Bawari M, Babu GN, Ali MM, Misra UK. Effect of neonatal monosodium glutamate on lipid peroxidation in adult rat brain. Neuroreport 1995; 6: 650-2.

[100] Beas-Zarate C, Perez-Vega M, Gonzalez-Burgos I. Neonatal exposure to monosodium L-glutamate induces loss of neurons and cytoarchitectural alterations in hippocampal CA1 pyramidal neurons of adult rats. Brain Res 2002; 952: 275-81.

[101] Olney JW. Brain lesions, obesity, and other disturbances in mice treated with monosodium glutamate. Science 1969; 164: 719-21.

[102] Davies LP, Johnston GA. Uptake and release of D- and L-aspartate by rat brain slices. J Neurochem 1976; 26: 1007-14.

[103] Ishida AT, Fain GL. D-aspartate potentiates the effects of L-glutamate on horizontal cells in goldfish retina. Proc Natl Acad Sci U S A 1981; 78: 5890-4.

CHAPTER 4

Immunoexcitotoxicity as a Central Mechanism of Autism Spectrum Disorders

Russell L. Blaylock, MD

Institute for Theoretical Neuroscience, LLC, and Visiting Professor of Biology, Belhaven University, Ridgeland, MS 39157, USA

Abstract: Autism has undergone a tremendous amount of study, and a number of often seemingly unconnected disorders have been disclosed. Yet, despite an enormous amount of study, no central mechanism to explain the causation of this syndrome or why it affects only a subset of children has come forth. In this chapter I propose such a central mechanism that explains a great number of biochemical, histological, neurodevelopmental and systemic dysfunctions, as well as behavioral findings in autism spectrum disorders (ASD). Since the discovery of excitotoxicity by Olney in 1968, neuroscientists have determined that not only is glutamate a neurotransmitter, but it is the most abundant neurotransmitter in the brain, exceeding the more traditional neurotransmitters combined. Recent studies have also shown that glutamatergic receptors (GluRs) interact with other receptors, not only neurotransmitters, but also immune receptors, in a way that can alter their sensitivity. Chronic brain inflammation is known to dramatically enhance the sensitivity of N-methyl-D-aspartic acid (NMDA) and α-amino-3-hydroxy-5-methyl-4-isoxazole propionic acid (AMPA) type GluRs and interfere with glutamate removal from the extraneuronal space, where it can trigger excitotoxicity and abnormal synaptic and dendritic physiology over a prolonged period. Importantly, neuroscience studies have clearly shown that sequential systemic immune stimulation can not only activate the brain's immune system, microglia and astrocytes, but that there occurs an amplified response to both subsequent stimulation, either systemic or within the CNS. The ASD child is exposed to such sequential immune stimulation via a growing number of vaccines, recurrent infections, chemical toxins and persistent viral infections.

INTRODUCTION

In 1957, Lucas and Newhouse made the observation that feeding monosodium glutamate to newborn rats caused widespread destruction of the inner ganglion layer of the retina [1]. Ten years later, Dr. John Olney repeated the study and discovered that not only was the retina affected but also select nuclei, primarily the arcuate nucleus of the hypothalamus, of newborn rats were also severely damaged [2]. Exposing neurons to elevated levels of glutamate triggered a delayed neuronal death, characterized by excessive neuronal firing rates and release of apoptosis factors. Based on the excitation of the exposed neurons, he coined the name excitotoxin to describe the agents and excitotoxicity to describe the process.

Since this early discovery, researchers have carefully worked out the mechanism of excitotoxicity and discovered a series of glutamate receptors (GluRs) with complex structures and intimate interactions among themselves and other neurotransmitters and immune receptors. Over 50% of the brain's neurotransmission is via glutamatergic neurotransmission and 90% of cortical neurons utilize glutamate as a neurotransmitter.

During embryonal life taurine is the most abundant neurotransmitter, whereas postnatally, glutamate becomes dominant [3]. Activation of GluRs plays a critical role in neuronal and glial cell migration, differentiation and maturation and is essential for pruning of excessive synaptic connection and dendritic processes during development. Disruption of glutamate content in the brain, either too low or too high, can alter the brain's cortical laminar development, leading to disorganized neural elements in the various layers of the cortex and abnormal organization of the hypothalamus [4].

Recently, a number of mechanisms linking immune pro-inflammatory activation with excitotoxicity have been shown. Studies have demonstrated that excitotoxicity can trigger immune activation and immune activation can, as well, trigger excitotoxicity either focally or widespread throughout the brain. It appears that the two processes are

---

*Address correspondence to: **Russell L. Blaylock** Institute for Theoretical Neuroscience, LLC, and Visiting Professor of Biology, Belhaven University, Ridgeland, MS 39157, USA; E-mail: Blay6307@bellsouth.net

always linked and one usually does not occur without the other, even though, in special situations, excitotoxicity is the dominant process.

While one speaks of immunoexcitotoxicity, in fact, it may better be characterized as immuno-glutamatergic dysfunction. The term immunoexcitotoxicity was coined by the author in an article written to explain a mechanism responsible for the development of ASD. Intimately linked with this process is the activation of brain microglia and infiltration of the brain by monocytes/macrophage cells from the periphery.

FETAL IMMUNE RESPONSES AND MICROGLIAL DEVELOPMENT

Microglia make their first appearance around 4.5 weeks of gestation in the human brain, mainly appearing near the meninges and choroid plexus [5]. From the second trimester onward, microglia are widely distributed throughout the developing brain and appear in a ramified state, with significant downregulation of their surface immune receptors. They are more numerous in the white matter at this stage [6]. Later, during gestation, they assume an amoeboid morphology, again not expressing immune mediators to any degree, yet they have the ability to do so under challenge.

Lee and co-workers, for example, examined production of three cytokines, interleukins (IL) IL-1ß, IL-6 and tumor necrosis factor-α (TNF-α) in 2nd trimester human fetal microglia and found that lipopolysaccharide (LPS) stimulation increased upregulation of mRNA for IL-1ß, IL-6 and TNF-α [7]. Fetal responses to immune mediators, while basically the same as in the adult brain can have unique effects, depending on the stage of development and the presence of other stimuli, such as excess release of glutamate and other excitatory molecules.

Using human fetal neurons in culture, Chao and co-workers found that IL-1ß and TNF-α were not neurotoxic, but in combination were [8]. Part of the neurotoxicity can be explained by nitric oxide (NO) release from surrounding astrocytes. In addition, in combination, they were found to inhibit glutamate uptake and suppress glutamine synthetase (GS) function, both of which serve to raise glutamate to excitotoxic levels.

IL-1ß is the major stimulus for microglial activation and once activated microglia secrete other pro-inflammatory cytokines. IL-1ß appears to activate microglia by way of MAPK signaling. It has been shown that in the presence of excitotoxic levels of glutamate, IL-1ß markedly enhances neuronal damage [9]. High levels of TNF-α have also been shown to inhibit neurite outgrowth and branching of hippocampal neurons [10].

During development, microglia enter the brain mainly via the meninges and choroid plexus [11]. There is evidence that the circumventricular organs (CVO) represent sites of entry in the adult brain, but less is known concerning macrophage entry during development. These are areas of slow blood flow and an absent blood-brain barrier (BBB).

Upon entry, the microglia assume an amoeboid morphology, allowing migration along axons to the cortical and subcortical areas, moving along tangential and radial migrational routes [12,13]. Movement is accomplished by extension and retraction of the microglial lamellipodia in the direction of the migration.

Microglia Activation

Microglia are considered to be the resident immune cell of the CNS and normally exist in a resting state called ramified. Actually, they are not resting and recent studies using confocal microscopy of living microglia demonstrate a constant activity, with extension and retraction of pseudopodia. It is also known that in a ramified state the microglia secrete a number of trophic growth factors.

Characteristic of microglia is the presence of pattern recognition receptors (PRR), which are constitutively expressed to identify and bind pathogens, utilizing pathogen-associated molecular patterns (PAMPS) [14]. Other, non-microbial stimuli can also activate these receptors and initiate microglial activation. Recently, researchers have identified toll-like receptors (TLR) that recognize these stimuli [15]. Thus far 12 TLR have been identified in immune cells, with TLR 1-9 being identified in microglia [16]. These receptors recognize viruses, fungi, bacteria, parasites and self-proteins. The TLR-4 is activated by LPS and is rapidly upregulated with brain inflammation [17,18]. TLR not only recognize microbial components, evoke inflammation and immune responses, but also regulate the magnitude and duration of the immune reaction. TLR mediate responses to the host molecules, including ROS. An unbalanced relationship between TNF-α, IL-10, and TLR signaling would fail to mitigate inflammation and oxidative stress.

Activation of TLR 2, TLR 4 and TLR 9 induce microglial NO production and TLR-4 can also have beneficial effects such as repair and remyelination. Microglia also contain scavenger receptors, which can recognize modified lipoproteins and various polyanionic ligands. They also play a role in phagocytosis. Class B scavenger receptor CD36 mediates free radical production and tissue injury in cerebral ischemia [19]. Activation of these receptors can result in ligand internalization and/or production of superoxide by microglia. This can increase peroxynitrite generation as superoxide combines with NO, which is also produced by activated microglia.

Microglia also have PRR that recognize advanced glycation end products via RAGE receptors [20]. Most important is the macrophage antigen complex-1 (MAC1) receptor, integrin CD11b/CD18, which functions both as an adhesion molecule and as a PRR [21]. MAC1 is essential for initiation of phagocytosis in response to a diverse number of stimuli. It is also associated with respiratory burst activation from both neutrophils and macrophages. There is evidence that it plays a key role in microglial oxidative stress production, with elevated production of the superoxide radical.

Activation of PRR can be quite complex and newer evidence indicates that neurotoxins might interact with multiple PRR as receptor complexes [14]. For example, MAC1, SR-A1/2 and Fcγ receptors are known to mediate phagocytosis of degenerating myelin by macrophages and microglia and that axonal injury initiates MAC1 and SR-A1/2 but not Fcγ receptors [22]. Recognition of various ligands by the PRR commonly initiate the generation and release of superoxide. Superoxide is produced by activation of NADPH oxidase, which is assembled and translocated to the membrane of the microglia during activation. While several PRR can activate NADPH oxidase, MAC1 may be crucial [23]. It is known that NADPH oxidase is crucial to a number of neurotoxic and neurodestructive processes and is intimately connected to glutamate excitotoxicity. It is also crucial to microglial signaling associated with changes in microglial morphology and proliferation [24, 25].

It is also known that intracellular reactive oxygen species (ROS)/ reactive nitrogen species (RNS) dramatically amplify the inflammatory response of microglia [26]. Alterations in ROS/RNS levels from NADPH oxidase activation have been shown to prime microglia, thus making them hyperresponsive to subsequent stimulation [27]. Both pro-inflammatory cytokines/chemokines and excitotoxicity dramatically increase intracellular ROS/RNS, thus creating a self-generating cascade.

Hypoxia/ischemia in conjunction with excitotoxicity has been shown to produce a robust microglial activation response in the neonatal rat brain [28]. Lesioning with *N*-methyl-D-aspartic acid (NMDA), an agonist for the NMDA receptor, has been shown to rapidly elevate IL-1ß and TNF-α gene expression [29]. Both have been shown to induce expression of macrophage inflammatory protein MIP-1α and MIP-1ß in human fetal microglia.

Another link between immune activation and excitotoxicity occurs in conjunction with kynurenine metabolism during inflammation. During inflammatory conditions in the brain, one sees an increase in kynurenine levels in response to interferon-C (INF-C), which upregulates indolamine-2,3-dioxygenase (IDO) [30]. This, in turn, increases the production of quinolinic acid (QUIN). QUIN, a tryptophan metabolite, is a powerful excitotoxins, acting through NMDA receptors.

The highest level of QUIN production during inflammatory states is from macrophages, with less from microglia. Fetal brain cultures have been shown to produce less QUIN because of the lower percentage of microglia [30]. QUIN elevations have been observed in a number of human inflammatory brain conditions [31-34].

Interestingly, in mice and gerbils, one sees large increases in IDO and QUIN in the brain after systemic inflammation or systemic immune stimulation using endotoxin, pokeweed mitogen or interferon [35]. Ironically, in this study, they did not see macrophage infiltrates, meaning that most of the QUIN was generated from microglia. In this same study they demonstrated QUIN production in human fetal microglia as well, but at significantly lower levels than adult microglia and macrophages. This is most likely due to the extremely low levels of kynurenine 3-hydroxylase and kynureninase in fetal microglia.

Importantly, studies have shown that brain levels of QUIN can be elevated without concomitant elevation of CSF or blood QUIN levels [36,37]. Studies of autistic children using CSF and blood, found QUIN levels to be low, again emphasizing that blood and CSF may not always accurately reflect brain levels of toxins [38]. It could also indicate a loss of enzymes necessary to convert tryptophan into QUIN.

Microglia also contain receptors for several neurotransmitters (glutamate and choline), glutamate transport proteins, interferons, cytokines, chemokines, prostaglandins, purinergic receptors (P2), hormone receptors (estrogen and androgens), thrombin receptors, Fcy receptors, complement receptors and MHC class II receptors [39]. They also secrete a great number of active substances, including NO, ROS/RNS, LPP, cytokines, chemokines, interferons, complement, prostaglandins, acute phase proteins, and proteases. Importantly, microglial activation does not always mean initiation of a neurodestructive mode. There appears to be at least three states of activation: predominantly neuroprotective and trophic, predominantly neurodestructive and an intermediate stage [40]. The particular activation state depends on the activation signals as well as the state of the brain at the time of activation. Also critical, appears to be the length of time the microglia are activated. During most pathological states, especially infectious ones, activation is relatively short lived and then switches to a neurotrophic mode. It is important to appreciate that microglia also play a major role in brain repair and neuroprotection.

Because microglia lie only nano1meters from neurons, they are able to continuously monitor neuronal homeostasis and activity and in doing so they regulate glutamate levels, clear immune antigen particulate matter and supply trophic molecules such as ***brain-derived neurotrophic factor*** (BDNF), neurotrophic factor 3, basic fibroblast growth factor (bFGF), hepatocyte growth factor and plasminogen [41]. There is considerable evidence that microglia play a critical role in synaptic remodeling in the normally developing brain. Proteases play a significant role in this remodeling and include matrix metalloproteases, elastase and plasminogen activator. Microglia by removing pruned elements, such as dying neurons, dendrites and axons, play an essential role in the ongoing development of the brain.

Human microglia constitutively express transcripts for mRNA controlling a wide spectrum of cytokines, both inflammatory and anti-inflammatory. They also secrete a number of chemokines and other chemotactic molecules, such as MIP-1, MIP-1ß and monocytes chemotactic protein-1 (MCP-1). In an activated state, microglia can secrete inflammatory cytokines (IL-1α, IL1ß, IL-6, IL-8, IL-18 and TNF-α), immunomodulatory cytokines (IL-5, IL-12 and IL-15), and anti-inflammatory cytokines (TGFß, IL-4, IL-10, IL-13). By acting through cell signaling, cytokines can act synergistically or as antagonists. Interleukin-1 and IL-18 appear to be major players in neurodestructive reactions and IL-1 becomes especially neurodestructive in the presence of TNF-α [42, 43].

Cytokines can also affect neurotransmission by altering the release of various neurotransmitters and by altering brain neurophysiology [44]. Most studies were done on hypothalamic and hippocampal structures, but they clearly demonstrate that cytokines can affect neurotransmitter release within minutes of application. The response is dose dependent. For example, high dose IL-2 was shown to suppress the release of acetylcholine and lower doses enhanced release from hippocampal slices [45,46]. This has been shown for acetylcholine, noradrenalin, dopamine, serotonin, GABA and glutamate. Sensitivity to IL-2 is also region specific, since some areas of the brain respond strongly, while other respond poorly.

Like glutamate, there is a timed rise and fall in brain cytokines, with significant elevations in pro-inflammatory IL-1, TNF-α and IL-6 during mid- to late gestation, extending into early postnatal development, after which levels begin to rapidly decline [47]. These cytokines have been shown to also play a role in migration, differentiation and maturation of neurons and glia during development [48]. In conjunction with excitatory amino acids, they may play a role in pruning of excessive synaptic, dendritic and axonal connections.

The important role played by immune factors is emphasized by the studies showing that immune stimulation during mid- to later term pregnancy and possible early postnatal development, can dramatically increase the risk of schizophrenia and autism in the offspring [49]. It has been estimated that 14% to 21% of the cases of schizophrenia are caused by maternal infection [50]. The greatest risk is during the 2nd trimester, with a 3-fold to 7-fold increased risk.

Convincing evidence indicates that it is the elevation in immune cytokines, rather than transfer of infectious agents, that is responsible. Using a mouse model, researchers found that by using poly (I:C) a double-stranded RNA molecule or LPS instead of an infectious virus, they could reproduce the clinical syndrome in offspring as well as the histological changes in the brain [51-54]. Further study pinpointed IL-6 as the culprit. Only injections of IL-6, and not IL-1ß, TNF-α or IFN-γ, cause the effect in the offspring [55]. Blocking IL-6 completely abrogated the abnormal behaviors. Studies using radiolabeled IL-6 have shown that it enters the rat fetus only during mid, but not late gestation, which correlates with the clinical findings in humans. It is known that IL-6 plays a major role in brain development, learning and working memory [56].

EXCITOTOXICITY AND NEURODEVELOPMENT

The other major reaction associated with microglial activation is the release of glutamate and other excitatory amino acids from the microglia. By way of intercellular communication utilizing IL-1ß, microglia influence astrocytic activation. Astrocytes are the major site of storage and generation of glutamate, and possibly cytokines.

Brain glutamate levels are controlled by a number of mechanisms, including the glutamate transport proteins (EAATS 1-5), activation states of the enzyme glutaminase, (which converts glutamine into glutamate), and an array of metabolic pathways for glutamate removal. The latter include glutamic acid decarboxylase (GAD) (GABA generation), GS (conversion of glutamate into glutamine) and glutamic acid dehydrogenase (GDH) (entry into Kreb's cycle via α-ketogluterate). Because of the extreme toxicity of extraneuronal glutamate, its levels are carefully monitored and regulated. In addition to controlling glutamate toxicity, the brain must also prevent diffusion into surrounding synapses, which can lead to interference and noise; that is, careful control of extraneuronal glutamate levels sharpens the signal.

Glutamate transmission is via a number of GluRs, which include three ionic GluRs and three groups of metabotropic receptors, each composed of an arrangement of eight subtype components. The principle ionic GluRs are named according to their principle agonist, NMDA, AMPA, and kainate. The NMDA receptors are composed of an arrangement of receptor subtypes, which include NR1, NR2$_{A-D}$ and NR3 subunits [57]. All NMDA receptors contain the NR1 subunit and three of the other subunit types. The AMPA receptors contain an assortment of subunits GluR1-4; the kainate receptors contain GluR5-7 subunits and KA-1 and KA-2 subunits. Function of the various NMDA receptors depends on the particular subunit composition within each of the NMDA receptors. NMDA receptors are found mainly postsynaptically, but have been identified presynaptically as well, primarily with GABAergic neurons [58].

AMPA receptors are faster conducting than NMDA receptors and are activated before the NMDA receptor. Synapses not containing AMPA receptors are inactive and referred to as silent synapses. AMPA receptors exist in axon membranes extrasynaptically as well as in oligodendrocytes and excess activation can result in excitotoxic damage to axons [59]. In general, they do not enhance calcium entry into the neuron, but those lacking the GluR2 have enhanced calcium permeability [60]. Because of the extensive presence of AMPA/kainate receptors throughout the cerebral and cerebellar white matter, excessive activation of these receptors can play a major role in white matter injury and a loss of connectivity seen in ASD.

The metabotropic GluRs (mGluRs) are significantly more complex and also play a major role in brain development and function. These receptors are divided into three functional groups I-III. Group I mGluRs contain mGluR1 and 5 and signal through phospholipase C (PLC), 1,2 – diacylglycerol (DAG) and inositol 1,4,5-trisphosphate (IP$_3$), initiating release of calcium from endoplasmic reticulum stores. Group II mGluRs contains mGluR2 and 3 and Group III mGluR4,6,7,8 subunits. Both Group II and III negatively regulate adenylyl cyclase (AC) and reduce cAMP.

Functional GluRs appear early during gestation, and by the second trimester are expressed throughout the brain [61]. Glutamate has been shown to play a major role in neural and glial migration, maturation and architectonic arrangement of the brain. Excessive glutamate stimulation during critical stages of brain development can cause significant disruption of brain architecture. Marret and co-workers [4], using the NMDA receptor agonist ibotenate found that excitotoxicity during brain development could cause arrested neural migration with the development of intracortical and molecular heterotopias, subcortical and intracortical arrest of migration and ectopias in the molecular layer of the neocortex. At higher doses they found periventricular and band heterotopias. The effects on brain architecture and migration patterns were dose dependent. Excitotoxicity produced arrest at all levels of the radial migratory corridors, that is, the germinative zone, white matter, cortical plate and molecular layer.

Many of these changes are seen in the more severely affected ASD cases, particularly involving the brainstem and cerebellum, but also minicolumn architecture in the frontal cortex [62, 63]. The final effect of the glutamate excess will depend on the concentration, stage of development, competence of anti-oxidant defense mechanisms and state of immune activation within the CNS.

Komuro and Rakic demonstrated that fluctuations in intracellular calcium secondary to voltage-gated NMDA receptor activation controls the speed of granule cell migration, with calcium peaks speeding up migration and troughs slowing migration [61]. It was also found that termination of migration was controlled by a fall in intracellular calcium [64].

It has been shown that glutamate transport proteins play a major role in brain development and that fluctuations in expression are vital to the formation of brain architecture [65]. These transport proteins are quite redox sensitive and are suppressed by ROS/RNS, 4-hydroxynonenal (4-HNE) and pro-inflammatory cytokines. Certain toxins can cause a reversal of transport, thus increasing extracellular glutamate levels.

The mGluRs also play a major role in brain development and do so in very complex ways. For example, specific mGluRs subtypes are necessary for pruning of excessive innervation of Purkinje cells by excitatory climbing fibers [66]. In the adult cerebellum there is a one to one ratio of climbing fibers to Purkinje cells. Loss of this metabotropic influence results in multiply innervated Purkinje cells and dysfunction of the cerebellum.

Studies of the development of thalamic innervation in the rat indicates that during the first postnatal week NMDA receptors make up most of the corticothalamic synaptic connections and that after two weeks its shifts to AMPA/kainate and metabotropic glutamate innervation [67]. Kainate receptor gene *GluR5* mRNA expression peaks at birth in the thalamus and then declines during the first two postnatal weeks. It is absent in the adult dorsal thalamus. Only GluR7 of the kainate receptors remain until adulthood [68].

Metabotropic GluR1 appears in the thalamic relay nuclei during first postnatal week and increases throughout development, whereas mGluR5 is present at birth in both the midbrain and thalamus and peaks at P7, after which it declines dramatically [69]. Metabotropic GluRs are heterogenously distributed in the developing and adult brain and demonstrate temporal expression profiles. Development of these particular receptors depends on corticothalamic stimulation, meaning that excessive stimulation, as with excitotoxicity, can alter receptor dynamics and morphology.

Metabotropic subtypes mGluR1, mGluR2 and mGluR4 mRNA expression is low at birth and progressively increases during postnatal development, whereas mGluR3 and mGluR5 are highly expressed at birth and decrease as the brain matures [69]. It is known that Group I mGluRs regulate proliferation, differentiation and survival of neuronal stem cells and progenitor cells. There is also evidence that mGluR5 may be subject to exaggerated activation and may be responsible for synaptic dysfunction in Fragile-X syndrome, which is a condition associated with elevated brain glutamate levels [70,71].

Another, less obvious effect of overstimulation of mGluRs, involves one of the brain's most important protectant molecules, melatonin. Norepinephrine receptors innervating pinealocytes stimulate the release of melatonin by enhancing cAMP production. Normally, pinealocytes also secrete glutamate and this suppresses further melatonin secretion, a possible feedback mechanism. Yamada and co-workers, using rat pinealocytes, demonstrated that suppression of melatonin secretion occurred when mGluR3 of Group II mGluRs were activated. Group II mGluRs negatively modulate cAMP, explaining the link to melatonin suppression [72]. Metabotropic GluRs require a higher concentration of glutamate to become activated than do ionotropic GluRs, which protects neurotransmission. In general, Group I mGluRs enhance excitotoxicity and Group II and III are mostly neuroprotective. Neuroprotection in most instances involves presynaptic mGluR innervation of GABAergic interneurons.

Several studies have found low melatonin levels in autistic patients and one study suggest that a genetic defect in the *ASMT* gene exists in autistics, which controls the final enzyme in melatonin synthesis [73, 74]. Melatonin not only plays a critical role in the sleep cycle, it is also a powerful antioxidant and increases brain antioxidant enzyme levels [75]. Other studies have shown that melatonin offers several layers of protection against mitochondrial dysfunction and ROS/RNS generation [76]. There is also evidence that it alters NMDA receptor subunit composition and hence function and promotes neurogenesis within the dentate gyrus [77]. Of particular importance in ASD is the finding that melatonin reduces microglial activation, stimulates oligodendroglial maturation (but not proliferation) and prevents demyelination [78].

Besides the effects of excess excitatory transmitters on brain development, they can also trigger destructive reactions that result in a loss of synapses, cause dendritic retraction, axonal injury, damage to astrocytes and cause the generation of high levels of RNS/ROS throughout the brain. A number of lipid peroxidation products are associated with excitotoxicity, but especially destructive are acrolein and 4-HNE. Of these destructive products of excitotoxicity, much attention has been directed at peroxynitrite, QUIN and 4-HNE. All are capable to suppressing the glutamate transport proteins (EAAT1-5), thus triggering a cascade effect [79,80].

Peroxynitrite has been shown to enhance excitotoxicity by altering nitration of the NMDA receptor subunits so as to increase glutamate binding [81]. It has also been shown to stimulate aspartate release from cerebellar granule neurons [82]. Besides its effects on glutamate transport proteins, 4-HNE has been shown to interfere with mitochondrial function in synaptosomes and impair signal transduction of muscarinic acetylcholine receptors and mGluRs [80].

Since dramatic generation of numerous ROS/RNS and LPP molecules is an intimate part of pro-inflammatory toxicity and excitotoxicity, it is difficult to separate the effects attributed to oxidative stress from immunoexcitotoxicity. It should also be appreciated that elevated free radical levels in the brain trigger microglial activation and immunoexcitotoxicity.

Another mechanism of excitotoxic injury that is indirect, involves reductions in glutathione. The glutamate/cystine antiporter X_c- is an exchange system involving an exchange of intracellular glutamate for extracellular cystine. Intracellular cystine is broken down into cysteine, which is metabolized to glutathione. High levels of extraneuronal glutamate inhibit the exchange and can lower brain glutathione. Unlike systemic cells, brain neurons depend on the glutamate/cystine antiporter X_c- for glutathione production and not the transsulfuration pathway.

Reduced levels of glutathione greatly increase the vulnerability of neurons and astrocytes to excitotoxicity and oxidative stress. When combined with reduced mitochondrial energy production, low antioxidant enzymes, low glutathione and reduced secretion of melatonin, one can reasonably expect an acceleration of damage to brain's elements. This is consistent with the finding of high levels of oxidative stress in the autistic brain reported by several authors [83-86]. High levels of oxidative stress also impair a number of redox-sensitive enzymes, including glutathione dehydrogenase, GAD and GS, which can result in an elevation of brain glutamate and an excitotoxic imbalance in the glutamate/GABA ratio. When combined with inhibition of glutamate transport and even reverse glutamate transport from astrocytes, one can see high and prolonged elevations in glutamate.

Microglia contain receptors for each of the cytokines, which when activated can increase the secretion of the various cytokines. Evidence indicates that microglial activation takes place through MAPK signaling and the particular state of microglial activation depends on specific cellular signaling. For example, extracellular signal-regulated kinase (ERK) is most responsive to growth factors and phorbol esters, while c-Jun N-terminal kinase/stress activated protein kinase (JNK and p38MAP kinase) is activated by stress signals, such as LPS stimulation [87]. Interferon-γ stimulates microglial phagocytosis, without cytokine release or glutamate release [88].

It has also been shown that non-toxic concentrations of glutamate can induce oligodendrocyte death by sensitizing the cells to complement attack [89]. Complement toxicity is induced by activation of kainate receptors on oligodendrocytes only and not AMPA, NMDA or mGluRs. It is known that blocking GluRs significantly reduces the symptoms and histological damage in experimental autoimmune encephalomyelitis [90,91]. Blocking both pro-inflammatory cytokines and GluRs haltered further neurodestruction and significantly improved symptoms [92]. This emphasizes the importance of the interaction of immune receptor stimulation and GluR activation. White matter contains mostly AMPA and kainate type receptors.

The origin of the neuronal toxicity caused by high, sustained levels of TNF-α has been shown not to stem from TNF-α toxicity itself, but rather from microglial release of excessive glutamate from activated microglia [93]. TNF-α in pure cultures of neurons is non-toxic. In mixed neuronal and microglial cultures TNF-α becomes fully toxic. Likewise, blocking FasL only reduced toxicity slightly, whereas blocking GluRs significantly reduced toxicity.

Taken together, microglial activation can be initiated during a primary brain insult or from a systemic immune stimulus. Within the brain, microglial activation can be initiated by ROS/RNS, LPPs, cytokines, chemokines, interferons, excitatory molecules, and ATP. The phenotypic activation mode depends on temporal factors, synergistic action of triggers, concentrations of excitotoxins and pro-inflammatory cytokines and age. A number of these factors, such as ROS/RNS, LPPs, pro-inflammatory cytokines and prostaglandins can raise glutamate or other excitatory molecule levels to excitatory concentrations by a number of mechanisms, including inhibition of glutamate transport proteins, oxidation of regulatory enzymes (GAD, GD and GS), enhance release of glutamate from microglia and astrocytes, and by changing the sensitivity of the GluRs (Fig. **1**.)

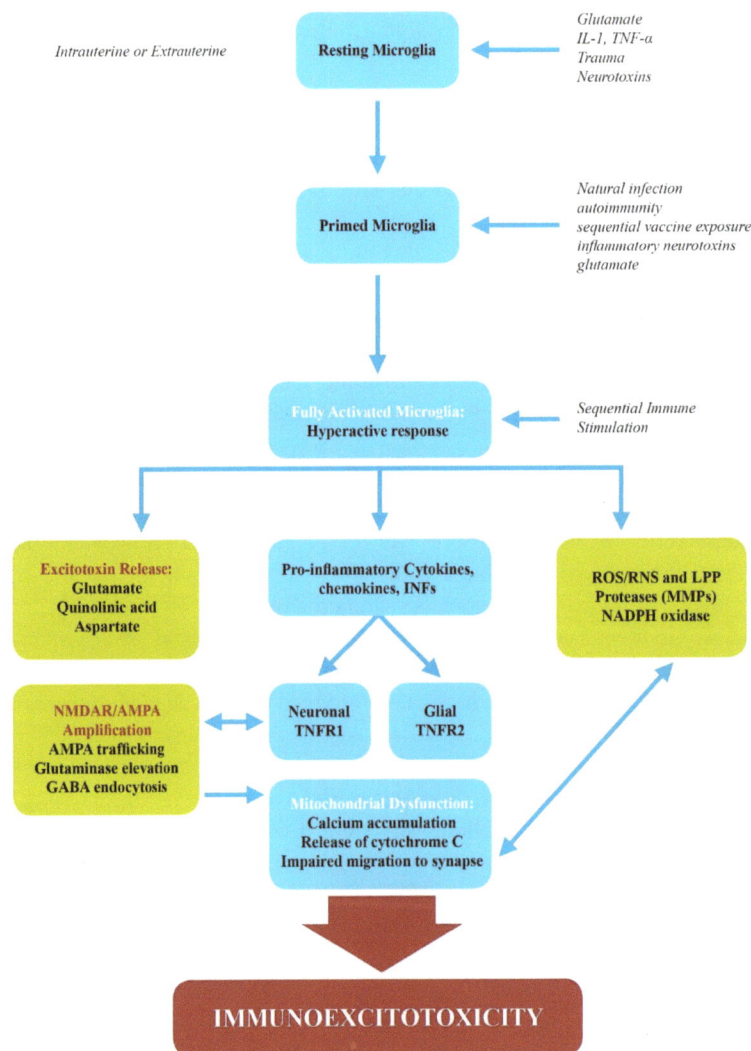

Figure 1: Immunoexcitotoxicity.

Under normal conditions, these processes are quickly terminated and microglial switch to a neurotrophic and neuroprotective mode (ramified), but under pathological conditions, microglia can become chronically activated and neurodestructive over very long periods. For example, it has been shown that a single exposure to 1-methyl-4-phenylpyridinium (MPP⁺) can activate midbrain microglia for a prolonged period [94, 95].

With chronic microglial activation and release of neurotoxic elements during periods of brain development, one would expect abnormalities in pathway development, maturation, migration and function as well as neurodegenerative effects on dendrites, axons and synapses.

IMMUNE-GLUTAMATERGIC INTERACTION

The process of immunoexcitotoxicity involves an interaction between immune cytokines, cytokine receptors and glutamatergic neurotransmission. Earlier studies indicated that there exist an interaction between pro-inflammatory cytokines and GluRs. For example, Panegytes and Hughes found that IL-1 receptor antagonist protected against kainic acid excitotoxicity [96]. Similar interactions, where IL-1 enhanced excitotoxicity, were described by others [8,9,29]. IL-1ß has been shown to prolong seizures by increasing glutamatergic neurotransmission as has IL-6 [97,98]. This not only worsened seizures but also lead to neurodegeneration.

In all these studies, there appears to be an interaction between certain cytokines and GluRs. More recent studies have elucidated a number of mechanisms that may explain this amplification effect. Stellwagen and co-workers demonstrated TNF-α-induced robust increase (2-fold) surface expression of AMPA receptors in cultured neurons [99]. IL-1ß also produced a smaller but significant increased trafficking of AMPA receptors to the synapse, but blocking IL-1ß did not reduce AMPA receptor surface expression. IL-6 and IL-10 had no part in this process. Of importance was their finding that GABA receptor endocytosis was also increased, which would further enhance excitotoxicity by changing the glutamate/GABA ratio.

Studies of autistic patients have demonstrated reduced $GABA_B$ receptors, the most abundant inhibitory receptor in the brain. GABBR1 was significantly decreased in the cerebellum, Broadman area 9 (BA9) and Broadman Area 40, while GABBR2 was decreased only in the cerebellum. Alteration in GABA receptor function upsets the excitatory/inhibitory balance and can increase neurodevelopmental disruption, neurodegeneration and seizure behavior. Autistic children have a very high incidence of seizures, especially in those displaying mental retardation and severe language disorders [100]. Others have demonstrated a decrease in GAD activity in ASD [101,102].

Trafficking of AMPA receptors stimulate transfer from the endoplasmic reticulum to the synaptic membrane. Once inserted, lateral diffusion determines its activity state. Receptors for TNF-α consist of TNFR1 (p55) and TNFR2 (p75). Both exist on glia and neurons, with the concentration of TNFR2 being higher on microglia [10]. Death domains exist on TNFR1, making it mostly neurodestructive when stimulated. TNFR2 is predominantly neuroprotective [103]. Trafficking of AMPA receptors is dependent on PI3K signaling and blocking this site completely prevents TNF-α-induced trafficking. Of importance is the finding that TNF-α preferentially increases synaptic expression of GluR2-lacking AMPA receptors, which makes the receptors calcium permeable. This loss of GluR2 has been observed in a number of neurological conditions [104,105].

During brain development there is a switch from GluR2-lacking (calcium permeable) to GluR2-containing AMPA receptors (calcium impermeable). The switch occurs at different times, depending on the cortical layer examined. On testing rat somatosensory cortex, it was found that pyramidal neurons in layer 2/3 switched on postnatal day P12 and P14 and that switching occurred between days P7 and P8 for stellate cells in layer four. Neurons and axons containing GluR2-lacking AMPA receptors are more sensitive to excitotoxicity [104,105], meaning that at certain pre and postnatal days the developing brain is more sensitive to the effects of excitotoxicity.

There is some debate as to which GluR subtype is involved with TNF-α enhancement of excitotoxicity. Zou and Crews, using organotypic hippocampal slices found NMDA receptors to be the predominantly involved receptors and not AMPA receptors [106]. The evidence indicates that enhancement of excitotoxicity is via two separate mechanisms—one by inhibiting glutamate transport proteins and another by stimulating AMPA receptors trafficking to the synaptic membrane surface.

Another mechanism, involves upregulation of the glutamate generating enzyme glutaminase by TNF-α stimulation [107]. The excess glutamate is released from the microglia by way of connexin 32 hemichannel of the gap junction. TLR-2, TLR-4 and TLR-9 signal through MyD88, an adaptor protein, and this, in turn, increases expression of glutaminase [108]. Glutaminase converts glutamine to glutamate and its upregulation can induce excitotoxicity.

It has also been shown that IL-ß, IL-18 and other factors, such as the chemokines MCP-1, MCP-3, MIP-1α, MIP-1β and RANTES, enhance the recruitment of microglia, which would indirectly increase available excitatory substances for release [109]. In some instances, enhancement of neurotoxicity occurs because of shared neurotoxic mechanisms that interact. For example, Floden and co-workers found that both pro-inflammatory cytokines, such as TNF-α, and glutamate synergistically stimulated iNOS expression with increased production of peroxynitrite, leading to neuronal death [110]. Taken together, there are a number of mechanisms by which pro-inflammatory cytokines can synergistically enhance GluR function and excitotoxicity.

INTERACTIONS OF REELIN WITH GLUTAMATE RECEPTORS

A number of studies have shown reelin is impaired in autism [111-113]. Reelin is a large protein found in the extracellular matrix, which operates through two pathways, ApoER2 and very low-density lipoprotein receptor (VLDLR) [114]. It also binds to the integrin family adhesion molecules [115].

During embryogenesis, it is secreted by Cajal-Retzius cells, which die after the first postnatal week [116]. Reelin is independent of GluR activity and neuronal activity, but plays an important role in ongoing maintenance of NMDA receptors [117,118]. It plays a significant role in brain development, being crucial for correct cytoarchitecture of the brain's laminated structures [119]. Postnatally reelin plays a vital role in synaptic plasticity, where it controls maturation and surface motility of NMDA receptors.

A recent study found that there are two forms of reelin-secreting cells, punctate and intense, with the intense form being secreted only by GABAergic neurons and the punctate by both GABAergic and non-GABAergic neurons [120]. During brain development NMDA receptors are predominately NR2B-containing and later switch to NR2A-containing assembly [121]. Reelin plays a vital role in this process. After brain development, reelin plays a critical role in maintaining NMDA receptor function, essential for plasticity.

SICKNESS BEHAVIOR AND ASD

Systemic infections are known to cause impairment in cognition, difficulty learning, impaired attention, fragmented sleep, irritability, reduced food and water intake, social withdrawal and depression; something referred to as sickness behavior [122]. Twenty years of research on this phenomenon indicates that it is caused by pro-inflammatory cytokine activation in the brain [123]. A number of recent studies have shown that systemic immune stimulation can rapidly activate brain microglia and if sustained, result in neurodegeneration. ASD children are known to have a number systemic infections and a higher incidence of inflammatory bowel diseases than the general population.

Churchill and co-workers found that systemic cytokines can elevate brain cytokine mRNA levels [124]. When IL-1ß was elevated systemically, brain IL-1ß, IL-6 and TNF-α mRNA levels in the nucleus of the tractus solitarius, hypothalamus, hippocampus, and somatosensory cortex also increased. It was also able to enhance IL-1ß, IL-6 and TNF-α levels in the amygdala, a structure commonly affected in ASD.

It is now accepted that IL-1ß is the main activator of microglia and elevations in systemic IL-1ß can bring about sickness behavior [125]. MAP kinase, ERK 1/2 and JNK are involved in the signaling required for this process [126].

Transfer of signals to the brain from the periphery occurs by a number of routes. Vagal and glossopharyngeal afferents activate glutamatergic neurons in the dorsal vagal complex and trigeminal neurons that activate macrophages and resident microglia in these areas [127]. Increased IL-1ß secretion then activates surrounding microglia and astrocytes. A humoral pathway involves TLRs residing in the CVO and choroid plexus which allow contact between systemic immune stimulants and resident microglia [128]. A system of saturable cytokine transporters exist at the BBB [129]. Finally, IL-1 receptors located on perivascular macrophages and endothelial cells of brain venules, when activated, can increase the production of intrinsic cytokines and prostaglandin E2 within the brain [130].

Basically, there appears to be two systems, a vagal afferent system that is rapidly acting and a slower transference system by way of the BBB. Both systems rely heavily on the CVO system [131]. LPS given systemically does not act via BBB cytokine transporters, but rather enters by way of the CVO. EM and functional studies of the subfornical organ (SFO) and area postrema (AP) indicates that they are areas of high blood volume and slow blood flow [132]. Transient times within the capillaries of labeled albumin in the SFO is 7 to 12X longer that capillaries of the BBB region. Studies have also demonstrated high concentrations of ramified microglia located within the area of the CVO.

Lacroix and co-workers studied the entry of immune cell activation within the brain following LPS stimulation i.v. and i. p. [133]. Using CD14 as an indicator of LPS induction of macrophage phagocytosis, they examined LPS injected Sprague-Dawley rats at 1,3,6 and 24 hours post-injection. The assay demonstrated that at 1 hour there was a profound increase expression of mRNA for CD14 in leptomeninges, choroid plexus and CVO, which peaked at 3 hours. Levels then declined at 6 hours and returned to baseline at 24 hours. Interestingly, a migratory pattern was seen from all CVOs into the deeper parenchymal brain 3 to 6 hours after LPS injection. At 6 hours, small positive cells were seen throughout the entire brain. Labeling identified these cells as CVO microglia and invasive macrophages.

Studies using EAE as a model concluded that the CVO sites are the major points of entry for macrophages and monocytes into the CNS [134]. The afferents entering the area postremia are primarily glutamatergic, and TNF-α has been shown to enhance these glutamate afferents. It has also been shown that vagal afferent signals regulate cytokine production via nicotinic acetylcholine receptors subunits α7 [135]. This pathway has been called the "cholinergic anti-inflammatory pathway". Nicotine not only suppresses TNF-α production via vagal terminals in the spleen, but also by way of α7 subunits on macrophages themselves.

The α7nicotinic-dependent pathway to the spleen has been shown to inhibit the critical proinflammatory mediator-high mobility group box-1 (HMGB-1) [136]. Interestingly, despite the fact that acetylcholine increased the phagocytosis of bacteria by intestinal macrophages, it at the same time reduced NFkB activation and pro-inflammatory cytokine generation, and increases IL-10 production [137].

While the BBB is considered to provide protection from systemic immune stimulation, newer studies have detected signs of occult breakdown of the BBB *in vivo* using iron oxide magnetic nanoparticles with MRI scanning in neuroinflammatory disorders that are not seen using conventional methods [138]. It has also been shown that chronic expression of monocytes chemoattractant protein-1 (MCP-1) alone can lead to a delayed encephalopathy (DESMO encephalitis) [139]. It appears that a slow accumulation of monocytes/macrophages at the BBB induces dysfunction of the barrier and a slowly destructive process. Interestingly, they found that DESMO was associated with impaired microglia, which no longer acted as APCs. The destructive effects were primarily from monocytes and macrophages.

A number of cytokines are involved in sickness behavior, including IL-1ß and TNF-α and it has been shown that systemic LPS, in doses that do not produce sepsis, can induce expression of IL-1ß and other pro-inflammatory cytokine mRNA and proteins in the brain [14,140-142].

Both systemic and central administration of IL-1ß and TNF-α to rats and mice can induce all of the signs and symptoms of sickness behavior in a time- and dose-dependent manner [143]. Animals injected with IL-1ß or TNF-α show little interest or no interest in their physical or social environment. That is, they show similar social withdrawal as seen in ASD patients. They also have altered cognition, increased slow-wave sleep and reduced motor activity. IL-6, when injected systemically has no behavioral effects, despite being able to induce fever, yet IL-6 is playing a role, since less impairment is seen with LPS treatment when IL-6 is removed [143,144]. IL-6 appears to enhance expression of IL-1ß and TNF-α. The anti-inflammatory cytokines, IL-10, IL-4 and TGF-ß regulate the intensity and duration of sickness behavior by downregulating cytokine production and attenuating its signals. Mice deficient in IL-10 respond to I.P. LPS with an exaggerated sickness behavior response [122].

Approximately twenty years ago, clinicians began to use specific cytokines and interferons to treat various cancers and hepatitis C [145]. They noticed that a significant number of these patients developed depression, anxiety, confusion and other behavioral problems, that most often dissipated with cessation of therapy.

The neurological effects of pro-inflammatory cytokines fall into two main categories: early-onset neurovegetative and late-onset psychological problems. The neurovegetative syndrome is characterized by fatigue, loss of appetite, disrupted sleep and other flu-like symptoms. The late-onset symptoms include mild cognitive impairment, depression, anxiety and irritability. Treatment with SSRI medications has little beneficial effects on these symptoms [146].

Renault and co-workers divided treatment-related late cytokine effects into three categories: organic personality syndrome, organic affective syndrome and delirium [147]. They described a set of symptoms that sounded remarkable like that suffered by at least some ASD patients—uncontrollable overreaction to minor frustration, marked irritability and a short temper. Some experience uncontrollable crying spells, clouded consciousness, disorientation, irritability and mood alterations. Patients taking INF-α often developed periods of severe agitation, became abusive and withdrew from others. Psychomotor retardation was seen in 47% to 80% of patients treated with INF-α; this included becoming socially withdrawn, memory loss, and mental fog. Other reports describe patients who exhibited periods of silence and who would without warning stare vacantly, even in mid-sentence [148]. While most cognitive changes were reversible with withdrawal of the cytokine, some reports describe persistent cognitive problems lasting as long as two years after treatment cessation. In some instances, individuals have become fully demented [149]. Slowing of thought processes, confusion and parkinsonian symptoms have been reported in patients give high-dose INF-γ [150]. Other pro-inflammatory cytokines have shown similar effects. For example, IL-2 use in patients to treat cancer and various infectious diseases has been shown to result in mental status changes, agitation, combativeness, hallucinations, difficulty concentrating and delusions [151].

It is clear from these reports in human patients that high levels of pro-inflammatory cytokines can alter the function of the brain and that this disruption of function is rapidly reversible in many cases, but not all. This may indicate either a physiological effect or temporary injury to synapses, axons and/or dendrites. In severe cases of autism, one may see more widespread neurodestructive effects and alterations of neurodevelopment, depending on intensity and/or duration of the inflammatory response.

MICROGLIAL PRIMING AND EFFECTS OF SEQUENTIAL IMMUNE STIMULATION

Microglial activation in the developing brain is mostly in the amoeboid phagocytic state, with no significant release of immune factors or excitatory amino acids as described earlier. Soon after birth, microglia assume a ramified state, similar to that seen in the adult brain. A number of stimuli, infectious and non-infectious, are able to activate resting microglia. The initial immune stimulation of microglia produces a priming effect, characterized by an upregulation of immune factors—that is, acting as an antigen presenting cells (APC). That activation of microglia itself is neurotoxic has been shown in a number of studies. LPS, for example, in a pure neuronal culture is not neurotoxic, but becomes toxic when microglia are added to the culture [152, 153].

On subsequent stimulation, the primed microglia initiates a hyperintense response in terms of the release of immune factors and excitatory amino acids. Some studies have indicated that this state of hyperresponsiveness can last for very long periods, eventually leading to neurodegeneration [154,155]. For example, priming with LPS soon after birth can lead to an enhanced response to LPS in adulthood, leading to a long-term loss of dopaminergic neurons. Activation of primed microglia can be initiated by a number of stimuli, other than antigenic stimulation, including trauma, excitotoxins, hypoxia/ischemia, oxidant stress, pesticide/herbicide exposure, and neurotoxic metals.

Another example of the priming effect is found in studies where brief exposure to MPTP (1-methyl-4-phenyl-1,2,3,6-tetrahydropyridine) in humans and primates results in permanently activated microglia [156,157]. It may be that the microglia themselves are not permanently activated, but rather primed and overreact when exposed to a subsequent stimuli, even many years later. There is good evidence that this is true [158]. For example, it has been shown that LPS exposure during critical periods of microglial development in utero is exacerbated when the animal is exposed to a subsequent dose of LPS as an adult, indicating that the microglia can remain primed for extremely long periods [155]. The finding of widespread microglia activation in the brains of autistic individuals from age 5 years to age 44 years, indicates the possibility of ongoing microglial activation as well [159].

Once microglia are activated, a number of cues can fully activate the microglia later in life, including exposure to neurotoxic metals, pesticides/herbicides, infections (bacterial, viral, parasitic or fungal), head trauma, periods of hypoxia/ischemia and even stress. In many ASD individuals, several of these activators are operational at any given time.

One often-ignored activator has been particulate matter from the environment, primarily from automobile and diesel exhaust. Several studies have demonstrated microglial activation in animals or cultures exposed to such particulate matter, especially the ultra-fine particles (<0.1 μM) [160-162]. The ultra-fine particles have been shown to enter the general circulation and disperse to a number of organs, including the brain [163].

Children and older autistics living in heavily polluted cities could be subject to a more accelerated immunoexcitotoxic reaction than those living in less polluted environments. The same can be said for those living in areas with high levels of atmospheric mercury.

One of the more common ways children are exposed to sequential systemic immune stimulation is by vaccination, which can begin *in utero* when mothers are vaccinated against influenza. As we have seen, immune stimulation in pregnant women, during the second trimester, increases substantially their risk of having offspring with a high risk of autism and schizophrenia.

Neuropsychiatric disorders are also much more common in children exposed to repeated episodes of immune stimulation at birth. This can be from either natural infections or vaccination. Most vaccine schedules have children getting multiple injections every two months during the first 2 years of life. Repetitive immune stimulation has been shown to trigger prolonged progressive immunoexcitotoxicity [164].

Several studies have also shown that there exist an interaction between systemic immune stimulation and a number of neurotoxins, with significant synergistic enhancement of toxicity of these agents [165]. For example, immune stimulation has been shown to enhance the toxicity of MPTP, rotenone and manganese [166-168]. The pesticide rotenone is one of the most commonly used pesticides and manganese levels in commonly consumed soy products is quite high, especially in soy baby formula.

It has been shown that pretreating mice with herbicide maneb and later challenging them with MPTP caused greater PD-associated neurodegeneration than if treated with either agent alone [169]. This should alert health officials to a real and present danger of the present vaccine schedule.

A number of vaccines are given together during a single office visit and this pattern is repeated sequentially until age six years. Giving six vaccines during a single office visit means the child is getting six full doses of immune adjuvant at once, a significant systemic immune exposure. Once the microglia are primed, either by natural infections or toxins, subsequent, intense systemic immune stimulation markedly enhances microglial responsiveness during the next exposure. Another consideration is the neurotoxic effects of vaccine additives, such as mercury and aluminum. Both of these metals have been shown to migrate to the brain following vaccination and can remain in the brain for quite prolonged periods. In addition to the direct neurotoxic effects of these metals, they can act as sources of local immune stimulation. Mercury, for example, accumulates primarily within the astrocytes and microglia, primary sites of immune stimulation. As with the ME7 discussed below, these metals can prime the microglia as well.

Charleston and co-workers found that exposing monkeys to methylmercury resulted in extensive microglial activation and that this activation persisted as long as 6 months after the mercury dosing was terminated, that is, during the entire length of observation [170]. Mercury also has a profound effect on glutamate uptake. Brookes found that concentrations of mercuric chloride as low as 0.5 μg resulted in a 50% reduction of glutamate transport into astrocytes [171]. Mercury has also been shown to induce reverse transport of glutamate from the astrocyte, resulting in elevations in brain glutamate levels.

Brain development is dependent on timed fluctuations in glutamate transporters [172]. Control of brain glutamate levels are also dependent on other removal systems, such as GDH, which in rats increase their expression at birth and rise to adult levels at P20 and P30 [173]. Mercury has been shown to suppress GDH at low concentrations [174]. Another important glutamate removal enzyme is GS, which is also quite sensitive to mercurial inhibition [175]. That low concentrations of methylmercury can result in dramatic elevations in brain glutamate has been shown by Juarez and co-workers by instilling methylmercury into the frontal cortex of freely moving awake rats using microdialysis techniques [176]. Interestingly, the lower dose (10 μM) produced a higher elevation of extraneuronal glutamate than did the higher dose (100 µM)—9.8-fold rise versus 2.4-fold rise respectively. Mercury also has a deleterious effect on a number of other neuroprotective systems, including glutathione and metallothionein. Since astrocytes are the major site for production of neuronal glutathione and are also the principle site for mercury accumulation, high levels of mercury can endanger this critical antioxidant/detoxification system.

As extraneuronal glutamate levels rise, one observes suppression of cystine exchange via the cystine/glutamate antiporter. This lowers neuronal glutathione levels and increases neuronal sensitivity to ROS/RNS and LPP. Taken together, mercury can lead to neurotoxicity by a number of mechanisms, such as free radical generation, suppression of glutathione generation, reversal of glutamate transport, suppression of critical glutamate-controlling enzymes and by acting as a source of microglial activation and resulting immunoexcitotoxicity.

One can see that a number of environmental events can magnify the immunoexcitotoxicity effect of recurrent infections, chronic infections, and sequential vaccination schedules.

Of particular concern are live-virus vaccines. A number of vaccines have been found to be contaminated with other pathogenic organisms, including pestivirus and mycoplasma. Katayama and co-workers, using reverse transcriptase-PCR technology, studied 51 autopsied adults and found that 45% were positive for measles virus mRNA in at least one of their organs, Of special concern, 20% were found to have measles virus in their brain [177]. Persistence of measles viral infection is seen more commonly with defective complement function.

Viral infections can trigger progressive neurodegeneration without infecting the brain with live organisms. For example, only low titers of the HIV virus are detected in the brain microglia and none in neurons. Rather than whole viruses, a fragment called gp41 inhabits the microglia and can fully activate microglia, leading to slowly progressive neurodestruction [178]. It has also been shown that gp120, another HIV-1 coat protein, increases NMDA receptor sensitivity and trafficking [179].

Broderick and co-workers demonstrated that elevating IL-1ß alters hippocampal norepinephrine and serotonin levels as well as increases glutamate levels in the brain [180]. Elevations in brain serotonin are linked to increased levels of QUIN associated with chronic states of inflammation. QUIN has been shown to activate microglia.

Long-term persistent immune activation and low-grade brain inflammation has been described in 3 children, younger than age 2 years, who recovered from herpes simplex encephalitis [181]. All of these children demonstrated abundant activated microglia on brain biopsy performed 3 to 10 years later because of epilepsy. Only one demonstrated a reactivated virus. Others have described similar long-term neurodegeneration in very small children following recovery from HSV [182].

Anderson and co-workers, using a hamster neurotropic strain of the measles virus, found that a non-infectious encephalopathy could occur with destruction of CA1 and CA3 segments of the hippocampus [183]. The neurodegeneration caused by this virus was prevented by blocking NMDA receptors with its antagonists, MK-801, indicating that the principle neurodegenerative mechanism was excitotoxicity.

Another viral disease demonstrating the phenomenon of priming is borna disease virus. This is a RNA neurotropic virus with an affinity for the limbic structures and cerebellum, which has been proposed as a model for autism because of the similarity of clinical and neuropathological effects on developing rats. Borna disease viral infections have been associated with psychiatric disorders in humans. Horning and co-workers injected borna disease virus intracerebrally in Lewis rat pups and assessed clinical, behavioral and histological changes up to 76 weeks after inoculation [184]. They found a transient reduction in neurotrophic factors and an upregulation of proinflammatory cytokine mRNA for IL-1α, IL-1ß, IL-6 and TNF-α. Histological examination of the brains indicated prominent neuronal loss in the cerebral cortex, dentate gyrus, Purkinje cell layer of cerebellum, deep cerebellar nuclei, ventral cochlear nucleus, and superior colliculus. No signs of neural migration deficiencies were seen. Widespread microglial activation was observed.

Overall, the model resembled ASD in a number of ways, including stereotypies, inhibited response to novel stimuli and abnormalities in motor development. An unpublished study of 61 autistic children found no evidence of borna disease viral infection. The weakness of the study was that they did not attempt repeated, sequential systemic immune stimulation to see if it would prolong microglial activation during the primary infection.

Cunningham and co-workers did approach this question in a recent study [158]. In a previous study they demonstrated that the prion diseased brain contains "primed" microglia and that subsequent central or systemic immune stimulation produces an amplified cytokine and inflammatory response. In this study they used a ME7 murine prion disease model of chronic neurodegeneration. Immune stimulation with LPS markedly enhanced microglial release of proinflammatory cytokines, pentraxin 3 and iNOS transcription. They chose 19 weeks post-innoculation with ME7 for the LPS systemic injections based on previous studies indicating that this was the peak time for producing sickness behavior exaggeration responses and regional distribution of activated microglia. They used controls injected with saline and injection of normal brain homogenates (NBH). The study demonstrated a greater intensity of IL-1 staining in the ME7 + LPS *i.p.* animals. The ME7 + LPS *i.p.* animals experienced a 3-fold higher increase in IL-1ß than that seen in the ME7 + NBH, and TNF-α was 1.7-fold higher in the former. IL-6 increased 3-fold higher and importantly, TGFß1 did not increase with systemic immune stimulation, indicating a proinflammatory state. Of interest is the finding that prion infection itself activated the microglia, but the microglia did not secrete proinflammatory cytokines, whereas systemic or intracerebral LPS stimulation markedly increased proinflammatory cytokine generation from activated microglia.

Microglia priming is not peculiar to prion infections, since similar priming and enhanced immune reactions with secondary immune stimulation are also seen with amyloid secretion and presenilin-1 [7,185]. In keeping with this mechanism, it has been shown that infections accelerate the progression of Alzheimer's disease [186]. The first human study to show microglial activation in autopsied brains of patients dying of sepsis, without brain infection, was by Lemstra and co-workers [187].

In the case of autism, priming could occur with systemic infections, vaccination or exposure to certain pro-inflammatory chemicals such as pesticides and herbicides. The ultimate outcome of the priming effect would depend on timing of exposure, intensity of initial reaction and number of subsequent immune stimulating events. Intrauterine immune-triggered priming would be expected to produce a different clinical picture than would postnatal exposures. This would be based on the stage of neurodevelopment, sensitivity of glutamate and cytokine receptors and subunit makeup of the receptors. As we have seen, during development receptor types and subunit composition can change and this alteration of receptors is region specific.

The initial stimulus may be, for example, a viral infection during mid-term pregnancy. This primes the microglia in the developing brain. Upon birth, either re-infection or the first exposure to a vaccination, such as hepatitis B vaccine, would trigger a more intense reaction from the primed microglia. The average child is exposed to sequential vaccination every two months during the first 6 years of life. This includes multiple vaccine exposure during each doctor visit—anywhere from 5 to 9 vaccines.

One should appreciate that this means that the systemic immune system is exposed to 5 doses of powerful immune adjuvants on a single day. This is sufficient to initiate activation of brain microglia and this can in turn lead to immunoexcitotoxicity. Subsequent vaccine-adjuvant dosing every two months may be sufficient to cause prolonged microglial activation, chronic neurological injury and abnormal brain development, especially within brain areas that are late to mature, such as the prefrontal cortex and precuneus.

CHRONIC COLITIS, FOOD INTOLERANCE AND MICROGLIAL ACTIVATION

Evidence that chronic systemic inflammation and concomitant chronic brain inflammation is occurring in autism is growing. It has been suggested that a number of gut–related events can act as a source of chronic immune activation, such as food allergies, food intolerance, and chronic gut infections [188-190].

Wakefield and co-workers were the first to note the link between certain vaccines, such as MMR, and a specific colitis in autistic children [191]. They proposed that chronic intestinal inflammation led to malabsorption as well as chronic systemic immune activation. A more recent study by Ashwood and Wakefield found that peripheral blood lymphocytes as well as mucosal CD3+ TNF-α and CD3+ IFN-γ cytokine responses were significantly increased in children with autism as compared to non-inflamed controls [192].

Malabsorption is common with various forms of colitis, including coeliac disease and can lead to significant losses in various nutrients needed for brain repair and antioxidant defense. Colitis in ASD differs from that seen with Crohn's disease in one very important aspect. Peripheral and mucosal IL-10 was found to be markedly lower in cases of ASD, colitis, thus increasing the inflammatory effect in ASD colitis.

Several studies have found a cross reactivity between food-derived proteins and peptides and neuron specific antigens. Vojdani and co-workers examined 9 different neuron-specific antigens and 3 cross-reactive peptides, which included cow's milk proteins, and found that autistic children had the highest IgG, IgM and IgA antibody reaction to all 9 neuronal antigens as well as cross-reaction to all 3 peptides [193]. In a follow-up study they assessed reactivity of sera from 50 autistic patients as compared to 50 healthy controls. They demonstrated that a significant number of autistic children expressed antibodies against gliadin and cerebellar Purkinje cells simultaneously [194].

A number of studies found reactions to common colon bacterial organisms in ASD children [195,196] For example, Candida overgrowth in the colon occurs frequently in ASD and penetration of the bowel wall can lead to chronic systemic inflammation, mainly by β-glucan stimulation—a component of the organism's cell wall.

In a study of 96 autistic children as compared to 449 healthy children, Black and co-workers found no greater incidence of gastrointestinal (GI) disease in the autistic children than found in healthy children [197]. Interestingly, the study only included cases with obvious GI symptomatology and they recognized that subclinical cases could have been overlooked. It is of interest that a number of studies have noted a lack of obvious GI symptoms in individuals sensitive to gluten with neurological manifestations, including full-blown cases of coeliac disease. One study found that only 13% of coeliac patients having cerebellar ataxia had symptoms of GI disease [198].

Gluten sensitivity is the most common cause for sporadic cerebellar ataxia, accounting for 41% of cases, many of which present with the neurological syndrome rather than gut dysfunction symptoms. Hu and co-workers examined 13 patients with coeliac disease and neurological involvement and found cognitive impairment in all [199]. A slow, progressive neurological onset was characteristic, with development of acalculia, confusion, personality change and amnesia.

Two of these patients underwent brain biopsy and two were examined at autopsy. One patient demonstrated frontopolar lobar degeneration with histological involvement of the frontal and temporal corticies and hippocampal dentate granular cell layer. Some suspect that the progression is secondary to excitotoxic neurodegeneration [14]. Medhi and co-workers using an induced colitis rat model found that rats with colitis showed significant increases in seizure score and a reduction in onset time as compared to controls [200]. Cerebral excitability was directly related to serum TNF-α level. This is in keeping with other studies that have shown a relation between other systemic inflammatory disorders, such as ulcerative colitis, psoriatic arthritis, and neurological disorders [201,202].

Taken together these studies clearly suggest that chronic systemic immune stimulation, with or without specific anti-neuronal antibodies, can lead to chronic brain inflammation and immunoexcitotoxicity.

SYSTEMIC IMMUNE STIMULATION AND ENHANCEMENT OF ENVIRONMENTAL NEUROTOXICITY

It has been shown that early immune stimulation can result in altered, hyperactive immune responses within the CNS during adulthood. Likewise, systemic immune stimulation can exacerbate preexisting neurological disorders, such as AD, Parkinson's disease and strokes. Early life immune stimulation can also influence later life events such as reactivity to stress, disease susceptibility, increased vulnerability to neurodegenerative cognitive disorders and neuropsychiatric disorders [184,203-205].

Bacterial infections are the leading cause of infection in newborns [206], but viral infections during pregnancy are more commonly associated with autism and schizophrenia. Several studies have shown a negative effect on long-term neurodevelopmental outcome with early-life immune stimulation [207,208].

Purisai and co-workers, using mice, demonstrated that priming microglia in the area of the substantia nigra led to significant dopaminergic neuronal loss with a subsequent exposure to the pesticide paraquat [209]. Single exposures to paraquat had no toxic effect on the neurons, but rather primed the microglia so that subsequent exposures to paraquat produced significant neuronal loss. That it was microglial activation that was causing the neuronal destruction was shown by using LPS to induce priming. Once primed by LPS, a single exposure to paraquat could induce neuronal death. Minocycline, which blocks microglial activation, completely prevented the priming effect by the paraquat. The fungicide maneb can have the same priming effect as paraquat.

Similar results were found in rats injected with the immune stimulant LPS prenatally and then exposed to neurotoxin rotenone 14 days later [210]. LPS given prenatally and with postnatal exposure to rotenone produced a 39% synergistic loss of dopaminergic neurons and a prolonged elevation of TNF-α.

In a real world situation, the developing baby is exposed to a number of pesticides/herbicides and fungicides both *in utero* and postnatally. They are also exposed to a number of vaccines given sequentially. When faced with natural infections as well, both bacterial and viral, one can see that conditions are ripe for duplicating the experimental conditions known to produce long-term immunoexcitotoxicity.

Immunoexcitotoxicity and ROS/RNS

There is growing evidence that many of the destructive effects of excitotoxicity are caused by oxidative stress [211,212] and that antioxidants can reduce excitotoxicity. Basically, there are two types of glutamate toxicity within the CNS, excitotoxicity and oxidative glutamate toxicity (non-receptors mediated). Oxidative glutamate toxicity is distinguished by requiring a higher concentration of glutamate and a longer duration of exposure to cause injury than that seen with classical excitotoxicity.

When glutamate levels rise extraneuronally it competes for cystine exchange and this reduces available intra-astrocytic cysteine needed for glutathione generation. Glutathion has a high turnover, so that a reduction in intracellular cystine quickly lowers glutathione levels [213]. Astrocytes are the major source for neuronal glutathione.

Single antioxidants are unable to completely neutralize classical excitotoxicity (about 50 to 70% protection) but can completely neutralize oxidative glutamate toxicity [214]. As glutathion levels fall, ROS/RNS levels rise, but the main effect on cell toxicity is thought to be secondary to lipid peroxidation.

In the case of glutamate excitotoxicity via NMDA receptors the rapid rise in intracellular Ca^{2+} levels lead to mitochondrial uptake with interference with mitochondrial membrane potential and a subsequent fall in energy production and a concomitant rise in free radical generation [215,216]. That calcium is the major player was demonstrated by using ruthenium red to inhibit Ca^{2+} entry into the mitochondria, which protected against excitotoxicity [217].

In the case of AMPA/kainate receptors, which have high conductance for K^+ and Na^+, it has been shown that these receptors have a higher threshold for producing toxicity and require longer times than NMDA receptors. Under such conditions AMPA/kainate receptors can activate calcium channels and raise intraneuronal Ca^{2+} levels, but less so than NMDA receptors [218]. As we have seen, under a number of pathological conditions, AMPA receptors will lose their GluR2 subunit and this increases Ca^{2+} conductance.

Examination of a number of studies suggests that the major oxidant species with excitotoxicity is peroxynitrite. It has been shown that NMDA receptor stimulation increases, in a dose-dependent manner, superoxide generation [218]. Likewise, increasing superoxide dismutase (SOD) inhibits glutamate toxicity, suggesting it, and not H_2O_2 or hydroxyl, is the main oxidant generated [219]. Superoxide itself is a mild oxidant, whereas peroxynitrite is a powerful radical. NO itself is not a major contributor to excitotoxicity and blocking it does not reduce appreciably excitotoxicity in most studies. Yet, in the presence of elevated levels of superoxide, NO rapidly reacts with superoxide to produce peroxynitrite, which then reacts with a number of amino acid residues. High levels of 3-nitrotyrosine are seen in most neurodegenerative disorders.

Proinflammatory cytokines activate microglia and increase ROS/RNS as well as upregulating iNOS and increase protein nitration [220]. One of the major sources of superoxide is activated microglia, with most superoxide being produced by NADPH oxidase, an inducible cytoplasmic enzyme [221,222]. Together, these studies indicate that microglial activation increases ROS/RNS and LPP by way of immune and excitotoxic mechanism, with significant interaction between the two processes.

CONCLUSION

We have seen that microglia and astrocytic activations are playing a central role in chronic inflammatory disorders of the brain and may play a central role in ASDs. Because of the high levels of sensitivity of microglial activation, one may see a number of additive and synergistic processes occurring in the life of a child, especially during early brain development. This can include natural infections, excessive, sequential vaccination, exposure to pesticides/herbicides and fungicides, use of soy formula, exposure to neurotoxic metals, especially mercury, lead, cadmium, manganese, aluminum and possibly fluoride, and exposure to obstetric drugs known to affect microglial activation.

Several studies were cited suggesting excitotoxicity as a process in abnormal brain development as well as neurodegeneration. It is known that brain development is critically dependent on temporal fluctuations in both immune cytokines and glutamate sensitivity. Abnormalities in these levels can produce varying degrees of brain maldevelopment and hence abnormal function. Elevated levels of glutamate have been found in some autistic children and in those with related conditions. Likewise, high levels of ROS/RNS and LPP have also been described and are intimately related to immunoexcitotoxicity.

The author was the first to suggest chronic microglial activation, with immunoexcitotoxicity, as a central mechanism in ASD. A considerable amount of data now supports this mechanism. The strongest confirmatory evidence was shown by Vargas and co-workers, who demonstrated chronic microglial activation in a number of autistic patients from age 5 years to 44 years. While they did not measure brain glutamate levels or sensitivity of the GluRs, there is compelling evidence that microglia activation with inflammation is always accompanied by excitotoxicity.

Microglia priming sets the stage for this process and subsequent immune stimulating events and/or exposure to glutamate, even much later in life, can trigger an exaggerated immunoexcitotoxicity reaction. The exact presentation of neurological dysfunction and degree of injury depends on a number of factors, the most important being timing of exposure, presence of closely spaced sequential immune stimulation, genetic alteration in glutamate receptor function and immune function and antioxidant defense status. It may also involve the status of the glutamate transport protein system, nutrient deficiencies and known excitotoxins in the diet of both mother and child. Concomitant exposure to neurotoxic substances, many of which are known to induce priming and microglial activation, can also worsen the immunoexcitotoxic reaction or prolong it.

REFERENCES

[1] Lucas DR, Newhouse JP. The toxic effect of sodium L-glutamate on the inner layers of the retina. AMA Arch Ophthalmol 1957; 58: 193-201.
[2] Olney JW. Brain lesions, obesity, and other disturbances in mice treated with monosodium glutamate. Science 1969; 164: 719-21.
[3] Benitez-Diaz P, Miranda-Contreras L, Mendoza-Briceno RV, Pena-Contreras Z, Palacios-Pru E. Prenatal and postnatal contents of amino acid neurotransmitters in mouse parietal cortex. Dev Neurosci 2003; 25: 366-74.
[4] Marret S, Gressens P, Evrard P. Arrest of neuronal migration by excitatory amino acids in hamster developing brain. Proc Natl Acad Sci U S A 1996; 93: 15463-8.

[5] Monier A, Adle-Biassette H, Delezoide AL, *et al.* Entry and distribution of microglial cells in human embryonic and fetal cerebral cortex. J Neuropathol Exp Neurol 2007; 66: 372-82.

[6] Navascues J, Calvente R, Marin-Teva JL, Cuadros MA. Entry, dispersion and differentiation of microglia in the developing central nervous system. An Acad Bras Cienc 2000; 72: 91-102.

[7] Lee SC, Liu W, Dickson DW, Brosnan CF, Berman JW. Cytokine production by human fetal microglia and astrocytes. Differential induction by lipopolysaccharide and IL-1 beta. J Immunol 1993; 150: 2659-67.

[8] Chao CC, Hu S, Ehrlich L, Peterson PK. Interleukin-1 and tumor necrosis factor-alpha synergistically mediate neurotoxicity: involvement of nitric oxide and of N-methyl-D-aspartate receptors. Brain Behav Immun 1995; 9: 355-65.

[9] Lawrence CB, Allan SM, Rothwell NJ. Interleukin-1beta and the interleukin-1 receptor antagonist act in the striatum to modify excitotoxic brain damage in the rat. Eur J Neurosci 1998; 10: 1188-95.

[10] Neumann H, Schweigreiter R, Yamashita T, *et al.* Tumor necrosis factor inhibits neurite outgrowth and branching of hippocampal neurons by a rho-dependent mechanism. J Neurosci 2002; 22: 854-62.

[11] Perry VH, Gordon S. Macrophages and the nervous system. Int Rev Cytol 1991; 125: 203-44.

[12] Cuadros MA, Rodriguez-Ruiz J, Calvente R, *et al.* Microglia development in the quail cerebellum. J Comp Neurol 1997; 389: 390-401.

[13] Cuadros MA, Navascues J. The origin and differentiation of microglial cells during development. Prog Neurobiol 1998; 56: 173-89.

[14] Block ML, Zecca L, Hong JS. Microglia-mediated neurotoxicity: uncovering the molecular mechanisms. Nat Rev Neurosci 2007; 8: 57-69.

[15] McKimmie CS, Fazakerley JK. In response to pathogens, glial cells dynamically and differentially regulate Toll-like receptor gene expression. J Neuroimmunol 2005; 169: 116-25.

[16] Olson JK, Miller SD. Microglia initiate central nervous system innate and adaptive immune responses through multiple TLRs. J Immunol 2004; 173: 3916-24.

[17] Bsibsi M, Ravid R, Gveric D, van Noort JM. Broad expression of Toll-like receptors in the human central nervous system. J Neuropathol Exp Neurol 2002; 61: 1013-21.

[18] Lehnardt S, Massillon L, Follett P, *et al.* Activation of innate immunity in the CNS triggers neurodegeneration through a Toll-like receptor 4-dependent pathway. Proc Natl Acad Sci U S A 2003; 100: 8514-9.

[19] Cho S, Park EM, Febbraio M, *et al.* The class B scavenger receptor CD36 mediates free radical production and tissue injury in cerebral ischemia. J Neurosci 2005; 25: 2504-12.

[20] Murphy JE, Tedbury PR, Homer-Vanniasinkam S, Walker JH, Ponnambalam S. Biochemistry and cell biology of mammalian scavenger receptors. Atherosclerosis 2005; 182: 1-15.

[21] Akiyama H, McGeer PL. Brain microglia constitutively express beta-2 integrins. J Neuroimmunol 1990; 30: 81-93.

[22] Reichert F, Rotshenker S. Complement-receptor-3 and scavenger-receptor-AI/II mediated myelin phagocytosis in microglia and macrophages. Neurobiol Dis 2003; 12: 65-72.

[23] Babior BM. Phagocytes and oxidative stress. Am J Med 2000; 109: 33-44.

[24] Qin L, Liu Y, Wang T, *et al.* NADPH oxidase mediates lipopolysaccharide-induced neurotoxicity and proinflammatory gene expression in activated microglia. J Biol Chem 2004; 279: 1415-21.

[25] Mander PK, Jekabsone A, Brown GC. Microglia proliferation is regulated by hydrogen peroxide from NADPH oxidase. J Immunol 2006; 176: 1046-52.

[26] Pawate S, Shen Q, Fan F, Bhat NR. Redox regulation of glial inflammatory response to lipopolysaccharide and interferongamma. J Neurosci Res 2004; 77: 540-51.

[27] Gao HM, Liu B, Zhang W, Hong JS. Critical role of microglial NADPH oxidase-derived free radicals in the *in vitro* MPTP model of Parkinson's disease. FASEB J 2003; 17: 1954-6.

[28] Ivacko JA, Sun R, Silverstein FS. Hypoxic-ischemic brain injury induces an acute microglial reaction in perinatal rats. Pediatr Res 1996; 39: 39-47.

[29] McManus CM, Brosnan CF, Berman JW. Cytokine induction of MIP-1 alpha and MIP-1 beta in human fetal microglia. J Immunol 1998; 160: 1449-55.

[30] Heyes MP, Achim CL, Wiley CA, *et al.* Human microglia convert l-tryptophan into the neurotoxin quinolinic acid. Biochem J 1996; 320 (Pt 2): 595-7.

[31] Heyes MP, Saito K, Crowley JS, *et al.* Quinolinic acid and kynurenine pathway metabolism in inflammatory and non-inflammatory neurological disease. Brain 1992; 115 (Pt 5): 1249-73.

[32] Heyes MP, Ellis RJ, Ryan L, *et al.* Elevated cerebrospinal fluid quinolinic acid levels are associated with region-specific cerebral volume loss in HIV infection. Brain 2001; 124: 1033-42.

[33] Guillemin GJ, Brew BJ, Noonan CE, Takikawa O, Cullen KM. Indoleamine 2,3 dioxygenase and quinolinic acid immunoreactivity in Alzheimer's disease hippocampus. Neuropathol Appl Neurobiol 2005; 31: 395-404.

[34]   Yamada A, Akimoto H, Kagawa S, Guillemin GJ, Takikawa O. Proinflammatory cytokine interferon-gamma increases induction of indoleamine 2,3-dioxygenase in monocytic cells primed with amyloid beta peptide 1-42: implications for the pathogenesis of Alzheimer's disease. J Neurochem 2009; 110: 791-800.

[35]   Heyes MP, Saito K, Chen CY, *et al.* Species heterogeneity between gerbils and rats: quinolinate production by microglia and astrocytes and accumulations in response to ischemic brain injury and systemic immune activation. J Neurochem 1997; 69: 1519-29.

[36]   Heyes MP, Brew BJ, Martin A, *et al.* Quinolinic acid in cerebrospinal fluid and serum in HIV-1 infection: relationship to clinical and neurological status. Ann Neurol 1991; 29: 202-9.

[37]   Heyes MP, Saito K, Major EO, *et al.* A mechanism of quinolinic acid formation by brain in inflammatory neurological disease. Attenuation of synthesis from L-tryptophan by 6-chlorotryptophan and 4-chloro-3-hydroxyanthranilate. Brain 1993; 116 (Pt 6): 1425-50.

[38]   Zimmerman AW, Jyonouchi H, Comi AM, *et al.* Cerebrospinal fluid and serum markers of inflammation in autism. Pediatr Neurol 2005; 33: 195-201.

[39]   Streit W. Microglia in Regenerating and Degenerating Central Nervous System. New York: Springer; 2001.

[40]   Nakajima S, Kohsaka S. Neuroprotective roles of microglia in central nervous system. In: Streit W, editor. Microglia in the Regenerating and Degenerating Central Nervous System. New York: Springer; 2001. p. 188-208.

[41]   Bessis A, Bechade C, Bernard D, Roumier A. Microglial control of neuronal death and synaptic properties. Glia 2007; 55: 233-8.

[42]   Fassbender K, Mielke O, Bertsch T, *et al.* Interferon-gamma-inducing factor (IL-18) and interferon-gamma in inflammatory CNS diseases. Neurology 1999; 53: 1104-6.

[43]   Fassbender K, Schneider S, Bertsch T, *et al.* Temporal profile of release of interleukin-1beta in neurotrauma. Neurosci Lett 2000; 284: 135-8.

[44]   Vitkovic L, Bockaert J, Jacque C. "Inflammatory" cytokines: neuromodulators in normal brain? J Neurochem 2000; 74: 457-71.

[45]   Hanisch UK, Seto D, Quirion R. Modulation of hippocampal acetylcholine release: a potent central action of interleukin-2. J Neurosci 1993; 13: 3368-74.

[46]   Seto D, Kar S, Quirion R. Evidence for direct and indirect mechanisms in the potent modulatory action of interleukin-2 on the release of acetylcholine in rat hippocampal slices. Br J Pharmacol 1997; 120: 1151-7.

[47]   Merrill JE. Tumor necrosis factor alpha, interleukin 1 and related cytokines in brain development: normal and pathological. Dev Neurosci 1992; 14: 1-10.

[48]   Dziegielewska KM, Moller JE, Potter AM, *et al.* Acute-phase cytokines IL-1beta and TNF-alpha in brain development. Cell Tissue Res 2000; 299: 335-45.

[49]   Patterson PH. Maternal infection: window on neuroimmune interactions in fetal brain development and mental illness. Curr Opin Neurobiol 2002; 12: 115-8.

[50]   Brown AS. Prenatal infection as a risk factor for schizophrenia. Schizophr Bull 2006; 32: 200-2.

[51]   Zuckerman L, Rehavi M, Nachman R, Weiner I. Immune activation during pregnancy in rats leads to a postpubertal emergence of disrupted latent inhibition, dopaminergic hyperfunction, and altered limbic morphology in the offspring: a novel neurodevelopmental model of schizophrenia. Neuropsychopharmacology 2003; 28: 1778-89.

[52]   Ashdown H, Dumont Y, Ng M, *et al.* The role of cytokines in mediating effects of prenatal infection on the fetus: implications for schizophrenia. Mol Psychiatry 2006; 11: 47-55.

[53]   Meyer U, Nyffeler M, Engler A, *et al.* The time of prenatal immune challenge determines the specificity of inflammation-mediated brain and behavioral pathology. J Neurosci 2006; 26: 4752-62.

[54]   Ozawa K, Hashimoto K, Kishimoto T, *et al.* Immune activation during pregnancy in mice leads to dopaminergic hyperfunction and cognitive impairment in the offspring: a neurodevelopmental animal model of schizophrenia. Biol Psychiatry 2006; 59: 546-54.

[55]   Bauer S, Kerr BJ, Patterson PH. The neuropoietic cytokine family in development, plasticity, disease and injury. Nat Rev Neurosci 2007; 8: 221-32.

[56]   Baier PC, May U, Scheller J, Rose-John S, Schiffelholz T. Impaired hippocampus-dependent and -independent learning in IL-6 deficient mice. Behav Brain Res 2009; 200: 192-6.

[57]   Ozawa S, Kamiya H, Tsuzuki K. Glutamate receptors in the mammalian central nervous system. Prog Neurobiol 1998; 54: 581-618.

[58]   Paquet M, Smith Y. Presynaptic NMDA receptor subunit immunoreactivity in GABAergic terminals in rat brain. J Comp Neurol 2000; 423: 330-47.

[59]   Stys P, Li S. Glutamate-induced white matter injury: excitotoxicity without synapses. The Neuroscientists 2000; 6: 230-3.

[60]   Washburn MS, Numberger M, Zhang S, Dingledine R. Differential dependence on GluR2 expression of three characteristic features of AMPA receptors. J Neurosci 1997; 17: 9393-406.

[61]   Komuro H, Rakic P. Distinct modes of neuronal migration in different domains of developing cerebellar cortex. J Neurosci 1998; 18: 1478-90.

[62]   Bauman M, Kemper TL. Histoanatomic observations of the brain in early infantile autism. Neurology 1985; 35: 866-74.

[63]   Casanova MF, van Kooten IA, Switala AE, *et al.* Minicolumnar abnormalities in autism. Acta Neuropathol 2006; 112: 287-303.

[64]   Kumada T, Komuro H. Completion of neuronal migration regulated by loss of Ca(2+) transients. Proc Natl Acad Sci U S A 2004; 101: 8479-84.

[65]   Sutherland ML, Delaney TA, Noebels JL. Glutamate transporter mRNA expression in proliferative zones of the developing and adult murine CNS. J Neurosci 1996; 16: 2191-207.

[66]   Hashimoto K, Ichikawa R, Takechi H, *et al.* Roles of glutamate receptor delta 2 subunit (GluRdelta 2) and metabotropic glutamate receptor subtype 1 (mGluR1) in climbing fiber synapse elimination during postnatal cerebellar development. J Neurosci 2001; 21: 9701-12.

[67]   Golshani P, Warren RA, Jones EG. Progression of change in NMDA, non-NMDA, and metabotropic glutamate receptor function at the developing corticothalamic synapse. J Neurophysiol 1998; 80: 143-54.

[68]   Bahn S, Volk B, Wisden W. Kainate receptor gene expression in the developing rat brain. J Neurosci 1994; 14: 5525-47.

[69]   Catania MV, De Socarraz H, Penney JB, Young AB. Metabotropic glutamate receptor heterogeneity in rat brain. Mol Pharmacol 1994; 45: 626-36.

[70]   Catania MV, D'Antoni S, Bonaccorso CM, *et al.* Group I metabotropic glutamate receptors: a role in neurodevelopmental disorders? Mol Neurobiol 2007; 35: 298-307.

[71]   Dolen G, Bear MF. Role for metabotropic glutamate receptor 5 (mGluR5) in the pathogenesis of fragile X syndrome. J Physiol 2008; 586: 1503-8.

[72]   Yamada H, Yatsushiro S, Ishio S, *et al.* Metabotropic glutamate receptors negatively regulate melatonin synthesis in rat pinealocytes. J Neurosci 1998; 18: 2056-62.

[73]   Kulman G, Lissoni P, Rovelli F, *et al.* Evidence of pineal endocrine hypofunction in autistic children. Neuro Endocrinol Lett 2000; 21: 31-4.

[74]   Melke J, Goubran Botros H, Chaste P, *et al.* Abnormal melatonin synthesis in autism spectrum disorders. Mol Psychiatry 2008; 13: 90-8.

[75]   Millan-Plano S, Piedrafita E, Miana-Mena FJ, *et al.* Melatonin and structurally-related compounds protect synaptosomal membranes from free radical damage. Int J Mol Sci 2010; 11: 312-28.

[76]   Jou MJ, Peng TI, Hsu LF, *et al.* Visualization of melatonin's multiple mitochondrial levels of protection against mitochondrial Ca(2+)-mediated permeability transition and beyond in rat brain astrocytes. J Pineal Res 2010; 48: 20-38.

[77]   Rennie K, De Butte M, Pappas BA. Melatonin promotes neurogenesis in dentate gyrus in the pinealectomized rat. J Pineal Res 2009; 47: 313-7.

[78]   Olivier P, Fontaine RH, Loron G, *et al.* Melatonin promotes oligodendroglial maturation of injured white matter in neonatal rats. PLoS One 2009; 4: e7128.

[79]   Trotti D, Rossi D, Gjesdal O, *et al.* Peroxynitrite inhibits glutamate transporter subtypes. J Biol Chem 1996; 271: 5976-9.

[80]   Keller JN, Mark RJ, Bruce AJ, *et al.* 4-Hydroxynonenal, an aldehydic product of membrane lipid peroxidation, impairs glutamate transport and mitochondrial function in synaptosomes. Neuroscience 1997; 80: 685-96.

[81]   Zanelli SA, Ashraf QM, Delivoria-Papadopoulos M, Mishra OP. Peroxynitrite-induced modification of the N-methyl-D-aspartate receptor in the cerebral cortex of the guinea pig fetus at term. Neurosci Lett 2000; 296: 5-8.

[82]   Moro MA, Leza JC, Lorenzo P, Lizasoain I. Peroxynitrite causes aspartate release from dissociated rat cerebellar granule neurones. Free Radic Res 1998; 28: 193-204.

[83]   Chauhan A, Chauhan V. Oxidative stress in autism. Pathophysiology 2006; 13: 171-81.

[84]   Kern JK, Jones AM. Evidence of toxicity, oxidative stress, and neuronal insult in autism. J Toxicol Environ Health B Crit Rev 2006; 9: 485-99.

[85]   McGinnis WR. Could oxidative stress from psychosocial stress affect neurodevelopment in autism? J Autism Dev Disord 2007; 37: 993-4.

[86]   Sajdel-Sulkowska EM, Xu M, Koibuchi N. Increase in cerebellar neurotrophin-3 and oxidative stress markers in autism. Cerebellum 2009; 8: 366-72.

[87]   Kyriakis JM, Banerjee P, Nikolakaki E, *et al.* The stress-activated protein kinase subfamily of c-Jun kinases. Nature 1994; 369: 156-60.

[88]   Quan Y, Moller T, Weinstein JR. Regulation of Fcgamma receptors and immunoglobulin G-mediated phagocytosis in mouse microglia. Neurosci Lett 2009; 464: 29-33.

[89] Alberdi E, Sanchcz-Gomez MV, Torre I, *et al.* Activation of kainate receptors sensitizes oligodendrocytes to complement attack. J Neurosci 2006; 26: 3220-8.

[90] Zhu B, Luo L, Moore GR, Paty DW, Cynader MS. Dendritic and synaptic pathology in experimental autoimmune encephalomyelitis. Am J Pathol 2003; 162: 1639-50.

[91] Sulkowski G, Dabrowska-Bouta B, Kwiatkowska-Patzer B, Struzynska L. Alterations in glutamate transport and group I metabotropic glutamate receptors in the rat brain during acute phase of experimental autoimmune encephalomyelitis. Folia Neuropathol 2009; 47: 329-37.

[92] Kanwar JR, Kanwar RK, Krissansen GW. Simultaneous neuroprotection and blockade of inflammation reverses autoimmune encephalomyelitis. Brain 2004; 127: 1313-31.

[93] Taylor DL, Jones F, Kubota ES, Pocock JM. Stimulation of microglial metabotropic glutamate receptor mGlu2 triggers tumor necrosis factor alpha-induced neurotoxicity in concert with microglial-derived Fas ligand. J Neurosci 2005; 25: 2952-64.

[94] Yasuda Y, Shimoda T, Uno K, *et al.* The effects of MPTP on the activation of microglia/astrocytes and cytokine/chemokine levels in different mice strains. J Neuroimmunol 2008; 204: 43-51.

[95] Schintu N, Frau L, Ibba M, *et al.* Progressive dopaminergic degeneration in the chronic MPTPp mouse model of Parkinson's disease. Neurotox Res 2009; 16: 127-39.

[96] Panegyres PK, Hughes J. The neuroprotective effects of the recombinant interleukin-1 receptor antagonist rhIL-1ra after excitotoxic stimulation with kainic acid and its relationship to the amyloid precursor protein gene. J Neurol Sci 1998; 154: 123-32.

[97] Vezzani A, Conti M, De Luigi A, *et al.* Interleukin-1beta immunoreactivity and microglia are enhanced in the rat hippocampus by focal kainate application: functional evidence for enhancement of electrographic seizures. J Neurosci 1999; 19: 5054-65.

[98] Samland H, Huitron-Resendiz S, Masliah E, *et al.* Profound increase in sensitivity to glutamatergic- but not cholinergic agonist-induced seizures in transgenic mice with astrocyte production of IL-6. J Neurosci Res 2003; 73: 176-87.

[99] Stellwagen D, Beattie EC, Seo JY, Malenka RC. Differential regulation of AMPA receptor and GABA receptor trafficking by tumor necrosis factor-alpha. J Neurosci 2005; 25: 3219-28.

[100] Canitano R. Epilepsy in autism spectrum disorders. Eur Child Adolesc Psychiatry 2007; 16: 61-6.

[101] Fatemi SH, Halt AR, Stary JM, *et al.* Glutamic acid decarboxylase 65 and 67 kDa proteins are reduced in autistic parietal and cerebellar cortices. Biol Psychiatry 2002; 52: 805-10.

[102] Yip J, Soghomonian JJ, Blatt GJ. Decreased GAD67 mRNA levels in cerebellar Purkinje cells in autism: pathophysiological implications. Acta Neuropathol 2007; 113: 559-68.

[103] Tartaglia LA, Weber RF, Figari IS, *et al.* The two different receptors for tumor necrosis factor mediate distinct cellular responses. Proc Natl Acad Sci U S A 1991; 88: 9292-6.

[104] Grooms SY, Opitz T, Bennett MV, Zukin RS. Status epilepticus decreases glutamate receptor 2 mRNA and protein expression in hippocampal pyramidal cells before neuronal death. Proc Natl Acad Sci U S A 2000; 97: 3631-6.

[105] Leonoudakis D, Zhao P, Beattie EC. Rapid tumor necrosis factor alpha-induced exocytosis of glutamate receptor 2-lacking AMPA receptors to extrasynaptic plasma membrane potentiates excitotoxicity. J Neurosci 2008; 28: 2119-30.

[106] Zou JY, Crews FT. TNF alpha potentiates glutamate neurotoxicity by inhibiting glutamate uptake in organotypic brain slice cultures: neuroprotection by NF kappa B inhibition. Brain Res 2005; 1034: 11-24.

[107] Takeuchi H, Jin S, Wang J, *et al.* Tumor necrosis factor-alpha induces neurotoxicity via glutamate release from hemichannels of activated microglia in an autocrine manner. J Biol Chem 2006; 281: 21362-8.

[108] Pais TF, Figueiredo C, Peixoto R, Braz MH, Chatterjee S. Necrotic neurons enhance microglial neurotoxicity through induction of glutaminase by a MyD88-dependent pathway. J Neuroinflammation 2008; 5: 43.

[109] Zeng H, Zhy X, Zhang C, *et al.* Identification of sequential events and factors associated with microglial activation, migration, and cytotoxicity in retinal degeneration in rd mice. Invest Ophthalmol Vis Sci 2005; 46: 2992-9.

[110] Floden AM, Li S, Combs CK. Beta-amyloid-stimulated microglia induce neuron death via synergistic stimulation of tumor necrosis factor alpha and NMDA receptors. J Neurosci 2005; 25: 2566-75.

[111] Fatemi SH, Snow AV, Stary JM, *et al.* Reelin signaling is impaired in autism. Biol Psychiatry 2005; 57: 777-87.

[112] Serajee FJ, Zhong H, Mahbubul Huq AH. Association of Reelin gene polymorphisms with autism. Genomics 2006; 87: 75-83.

[113] Kelemenova S, Ostatnikova D. Neuroendocrine pathways altered in autism. Special role of reelin. Neuro Endocrinol Lett 2009; 30: 429-36.

[114] Herz J, Chen Y. Reelin, lipoprotein receptors and synaptic plasticity. Nat Rev Neurosci 2006; 7: 850-9.

[115] Dulabon L, Olson EC, Taglienti MG, *et al.* Reelin binds alpha3beta1 integrin and inhibits neuronal migration. Neuron 2000; 27: 33-44.

[116] Drakew A, Frotscher M, Deller T, Ogawa M, Heimrich B. Developmental distribution of a reeler gene-related antigen in the rat hippocampal formation visualized by CR-50 immunocytochemistry. Neuroscience 1998; 82: 1079-86.

[117] Lacor PN, Grayson DR, Auta J, *et al.* Reelin secretion from glutamatergic neurons in culture is independent from neurotransmitter regulation. Proc Natl Acad Sci U S A 2000; 97: 3556-61.

[118] Groc L, Choquet D, Stephenson FA, *et al.* NMDA receptor surface trafficking and synaptic subunit composition are developmentally regulated by the extracellular matrix protein Reelin. J Neurosci 2007; 27: 10165-75.

[119] Tissir F, Goffinet AM. Reelin and brain development. Nat Rev Neurosci 2003; 4: 496-505.

[120] Campo CG, Sinagra M, Verrier D, Manzoni OJ, Chavis P. Reelin secreted by GABAergic neurons regulates glutamate receptor homeostasis. PLoS One 2009; 4: e5505.

[121] Stocca G, Vicini S. Increased contribution of NR2A subunit to synaptic NMDA receptors in developing rat cortical neurons. J Physiol 1998; 507 (Pt 1): 13-24.

[122] Dantzer R, O'Connor JC, Freund GG, Johnson RW, Kelley KW. From inflammation to sickness and depression: when the immune system subjugates the brain. Nat Rev Neurosci 2008; 9: 46-56.

[123] Dantzer R, Kelley KW. Twenty years of research on cytokine-induced sickness behavior. Brain Behav Immun 2007; 21: 153-60.

[124] Churchill L, Taishi P, Wang M, *et al.* Brain distribution of cytokine mRNA induced by systemic administration of interleukin-1beta or tumor necrosis factor alpha. Brain Res 2006; 1120: 64-73.

[125] Rothwell N, Allan S, Toulmond S. The role of interleukin 1 in acute neurodegeneration and stroke: pathophysiological and therapeutic implications. J Clin Invest 1997; 100: 2648-52.

[126] Nadjar A, Combe C, Busquet P, Dantzer R, Parnet P. Signaling pathways of interleukin-1 actions in the brain: anatomical distribution of phospho-ERK1/2 in the brain of rat treated systemically with interleukin-1beta. Neuroscience 2005; 134: 921-32.

[127] Romeo HE, Tio DL, Rahman SU, Chiappelli F, Taylor AN. The glossopharyngeal nerve as a novel pathway in immune-to-brain communication: relevance to neuroimmune surveillance of the oral cavity. J Neuroimmunol 2001; 115: 91-100.

[128] Quan N, Whiteside M, Herkenham M. Time course and localization patterns of interleukin-1beta messenger RNA expression in brain and pituitary after peripheral administration of lipopolysaccharide. Neuroscience 1998; 83: 281-93.

[129] Banks WA. The blood-brain barrier in psychoneuroimmunology. Neurol Clin 2006; 24: 413-9.

[130] Schiltz JC, Sawchenko PE. Distinct brain vascular cell types manifest inducible cyclooxygenase expression as a function of the strength and nature of immune insults. J Neurosci 2002; 22: 5606-18.

[131] Dantzer R, Konsman JP, Bluthe RM, Kelley KW. Neural and humoral pathways of communication from the immune system to the brain: parallel or convergent? Auton Neurosci 2000; 85: 60-5.

[132] Gross PM, Sposito NM, Pettersen SE, Fenstermacher JD. Differences in function and structure of the capillary endothelium in gray matter, white matter and a circumventricular organ of rat brain. Blood Vessels 1986; 23: 261-70.

[133] Lacroix S, Feinstein D, Rivest S. The bacterial endotoxin lipopolysaccharide has the ability to target the brain in upregulating its membrane CD14 receptor within specific cellular populations. Brain Pathol 1998; 8: 625-40.

[134] Schulz M, Engrlhardt B. The circumventricular organs participate in the immunopathogenesis of experimental autoimmune encephalitis. Cerebrospinal Fluid Res 2005; 30: 8-12.

[135] Wang H, Yu M, Ochani M, *et al.* Nicotinic acetylcholine receptor alpha7 subunit is an essential regulator of inflammation. Nature 2003; 421: 384-8.

[136] Wang H, Liao H, Ochani M, *et al.* Cholinergic agonists inhibit HMGB1 release and improve survival in experimental sepsis. Nat Med 2004; 10: 1216-21.

[137] van der Zanden EP, Snoek SA, Heinsbroek SE, *et al.* Vagus nerve activity augments intestinal macrophage phagocytosis via nicotinic acetylcholine receptor alpha4beta2. Gastroenterology 2009; 137: 1029-39, 39 e1-4.

[138] Tysiak E, Asbach P, Aktas O, *et al.* Beyond blood brain barrier breakdown - *in vivo* detection of occult neuroinflammatory foci by magnetic nanoparticles in high field MRI. J Neuroinflamm 2009; 6: 20.

[139] Huang D, Wujek J, Kidd G, *et al.* Chronic expression of monocyte chemoattractant protein-1 in the central nervous system causes delayed encephalopathy and impaired microglial function in mice. FASEB J 2005; 19: 761-72.

[140] van Dam AM, Brouns M, Louisse S, Berkenbosch F. Appearance of interleukin-1 in macrophages and in ramified microglia in the brain of endotoxin-treated rats: a pathway for the induction of non-specific symptoms of sickness? Brain Res 1992; 588: 291-6.

[141] Gatti S, Bartfai T. Induction of tumor necrosis factor-alpha mRNA in the brain after peripheral endotoxin treatment: comparison with interleukin-1 family and interleukin-6. Brain Res 1993; 624: 291-4.

[142] Laye S, Parnet P, Goujon E, Dantzer R. Peripheral administration of lipopolysaccharide induces the expression of cytokine transcripts in the brain and pituitary of mice. Brain Res Mol Brain Res 1994; 27: 157-62.

[143] Dantzer R. Cytokine-induced sickness behavior: mechanisms and implications. Ann N Y Acad Sci 2001; 933: 222-34.

[144] Sparkman NL, Buchanan JB, Heyen JR, *et al.* Interleukin-6 facilitates lipopolysaccharide-induced disruption in working memory and expression of other proinflammatory cytokines in hippocampal neuronal cell layers. J Neurosci 2006; 26: 10709-16.

[145] Denicoff KD, Rubinow DR, Papa MZ, *et al.* The neuropsychiatric effects of treatment with interleukin-2 and lymphokine-activated killer cells. Ann Intern Med 1987; 107: 293-300.

[146] Capuron L, Gumnick JF, Musselman DL, *et al.* Neurobehavioral effects of interferon-alpha in cancer patients: phenomenology and paroxetine responsiveness of symptom dimensions. Neuropsychopharmacology 2002; 26: 643-52.

[147] Renault PF, Hoofnagle JH, Park Y, *et al.* Psychiatric complications of long-term interferon alfa therapy. Arch Intern Med 1987; 147: 1577-80.

[148] Meyers CA, Scheibel RS, Forman AD. Persistent neurotoxicity of systemically administered interferon-alpha. Neurology 1991; 41: 672-6.

[149] Adams F, Quesada JR, Gutterman JU. Neuropsychiatric manifestations of human leukocyte interferon therapy in patients with cancer. JAMA 1984; 252: 938-41.

[150] Kurzrock R, Quesada JR, Talpaz M, *et al.* Phase I study of multiple dose intramuscularly administered recombinant gamma interferon. J Clin Oncol 1986; 4: 1101-9.

[151] Turowski R, Triozzi P. Central nervous system toxicities of cytokine therapy. In: Plotnikoff N, Faith R, Murgo A, Good R, editors. Cytokines, Stress and Immunity. Baco Raton: CRC Press; 1998. p. 93-114.

[152] Gao HM, Jiang J, Wilson B, *et al.* Microglial activation-mediated delayed and progressive degeneration of rat nigral dopaminergic neurons: relevance to Parkinson's disease. J Neurochem 2002; 81: 1285-97.

[153] Gibbons HM, Dragunow M. Microglia induce neural cell death via a proximity-dependent mechanism involving nitric oxide. Brain Res 2006; 1084: 1-15.

[154] Carvey PM, Chang Q, Lipton JW, Ling Z. Prenatal exposure to the bacteriotoxin lipopolysaccharide leads to long-term losses of dopamine neurons in offspring: a potential, new model of Parkinson's disease. Front Biosci 2003; 8: s826-37.

[155] Ling Z, Zhu Y, Tong C, *et al.* Progressive dopamine neuron loss following supra-nigral lipopolysaccharide (LPS) infusion into rats exposed to LPS prenatally. Exp Neurol 2006; 199: 499-512.

[156] Langston JW, Forno LS, Tetrud J, *et al.* Evidence of active nerve cell degeneration in the substantia nigra of humans years after 1-methyl-4-phenyl-1,2,3,6-tetrahydropyridine exposure. Ann Neurol 1999; 46: 598-605.

[157] McGeer PL, Schwab C, Parent A, Doudet D. Presence of reactive microglia in monkey substantia nigra years after 1-methyl-4-phenyl-1,2,3,6-tetrahydropyridine administration. Ann Neurol 2003; 54: 599-604.

[158] Cunningham C, Wilcockson DC, Campion S, Lunnon K, Perry VH. Central and systemic endotoxin challenges exacerbate the local inflammatory response and increase neuronal death during chronic neurodegeneration. J Neurosci 2005; 25: 9275-84.

[159] Vargas DL, Nascimbene C, Krishnan C, Zimmerman AW, Pardo CA. Neuroglial activation and neuroinflammation in the brain of patients with autism. Ann Neurol 2005; 57: 67-81.

[160] Calderon-Garciduenas L, Maronpot RR, Torres-Jardon R, *et al.* DNA damage in nasal and brain tissues of canines exposed to air pollutants is associated with evidence of chronic brain inflammation and neurodegeneration. Toxicol Pathol 2003; 31: 524-38.

[161] Block ML, Wu X, Pei Z, *et al.* Nanometer size diesel exhaust particles are selectively toxic to dopaminergic neurons: the role of microglia, phagocytosis, and NADPH oxidase. FASEB J 2004; 18: 1618-20.

[162] Calderon-Garciduenas L, Reed W, Maronpot RR, *et al.* Brain inflammation and Alzheimer's-like pathology in individuals exposed to severe air pollution. Toxicol Pathol 2004; 32: 650-8.

[163] Takenaka S, Karg E, Roth C, *et al.* Pulmonary and systemic distribution of inhaled ultrafine silver particles in rats. Environ Health Perspect 2001; 109 Suppl 4: 547-51.

[164] Blaylock RL, Strunecka A. Immune-glutamatergic dysfunction as a central mechanism of the autism spectrum disorders. Curr Med Chem 2009; 16: 157-70.

[165] Block ML, Hong JS. Microglia and inflammation-mediated neurodegeneration: multiple triggers with a common mechanism. Prog Neurobiol 2005; 76: 77-98.

[166] Gao HM, Hong JS, Zhang W, Liu B. Synergistic dopaminergic neurotoxicity of the pesticide rotenone and inflammogen lipopolysaccharide: relevance to the etiology of Parkinson's disease. J Neurosci 2003; 23: 1228-36.

[167] Gao HM, Liu B, Zhang W, Hong JS. Synergistic dopaminergic neurotoxicity of MPTP and inflammogen lipopolysaccharide: relevance to the etiology of Parkinson's disease. FASEB J 2003; 17: 1957-9.

[168] Filipov NM, Seegal RF, Lawrence DA. Manganese potentiates *in vitro* production of proinflammatory cytokines and nitric oxide by microglia through a nuclear factor kappa B-dependent mechanism. Toxicol Sci 2005; 84: 139-48.

[169] Takahashi RN, Rogerio R, Zanin M. Maneb enhances MPTP neurotoxicity in mice. Res Commun Chem Pathol Pharmacol 1989; 66: 167-70.

[170] Charleston JS, Body RL, Bolender RP, *et al.* Changes in the number of astrocytes and microglia in the thalamus of the monkey Macaca fascicularis following long-term subclinical methylmercury exposure. Neurotoxicology 1996; 17: 127-38.

[171] Brookes N. Specificity and reversibility of the inhibition by HgCl2 of glutamate transport in astrocyte cultures. J Neurochem 1988; 50: 1117-22.

[172] Furuta A, Rothstein JD, Martin LJ. Glutamate transporter protein subtypes are expressed differentially during rat CNS development. J Neurosci 1997; 17: 8363-75.

[173] Kugler P, Schleyer V. Developmental expression of glutamate transporters and glutamate dehydrogenase in astrocytes of the postnatal rat hippocampus. Hippocampus 2004; 14: 975-85.

[174] Chmielnicka J, Komsta-Szumska E, Sulkowska B. Activity of glutamate and malate dehydrogenases in liver and kidneys of rats subjected to multiple exposures of mercuric chloride and sodium selenite. Bioinorg Chem 1978; 8: 291-302.

[175] Allen JW, Mutkus LA, Aschner M. Mercuric chloride, but not methylmercury, inhibits glutamine synthetase activity in primary cultures of cortical astrocytes. Brain Res 2001; 891: 148-57.

[176] Juarez BI, Martinez ML, Montante M, *et al.* Methylmercury increases glutamate extracellular levels in frontal cortex of awake rats. Neurotoxicol Teratol 2002; 24: 767-71.

[177] Katayama Y, Kohso K, Nishimura A, *et al.* Detection of measles virus mRNA from autopsied human tissues. J Clin Microbiol 1998; 36: 299-301.

[178] Adamson DC, Kopnisky KL, Dawson TM, Dawson VL. Mechanisms and structural determinants of HIV-1 coat protein, gp41-induced neurotoxicity. J Neurosci 1999; 19: 64-71.

[179] Viviani B, Gardoni F, Bartesaghi S, *et al.* Interleukin-1 beta released by gp120 drives neural death through tyrosine phosphorylation and trafficking of NMDA receptors. J Biol Chem 2006; 281: 30212-22.

[180] Broderick PA. Interleukin 1alpha alters hippocampal serotonin and norepinephrine release during open-field behavior in Sprague-Dawley animals: differences from the Fawn-Hooded animal model of depression. Prog Neuropsychopharmacol Biol Psychiatry 2002; 26: 1355-72.

[181] Lellouch-Tubiana A, Fohlen M, Robain O, Rozenberg F. Immunocytochemical characterization of long-term persistent immune activation in human brain after herpes simplex encephalitis. Neuropathol Appl Neurobiol 2000; 26: 285-94.

[182] De Tiege X, De Laet C, Mazoin N, *et al.* Postinfectious immune-mediated encephalitis after pediatric herpes simplex encephalitis. Brain Dev 2005; 27: 304-7.

[183] Andersson T, Schultzberg M, Schwarcz R, *et al.* NMDA-Receptor Antagonist Prevents Measles Virus-induced Neurodegeneration. Eur J Neurosci 1991; 3: 66-71.

[184] Hornig M, Weissenbock H, Horscroft N, Lipkin WI. An infection-based model of neurodevelopmental damage. Proc Natl Acad Sci U S A 1999; 96: 12102-7.

[185] Sly LM, Krzesicki RF, Brashler JR, *et al.* Endogenous brain cytokine mRNA and inflammatory responses to lipopolysaccharide are elevated in the Tg2576 transgenic mouse model of Alzheimer's disease. Brain Res Bull 2001; 56: 581-8.

[186] Dunn N, Mullee M, Perry VH, Holmes C. Association between dementia and infectious disease: evidence from a case-control study. Alzheimer Dis Assoc Disord 2005; 19: 91-4.

[187] Lemstra AW, Groen in't Woud JC, Hoozemans JJ, *et al.* Microglia activation in sepsis: a case-control study. J Neuroinflammation 2007; 4: 4.

[188] O'Banion D, Armstrong B, Cummings RA, Stange J. Disruptive behavior: a dietary approach. J Autism Child Schizophr 1978; 8: 325-37.

[189] Lucarelli S, Frediani T, Zingoni AM, *et al.* Food allergy and infantile autism. Panminerva Med 1995; 37: 137-41.

[190] Kidd PM. Autism, an extreme challenge to integrative medicine. Part 2: medical management. Altern Med Rev 2002; 7: 472-99.

[191] Wakefield AJ, Murch SH, Anthony A, *et al.* Ileal-lymphoid-nodular hyperplasia, non-specific colitis, and pervasive developmental disorder in children. Lancet 1998; 351: 637-41.

[192] Ashwood P, Wakefield AJ. Immune activation of peripheral blood and mucosal CD3+ lymphocyte cytokine profiles in children with autism and gastrointestinal symptoms. J Neuroimmunol 2006; 173: 126-34.

[193] Vojdani A, Campbell AW, Anyanwu E, *et al.* Antibodies to neuron-specific antigens in children with autism: possible cross-reaction with encephalitogenic proteins from milk, Chlamydia pneumoniae and Streptococcus group A. J Neuroimmunol 2002; 129: 168-77.

[194] Vojdani A, O'Bryan T, Green JA, *et al.* Immune response to dietary proteins, gliadin and cerebellar peptides in children with autism. Nutr Neurosci 2004; 7: 151-61.

[195] Hida S, Miura NN, Adachi Y, Ohno N. Effect of Candida albicans cell wall glucan as adjuvant for induction of autoimmune arthritis in mice. J Autoimmun 2005; 25: 93-101.

[196] Hida S, Nagi-Miura N, Adachi Y, Ohno N. Beta-glucan derived from zymosan acts as an adjuvant for collagen-induced arthritis. Microbiol Immunol 2006; 50: 453-61.

[197] Black C, Kaye JA, Jick H. Relation of childhood gastrointestinal disorders to autism: nested case-control study using data from the UK General Practice Research Database. BMJ 2002; 325: 419-21.

[198] Hadjivassiliou M, Grunewald R, Sharrack B, *et al.* Gluten ataxia in perspective: epidemiology, genetic susceptibility and clinical characteristics. Brain 2003; 126: 685-91.

[199] Hu WT, Murray JA, Greenaway MC, Parisi JE, Josephs KA. Cognitive impairment and coeliac disease. Arch Neurol 2006; 63: 1440-6.

[200] Medhi B, Prakash A, Avti PK, Chakrabarti A, Khanduja KL. Intestinal inflammation and seizure susceptibility: understanding the role of tumour necrosis factor-alpha in a rat model. J Pharm Pharmacol 2009; 61: 1359-64.

[201] Geissler A, Andus T, Roth M, *et al.* Focal white-matter lesions in brain of patients with inflammatory bowel disease. Lancet 1995; 345: 897-8.

[202] Azzouz D, Gargouri A, Hamdi W, *et al.* Coexistence of psoriatic arthritis and collagenous colitis with inflammatory nervous system disease. Joint Bone Spine 2008; 75: 624-5.

[203] Nelson KB, Willoughby RE. Infection, inflammation and the risk of cerebral palsy. Curr Opin Neurol 2000; 13: 133-9.

[204] Shi L, Fatemi SH, Sidwell RW, Patterson PH. Maternal influenza infection causes marked behavioral and pharmacological changes in the offspring. J Neurosci 2003; 23: 297-302.

[205] Shi L, Smith SE, Malkova N, *et al.* Activation of the maternal immune system alters cerebellar development in the offspring. Brain Behav Immun 2009; 23: 116-23.

[206] Skogstrand K, Hougaard DM, Schendel DE, *et al.* Association of preterm birth with sustained postnatal inflammatory response. Obstet Gynecol 2008; 111: 1118-28.

[207] Marshall-Clarke S, Reen D, Tasker L, Hassan J. Neonatal immunity: how well has it grown up? Immunol Today 2000; 21: 35-41.

[208] Osrin D, Vergnano S, Costello A. Serious bacterial infections in newborn infants in developing countries. Curr Opin Infect Dis 2004; 17: 217-24.

[209] Purisai MG, McCormack AL, Cumine S, *et al.* Microglial activation as a priming event leading to paraquat-induced dopaminergic cell degeneration. Neurobiol Dis 2007; 25: 392-400.

[210] Ling Z, Chang QA, Tong CW, *et al.* Rotenone potentiates dopamine neuron loss in animals exposed to lipopolysaccharide prenatally. Exp Neurol 2004; 190: 373-83.

[211] Bondy SC, LeBel CP. The relationship between excitotoxicity and oxidative stress in the central nervous system. Free Radic Biol Med 1993; 14: 633-42.

[212] Coyle JT, Puttfarcken P. Oxidative stress, glutamate, and neurodegenerative disorders. Science 1993; 262: 689-95.

[213] Bannai S, Tateishi N. Role of membrane transport in metabolism and function of glutathione in mammals. J Membr Biol 1986; 89: 1-8.

[214] Han D. In: Poli G, Gadenase E, Packer L, Ed. Free Radicals in Brain Pathophysiology. New York: Marcel Dekker, Inc;; 2000.

[215] Gunter KK, Gunter TE. Transport of calcium by mitochondria. J Bioenerg Biomembr 1994; 26: 471-85.

[216] Schinder AF, Olson EC, Spitzer NC, Montal M. Mitochondrial dysfunction is a primary event in glutamate neurotoxicity. J Neurosci 1996; 16: 6125-33.

[217] Dessi F, Ben-Ari Y, Charriaut-Marlangue C. Ruthenium red protects against glutamate-induced neuronal death in cerebellar culture. Neurosci Lett 1995; 201: 53-6.

[218] Patel M, Day BJ, Crapo JD, Fridovich I, McNamara JO. Requirement for superoxide in excitotoxic cell death. Neuron 1996; 16: 345-55.

[219] Chan PH, Chu L, Chen SF, Carlson EJ, Epstein CJ. Reduced neurotoxicity in transgenic mice overexpressing human copper-zinc-superoxide dismutase. Stroke 1990; 21: III80-2.

[220] Hensley K, Tabatabaie T, Stewart CA, Pye Q, Floyd RA. Nitric oxide and derived species as toxic agents in stroke, AIDS dementia, and chronic neurodegenerative disorders. Chem Res Toxicol 1997; 10: 527-32.

[221] Ano Y, Sakudo A, Kimata T, *et al.* Oxidative damage to neurons caused by the induction of microglial NADPH oxidase in encephalomyocarditis virus infection. Neurosci Lett 2010; 469: 39-43.

[222] Brown GC, Neher JJ. Inflammatory Neurodegeneration and Mechanisms of Microglial Killing of Neurons. Mol Neurobiol 2010; 41: 242-7.

CHAPTER 5

Immune Dysfunction in Autism Spectrum Disorders

Russell L. Blaylock, MD

Institute for Theoretical Neuroscience, LLC, and Visiting Professor of Biology, Belhaven University, Ridgeland, MS 39157, USA

Abstract: A great number of studies have been done examining immune function in children with autism spectrum disorders (ASD). Most of these studies have demonstrated immune dysfunction, especially involving cellular immunity. Important to the immunoexcitotoxicity hypothesis is the finding that macrophages and lymphocytes from ASD children have been shown to demonstrate an amplified release of pro-inflammatory cytokines with stimulation, especially in those having gastrointestinal (GI) symptoms. Because of the intimate connection between the gut and brain, hyperimmune responses from the gut, vial vagal afferents, can rapidly activate brain microglia, leading to an exaggerated innate immune response within the brain. It has also been shown that ASD children often react to food peptides, such as gliadin, gluten and casein as well as a number of bacterial and fungal antigens, all of which can exaggerate immunoexcitotoxicity. The finding of cross-reacting food antigen with brain components also indicates the presence of bystander damage and would trigger immunoexcitotoxicity as well.

INTRODUCTION

Accumulated evidence clearly indicates a major involvement of immune system dysfunction in the ASD [1, 2]. Because of the complex interaction between the immune system and the central nervous system (CNS) many of the details of this dysfunction have not been clearly worked out.

Several studies have shown that at least a subset of ASD children have affected genes that are related to immune function, such as *HLA, C4B, PTEN, REELIN*, and *MET* [3-6]. It appears that autoimmunity is playing a significant role in risk, but that only in combination with certain environmental triggers. The genetic influence involves at least 10 genes and perhaps more [7].

Warren and co-workers investigated activated T-cells in 26 autistic subjects and found that 14 had DR+T cells, indicating activation, but that none of these cells expressed interleukin-2 (IL-2), another indicator of activation in normal immune systems [3]. This combination has been seen with autoimmunity.

A high incidence of autoimmune diseases in family members has been noted by several authors [8-10]. In most cases a first-degree relative was identified in families with an autistic child. It is also of some interest that maternal immune abnormalities, including autoimmune disorders, allergies and asthma during the second trimester of pregnancy are associated with a two-fold higher incidence of autism in their offspring. This may be related to eventual microglial priming in the developing baby triggered by systemic immune activation in the mother during pregnancy.

A number of studies have found antibodies to brain structures in autistic individuals. Antibodies to serotonin receptors, myelin basic protein, neuron filament protein, cerebellar neurofilaments, nerve growth factor, α2-adrenergic-binding sites, anti-brain endothelial cell proteins and other less specific anti-brain protein antibodies have been described [11-15].

The question that remains is what is the relationship between these anti-brain antibodies and neurological symptoms in ASD? Are these merely reactions to the brain injury itself, as we see in cases of strokes and brain trauma, or are they responsible for many of the symptoms seen with these disorders?

*Address correspondence to: **Russell L. Blaylock** Institute for Theoretical Neuroscience, LLC, and Visiting Professor of Biology, Belhaven University, Ridgeland, MS 39157, USA; E-mail: Blay6307@bellsouth.net

Anna Strunecka (Ed)

Evidence for the latter come from a study in which serum from the mother of an autistic child was injected into gestating mice and produced alterations in exploration, motor coordination and changes in cerebellar magnetic resonance spectroscopy in the offspring [16]. The serum from this mother was found to bind with Purkinje cells and others neuronal cell types. It is known that antibody binding to neurotransmitter receptors, as seen with anti-serotonin receptor antibodies in ASD, can trigger action potentials and this could explain some of the symptoms triggered by anti-brain antibodies [17-19]. This can not only lead to progressive neurodegeneration and neurological symptoms but also seizures, which are seen in 30% to as high as 80% of ASD children depending on the accuracy of testing. To date, no studies have looked for anti-glutamate receptor antibodies in ASD.

In one interesting study, Singh and Rivas demonstrated antibodies to rat caudate nucleus in 49% of autism patients examined and in none of control patients [20]. Injury to the caudate has been associated with obsessive-compulsive behavior.

While blocking immune responses has shown some beneficial results in a subset of autistic children, this could represent a reduction in the immune arm of the immunoexcitotoxic reaction and not a reduction in autoimmune damage to neuronal structures.

While these studies appear to imply a common link to autism symptoms, there are a number of conflicting reports and in most studies only a portion of autistic patients demonstrated such antibodies. Again, this could represent a reaction to the primary injury itself and not an etiology. Even if autoimmunity is not the cause of autism, it may be responsible for some of the symptomatology either through direct interaction with neural elements or more likely by way of bystander injury.

ABNORMALITIES OF THE PERIPHERAL IMMUNE SYSTEM

Compelling evidence indicates that the immune arm of the immunoexcitotoxic reaction within the brain is not secondary to excessive stimulation of a normally functioning systemic immune system alone. There appears to be a number of abnormalities of the peripheral immune system in ASD, such as decreased lymphocyte numbers, reduced responsiveness of T cells to mitogens, abnormal release of IL-2 from activated lymphocytes, abnormalities in apoptosis mechanisms, dysregulated IgG and elevation in a number of cytokines.

A considerable amount of evidence indicates that immune mediators, such as cytokines and chemokines, play a vital role in neurodevelopment and that their rise and fall parallels that of glutamate. The cytokines appear to play a neurotrophic role as well as assisting in glutamate pruning during later stages of brain development.

Of particular importance is the finding of significant increased release of tumor necrosis factor-α (TNF-α) from lymphocytes isolated from autistic patients having GI symptoms as compared to age-matched control patients, even those with constipation [21]. Of special interest was the finding that stimulation of mucosal lymphocytes failed to trigger counter regulatory release of IL-10, thus increasing the intensity of the pro-inflammatory immune reaction [22, 1]. Since this inflammatory reaction is occurring within the territory of the vagus nerve, one would expect significant stimulation of brain microglia via the dorsal vagal complex.

Jyonouchi and co-workers have demonstrated peripheral hyper-reactivity of lymphocytes from autistic patients upon stimulation with specific dietary proteins [23]. This would activate brain microglia via the blood brain barrier (BBB) cytokine transport process, by way of the circumventricular organs (CVO) and the choroid plexus. Thus, all routes of systemic immune communication with the CNS are utilized.

Autism shares with schizophrenia a number of immune abnormalities, including lymphocyte abnormalities, dysregulated IgG, elevated pro-inflammatory cytokines and the presence of autoantibodies against brain [24, 25]. One also sees microglial activation in post-mortem studies of the schizophrenic brain, primarily in the prefrontal cortex and temporal lobes [26]. It has been noted that prior to specific diagnostic criteria for autism, most cases would have been diagnosed as juvenile schizophrenia. High levels of immune activation during mid-term pregnancy is associated with a high risk of both autism and schizophrenia, demonstrating another immunological link between the two disorders [27].

Autistic children are known to suffer from recurrent infections and to have difficulty recovering from these infections. Impaired ability to suppress viral and bacterial infection has been associated with ASD [1, 28]. Persistent viral infections, especially those latent in the brain, can not only act as microglial priming agents but also with periodic reactivation initiate the secretion of high levels of pro-inflammatory cytokines and excitotoxins, leading to a loss of synapses and dendritic retraction as well as neuronal loss. Likewise, retained viral fragments can act as sources of immunoexcitotoxicity as well and do so without live viral elements.

Measles virus is known to have profound immunosuppressive effects and is also frequently retained in brain tissue for a lifetime [29]. The presence of preexisting brain pathology or retained immunogenic elements has been shown to greatly magnify systemic-triggered microglial release of immunoexcitotoxic factors. For example, Cunningham and co-workers found that priming brain microglia with an antigen (ME7) followed by systemic lipopolysaccharide (LPS) challenge produced a 3-fold higher release of brain IL-1ß than when microglia were not primed [30]. IL-1ß is the prime regulator of microglial activation.

A number of environmental toxins are known to activate microglia and initiate both priming and hyperreactivity of previously primed microglia during subsequent systemic immune stimulation. These include organic mercury, aluminum, lead and possibly fluoride (as aluminofluoride complex) [31-33]. Vaccines can contain either mercury or more commonly, aluminum. Relatively high levels of aluminum are injected into children secondary to the large number of vaccines given over a relatively short period of time. (Some 36 to 38 vaccines by age six years). In addition, food and public drinking water are a major source of aluminum. Dental amalgam and atmospheric mercury, in some geographic areas, also remain a major source of mercury.

So, we see that not only is the autistic child subject to repeated episodes of immune stimulation systemically, but because of a dysfunctional immune system the normal immune counter regulatory system, primarily IL-10 and TGF-ß, one would expect a hyperimmune response to a number of antigens. The presence of early brain pathology from either intrauterine or early postnatal immunoexcitotoxicity, would mean that the response of the ASD child's brain would be much greater than a normal child. This may explain, along with genetic influences, why all children are not profoundly affected by environmental, infectious and vaccine-induced immune activation and immunoexcitotoxicity.

Some immune dysfunctions have been linked to particular behavioral expressions in autistic children. For example, Ashwood and co-workers examined the level of transforming growth factor-ß1 (TGF-ß1) in the plasma of 75 autistic children as compared to 96 aged-matched controls and children with other types of developmental disabilities adjusted for age and gender [34]. They found that TGF-ß1 was significantly lower in the plasma of autistic children than controls. Importantly, they found a direct correlation between the lowest TGF-ß1 and lower adaptive behaviors and worse behavioral symptoms.

There was an inverse correlation between TGF-ß1 plasma levels and measures of irritability, lethargy, stereotypy and hyperactivity. They found no difference between children with regressive autism and early onset autism as related to TGF-ß1 levels. These studies indicated that as TGF-ß1 levels fall, one sees a more atypical behavioral profile, which also involves communication and social interaction. They also found a significant association between low TGF-ß1 and irritability as measured by ABC in children who regressed.

As they demonstrate, TGF-ß1 has a very complex physiology. TGF-ß1 plays a major role in cell growth and differentiation, organ development, cell migration, apoptosis and plays a critical role in immune system regulation. During early immune response, TGF-ß1 can increase inflammation and later down-regulate T and B-cell development and function and control differentiation and activation of natural killer (NK) cells, dendritic cells, monocytes/macrophages, granulocytes, and mast cells [35]. Studies of autoimmune disease models indicate that deficiencies in TGF-ß1 can worsen symptoms and damage [36, 37].

Eric Hollander and co-workers in examining eighteen children with autism (14 boys and four girls) for the presence of monoclonal antibody D8/17 on their lymphocytes found that 77.8% of the autistic children were positive for the antigen versus 21.4% of control children [38]. This antibody has been linked to a greater susceptibility to rheumatic fever and also linked with obsessive-compulsive disorder (OCD) [39]. They also found that the severity of repetitive behaviors correlated positively with D8/17 positives, as measured by the Yale-Brown Obsessive-Compulsive Scale. Monoclonal antibody D8/17 identifies a B cell antigen. There was no link to communication or social scores.

A number of other autoimmune disorders are associated with neuropsychiatric symptoms and behavioral dysfunction, including Sydneham's chorea, PANDAS and SLE, demonstrating a link between autoimmune antibodies, immune cells and brain-behavioral function. Just how much of autistic behavior is linked to peripheral autoimmunity has not been determined, but a stronger link can be made between brain immunoexcitotoxicity and repetitive, sequential systemic immune stimulation in the face of a dysfunctional immune system.

Cytokines and ASD

A number of studies have shown abnormalities in cytokines and chemokine levels in ASD. One study reported elevations in interferon-α (INF-α) in 10 autistic children and no elevation in four adult controls [40]. Another study found increased levels of IL-12 and INF-γ but found no changes in IL-6, TNF-α, sICAM and INF-α [41]. The author suggested that this might indicate Th1 dominant autoimmunity.

Sweeten and co-workers, in a series of studies, found that autistic patients had elevated measures of nitric oxide (NO) production with otherwise normal cytokine levels. It was assumed that the source of the NO was INF-γ, based on a previous finding of elevated levels of neopterin in the blood of autistic children [10, 42]. Immunoexcitotoxicity could also account for elevations in NO, primarily from calcium-activated iNOS [43]. They also found significantly elevated numbers of monocytes in 31 autistic children examined but not other leukocytes.

Stimulation studies indicate abnormalities in the production of certain cytokines. In one such study involving 20 autistic subjects as compared to 20 age-matched controls, researchers found impaired production of IFN-γ and IL-2 in CD4+ and CD8+ lymphocytes with stimulation, but higher levels of IL-4, indicating a Th2 bias [44]. Of particular interest in terms of sickness-like behavior is the finding of amplified release of TNF-α and IL-1ß in monocytes from autistic subjects with stimulation by LPS. Both of these cytokines, primarily IL-1ß, is thought to be the primary cytokine messenger for activation of brain microglia upon systemic immune stimulation. They also play a significant role in inter-microglia communication and microglial communication with astrocytes.

This finding of immune cell hyperactivity with stimulation would help explain why only a subset of children are at increased risk from sequential immune stimulation, especially when combined with the finding that autistic children have reduced counter regulation mechanism, IL-10 and TGF-ß1, in the face of inflammation. With each infection, exposure to immune-activating environmental toxins and each exposure to dietary excitotoxins, one would expect to see an intense systemic immune-driven stimulation of microglial activation.

Recent studies have found that immune reactions within the autistic brain are also amplified. Li and co-workers found that ASD patients displayed an increased innate and adaptive immune response via the TH1 pathway [45]. Using Multiplex Bead Immunoassays they found that TNF-α, IL-6, GM-CSF, INF-γ and IL-8 were significantly increased in the ASD brain. Importantly, they also found that regulatory IL-10 was not increased in a compensatory way as should be seen in the normal brain. Again, this shows an amplification of immune responses in the ASD brain with an inability to attenuate these immune reactions. That is, both the systemic and the brain immune response to stimulation are amplified.

Garbett and co-workers found an increased expression of a considerable number of immune system-related genes in superior temporal lobes of autistic brains [46]. They note that these dysregulated gene pathways might indicate an inability to attenuate a cytokine activation signal in these autistic brains. That is, unlike in the case of natural infections or in sickness behavior in normal children, the autistic person would have great difficulty shutting down their microglial activation once it was initiated. This would result in either a prolonged activation or intense intermittent activation with each immune stimulus. With each immune challenge, the activated microglia release a high level of immunoexcitotoxic mediators, including cytokines, chemokines, interferons, prostaglandins-2 (PGE2), reactive oxygen species/ reactive nitrogen species (ROS/RNS), lipoperoxidation products (LPP), glutamate, aspartate, and quinolinic acid (QUIN). This seems to be what we are seeing in the Vargas study [47].

Several studies have also shown altered expression of glutamate system regulating genes in the brain of autistic subjects. Purcell and co-workers using DNA microarray studies of postmortem cerebellum from autistic subjects found upregulation of AMPA glutamate genes and glutamate binding proteins [48]. It is important to recall that GluR transport proteins under certain conditions can operate in reverse, increasing extraneuronal glutamate levels. Garbett and co-workers also found abnormalities in glutamate system genes [46]. Rather than upregulation, they found downregulation of glutamatergic transcripts, including those for GluRs and glutamate transporters. This may, as they point out, be a regional difference in brain areas being tested, cerebellum versus temporal lobe.

Reduced glutamic acid decarboxylase transcripts (GAD1 and GAD2) were seen in the temporal lobe in this study and have been previously reported by Yip and co-workers in cerebellar Purkinje cells, and in the parietal cortex and cerebellum by others [49, 50]. Lower GABAergic function would raise the excitotoxic ratio. Because of the high levels of ROS/RNS and LPPs, one sees suppression of two other glutamate regulatory enzymes, GDH and GS, which would also elevate extraneuronal glutamate levels.

Ashwood suggest that cytokine production, based on available studies, implies a more complex pattern than originally thought and that this may, in part be secondary to patient selection of various phenotypic forms of autism [2].

ROLE OF ENVIRONMENTAL NEUROTOXINS

A number of environmental toxins are known to affect neurodevelopment, immune function and excitotoxicity. Of greatest interest is the effect of organophosphate compounds on the developing nervous system.

In one interesting study, it was found that the children of mothers who lived within 500 meters of field sites with the highest organochlorine poundage (dicofol and endosulfan) had a risk factor for autism 6.1 times higher than mothers not living near these fields [51]. The highest incidence was with the highest exposures during the 8 weeks immediately following cranial neural tube closing.

It has been proposed that autism is associated with an imbalance between excitatory and inhibitory neurotransmission during embryogenesis. Organochloride pesticides are known to inhibit γ-amino butyric acid (GABA) receptors. Even though many of the organochloride pesticides have been banned, because of their prolonged half-life and bioaccumulation, risk are still of concern. Heptachlor, dieldrin, and toxaphene are among these banned organochloride pesticides. Of the polychloroalkane insecticides still being used are endosulfane and lindane.

Another class of insecticides is known to interfere with GABA neurotransmission and include the 4-alkyl-1-phenylpyrazoles and while they do not tend to persist in the environment, they are heavily used in homes and for commercial pest control. Fipronil is used to control fleas and ticks on pets and is a non-competitive antagonist of GABRs. Pessah and co-workers note that a group of widely used insecticides all interact with mammalian GABRß3 [52]. This is of special importance for the autistic child since studies have shown impaired GABRß3 expression in autism, Rett Syndrome, and Angleman Syndrome. Dyregulation of GABRs increases risk of excitotoxicity. A large number of environmental chemicals, including pesticides, herbicides and fungicides, are known to alter intracellular calcium function. Dioxins mediate their toxicity via the arylhydrocarbon receptor and non-coplanar polychlorinated biphenyls via ryanodine receptors [53]. Disruptions of calcium levels within neurons can increase the activity of the excitotoxic cascade and elevations in mitochondrial calcium can suppress energy production and increase free radical generation.

Recently Ashwood and co-workers defined abnormal cytokine and chemokine responses of monocytes to the flame retardant 2,2′,4,4′-tetrabrominated biphenyl (BDE-47) [54]. They exposed peripheral blood mononuclear cells (PBMC) isolated from 19 children with autism and 18 aged-matched typically developing (TD) controls to either 100 nM or 500 nM of BDE-47 and measured the induced release of cytokines and chemokines.

Normally LPS stimulated monocytes/macrophages produce IL-1ß, IL-6, IL-8, IL-10, IL-12p40, TNF-α, GM-CSF, macrophage inflammatory protein (MIP) MIP-1α and MIP-1ß. Lymphocytes release IL-2, IL-4, IL-5, IFN-γ, RANTES, and eotaxin. LPS signaling is via CD-14-TLR-4 complexes in both monocytes and macrophages. Microglia contain a number of toll-like receptors (TLR), including TLR4.

Unstimulated PBMC on exposure to 100 nM or 500 nM of BDE-47 demonstrated no cytokine releasing effects on ASD cells or TD control cells. They did see a fall in chemokine IL-8 release in both ASD and TD controls with exposure to 100 nM of BDE-47 but not 500 nM and a significant increase release of MIP-1α at the 100 nM and 500 nM dose in both ASD and TD control PBMC cells in culture. Cells pretreated with 500 nM of BDE-47 but not stimulated with LPS demonstrated an increase in MIP-1ß release in ASD cells but not TD controls.

Cells pretreated with BDE-47 and then stimulated with LPS demonstrated significant differences in immune profiles between ASD cells and TD controls. At 100 nM dose, TD controls demonstrated a significant decrease in granulocyte macrophage colony-stimulating factor (GM-CSF), IL-12p40, TNF-α, IL-6, MIP-1α, and MIP-1ß. In the ASD cells, IL-1ß was significantly increased when preincubated with 100 nM of BDE-47. At the 500 nM dose similar responses were seen in both the ASD and TD control patients. They also found an increase release of chemokine IL-8 with LPS stimulation in the ASD cells [54]. This study and those showing abnormal IL-1ß, IL-12, and TNF-α release from PBMC in ASD patients in children with GI problems, again suggest abnormal cytokine and chemokine profiles in response to immune stimulation in ASD [41].

BDE-47 is from a class of flame retardants, polybrominated diphenyl ether (PBDE), that are used widely in the textile industry, building and the manufacturing of computers and televisions. These are also known to be some of the most prevalent environmental compounds found in human tissues [55]. They have a half-life of 1.8 years. We are seeing rapid increasing tissue levels of these toxins in human tissues [56].

Children are exposed to higher levels of PBDE in contaminated foods, indoor and outdoor air, ingestion of dust, and from direct dermal exposure [57]. It is also found in human milk, with levels having increased 60-fold over the last 25 years [56]. Interestingly, PBDE have been shown to lower hippocampal *brain-derived neurotrophic factor* (BDNF) during early brain development [58]. Some studies of BDNF in plasma of children with autism have not shown deficiencies of this neurotrophic substance [59]. Yet, Katoh-Semba and co-workers have shown that the normal early elevation in BDNF is delayed in autism and significantly lower in children aged 0-9 years old and Hashimoto and co-workers found lower BDNF in adults with autism [60, 61].

In essence, pregnant women and children are exposed to a great number of environmental toxins including neurotoxic metals, pesticides, herbicides, fungicides, and industrial chemicals that have been shown to adversely affect neurodevelopment, induce autoimmunity and disrupt immune system functions.

CONCLUSIONS

A number of immune dysfunctions and abnormalities have been described in autistic children, which increase their risk of autoimmune disorders and recurrent infections. They also have a higher risk of persistent infections, especially viruses. As a result, we see, not only a greater incidence of recurrent, sequential, infections, but also prolonged immune activation, both systemically and in the CNS.

As noted, the immune cells of the autistic person produce higher levels of pro-inflammatory cytokines upon stimulation, have abnormalities in peripheral immune cells and hyperactivity of brain microglia. Of great importance is the finding that counter regulatory anti-inflammatory cytokines, including IL-10 and TGF-ß1, are suppressed. This not only increases inflammatory reactions, but also may trigger persistent immune activation. The Vargas et al study suggests that the immune reaction, via activated microglia and astrocytes, is continuous over decades [47].

Because both excitatory amino acid and immune cytokines play such a critical role in neurodevelopment, prolonged immunoexcitotoxicity would be expected to result in many of the neuropathological and neurobehavioral findings we see in ASD. I suspect that the original priming event can occur either *in utero* or within the first several years after birth, as we see with schizophrenia, a closely related disorder. Repeated, closely spaced vaccinations would also be expected to produce the same neurological consequences that we see with using LPS experimentally in a similar manner [62, 63].

Other factors involved in prolonged and/or repeated systemic immune stimulation would be food intolerances, leaky gut syndrome, candidasis and vaccine-induced colitis, all of which would intensely activate the brain's microglia via the vagal afferents. Colitis models in animals have clearly demonstrated activation of brain microglia and evidence of immunoexcitotoxicity [64].

We also see that during critical phases of neurodevelopment the child is exposed to numerous environmental chemicals that can alter neurological function, activate microglia, trigger excitotoxic cascades, increase brain ROS/RNS and LPP levels, and cause dysfunction of the immune system.

Taken together, we now have sufficient evidence to say that immunoexitotoxicity risks are higher in children having a set of altered genes operational for glutamate function and immune function and especially when these children are exposed to closely spaced sequential immune stimulation, no matter the origin.

REFERENCES:

[1]   Ashwood P, Van de Water J. A review of autism and the immune response. Clin Dev Immunol 2004; 11: 165-74.

[2]   Ashwood P, Wills S, Van de Water J. The immune response in autism: a new frontier for autism research. J Leukoc Biol 2006; 80: 1-15.

[3]   Warren RP, Yonk J, Burger RW, Odell D, Warren WL. DR-positive T cells in autism: association with decreased plasma levels of the complement C4B protein. Neuropsychobiology 1995; 31: 53-7.

[4]   Polleux F, Lauder JM. Toward a developmental neurobiology of autism. Ment Retard Dev Disabil Res Rev 2004; 10: 303-17.

[5]   Campbell DB, Sutcliffe JS, Ebert PJ, *et al.* A genetic variant that disrupts MET transcription is associated with autism. Proc Natl Acad Sci U S A 2006; 103: 16834-9.

[6]   Campbell DB, Warren D, Sutcliffe JS, Lee EB, Levitt P. Association of MET with social and communication phenotypes in individuals with autism spectrum disorder. Am J Med Genet B Neuropsychiatr Genet 2010; 153B: 438-46.

[7]   Muhle R, Trentacoste SV, Rapin I. The genetics of autism. Pediatrics 2004; 113: e472-86.

[8]   Money J, Bobrow NA, Clarke FC. Autism and autoimmune disease: a family study. J Autism Child Schizophr 1971; 1: 146-60.

[9]   Comi AM, Zimmerman AW, Frye VH, Law PA, Peeden JN. Familial clustering of autoimmune disorders and evaluation of medical risk factors in autism. J Child Neurol 1999; 14: 388-94.

[10]   Sweeten TL, Bowyer SL, Posey DJ, Halberstadt GM, McDougle CJ. Increased prevalence of familial autoimmunity in probands with pervasive developmental disorders. Pediatrics 2003; 112: e420.

[11]   Todd RD, Hickok JM, Anderson GM, Cohen DJ. Antibrain antibodies in infantile autism. Biol Psychiatry 1988; 23: 644-7.

[12]   Tuchman RF, Rapin I, Shinnar S. Autistic and dysphasic children. I: Clinical characteristics. Pediatrics 1991; 88: 1211-8.

[13]   Singh VK, Warren RP, Odell JD, Warren WL, Cole P. Antibodies to myelin basic protein in children with autistic behavior. Brain Behav Immun 1993; 7: 97-103.

[14]   Singh VK, Warren R, Averett R, Ghaziuddin M. Circulating autoantibodies to neuronal and glial filament proteins in autism. Pediatr Neurol 1997; 17: 88-90.

[15]   Connolly AM, Chez MG, Pestronk A, *et al.* Serum autoantibodies to brain in Landau-Kleffner variant, autism, and other neurologic disorders. J Pediatr 1999; 134: 607-13.

[16]   Dalton P, Deacon R, Blamire A, *et al.* Maternal neuronal antibodies associated with autism and a language disorder. Ann Neurol 2003; 53: 533-7.

[17]   Kubota M, Takahashi Y. Steroid-responsive chronic cerebellitis with positive glutamate receptor delta 2 antibody. J Child Neurol 2008; 23: 228-30.

[18]   Levite M, Ganor Y. Autoantibodies to glutamate receptors can damage the brain in epilepsy, systemic lupus erythematosus and encephalitis. Expert Rev Neurother 2008; 8: 1141-60.

[19]   Kawashima H, Suzuki K, Yamanaka G, *et al.* Anti-glutamate receptor antibodies in pediatric enteroviral encephalitis. Int J Neurosci 2010; 120: 99-103.

[20]   Singh VK, Rivas WH. Prevalence of serum antibodies to caudate nucleus in autistic children. Neurosci Lett 2004; 355: 53-6.

[21]   Ashwood P, Wakefield AJ. Immune activation of peripheral blood and mucosal CD3+ lymphocyte cytokine profiles in children with autism and gastrointestinal symptoms. J Neuroimmunol 2006; 173: 126-34.

[22]   Ashwood P, Anthony A, Torrente F, Wakefield AJ. Spontaneous mucosal lymphocyte cytokine profiles in children with autism and gastrointestinal symptoms: mucosal immune activation and reduced counter regulatory interleukin-10. J Clin Immunol 2004; 24: 664-73.

[23]   Jyonouchi H, Geng L, Ruby A, Reddy C, Zimmerman-Bier B. Evaluation of an association between gastrointestinal symptoms and cytokine production against common dietary proteins in children with autism spectrum disorders. J Pediatr 2005; 146: 605-10.

[24]   Gaughran F. Immunity and schizophrenia: autoimmunity, cytokines, and immune responses. Int Rev Neurobiol 2002; 52: 275-302.

[25]   Jones AL, Mowry BJ, Pender MP, Greer JM. Immune dysregulation and self-reactivity in schizophrenia: do some cases of schizophrenia have an autoimmune basis? Immunol Cell Biol 2005; 83: 9-17.

[26]   Radewicz K, Garey LJ, Gentleman SM, Reynolds R. Increase in HLA-DR immunoreactive microglia in frontal and temporal cortex of chronic schizophrenics. J Neuropathol Exp Neurol 2000; 59: 137-50.

[27]   Patterson PH. Maternal infection: window on neuroimmune interactions in fetal brain development and mental illness. Curr Opin Neurobiol 2002; 12: 115-8.

[28]   Uhlmann V, Martin CM, Sheils O, *et al.* Potential viral pathogenic mechanism for new variant inflammatory bowel disease. Mol Pathol 2002; 55: 84-90.

[29]   Katayama Y, Hotta H, Nishimura A, Tatsuno Y, Homma M. Detection of measles virus nucleoprotein mRNA in autopsied brain tissues. J Gen Virol 1995; 76 (Pt 12): 3201-4.

[30]   Cunningham C, Deacon R, Wells H, *et al.* Synaptic changes characterize early behavioural signs in the ME7 model of murine prion disease. Eur J Neurosci 2003; 17: 2147-55.

[31]   Charleston JS, Body RL, Bolender RP, *et al.* Changes in the number of astrocytes and microglia in the thalamus of the monkey Macaca fascicularis following long-term subclinical methylmercury exposure. Neurotoxicology 1996; 17: 127-38.

[32]   Campbell A. The role of aluminum and copper on neuroinflammation and Alzheimer's disease. J Alzheimers Dis 2006; 10: 165-72.

[33]   Strunecka A, Patocka J, Blaylock RL, Chinoy N.J. Fluoride interactions: From molecules to disease. Curr Signal Transduct Ther 2007; 2: 190-213.

[34]   Ashwood P, Enstrom A, Krakowiak P, *et al.* Decreased transforming growth factor beta1 in autism: a potential link between immune dysregulation and impairment in clinical behavioral outcomes. J Neuroimmunol 2008; 204: 149-53.

[35]   Li MO, Sanjabi S, Flavell RA. Transforming growth factor-beta controls development, homeostasis, and tolerance of T cells by regulatory T cell-dependent and -independent mechanisms. Immunity 2006; 25: 455-71.

[36]   Li MO, Wan YY, Sanjabi S, Robertson AK, Flavell RA. Transforming growth factor-beta regulation of immune responses. Annu Rev Immunol 2006; 24: 99-146.

[37]   Marie JC, Liggitt D, Rudensky AY. Cellular mechanisms of fatal early-onset autoimmunity in mice with the T cell-specific targeting of transforming growth factor-beta receptor. Immunity 2006; 25: 441-54.

[38]   Hollander E, DelGiudice-Asch G, Simon L, *et al.* B lymphocyte antigen D8/17 and repetitive behaviors in autism. Am J Psychiatry 1999; 156: 317-20.

[39]   Murphy TK, Goodman WK, Fudge MW, *et al.* B lymphocyte antigen D8/17: a peripheral marker for childhood-onset obsessive-compulsive disorder and Tourette's syndrome? Am J Psychiatry 1997; 154: 402-7.

[40]   Stubbs G. Interferonemia and autism. J Autism Dev Disord 1995; 25: 71-3.

[41]   Singh VK. Plasma increase of interleukin-12 and interferon-gamma. Pathological significance in autism. J Neuroimmunol 1996; 66: 143-5.

[42]   Sweeten TL, Posey DJ, Shankar S, McDougle CJ. High nitric oxide production in autistic disorder: a possible role for interferon-gamma. Biol Psychiatry 2004; 55: 434-7.

[43]   Szydlowska K, Tymianski M. Calcium, ischemia and excitotoxicity. Cell Calcium 2010; 47: 122-9.

[44]   Gupta S, Aggarwal S, Rashanravan B, Lee T. Th1- and Th2-like cytokines in CD4+ and CD8+ T cells in autism. J Neuroimmunol 1998; 85: 106-9.

[45]   Li X, Chauhan A, Sheikh AM, *et al.* Elevated immune response in the brain of autistic patients. J Neuroimmunol 2009; 207: 111-6.

[46]   Garbett K, Ebert PJ, Mitchell A, *et al.* Immune transcriptome alterations in the temporal cortex of subjects with autism. Neurobiol Dis 2008; 30: 303-11.

[47]   Vargas DL, Nascimbene C, Krishnan C, Zimmerman AW, Pardo CA. Neuroglial activation and neuroinflammation in the brain of patients with autism. Ann Neurol 2005; 57: 67-81.

[48]   Purcell AE, Jeon OH, Zimmerman AW, Blue ME, Pevsner J. Postmortem brain abnormalities of the glutamate neurotransmitter system in autism. Neurology 2001; 57: 1618-28.

[49]   Fatemi SH, Halt AR, Stary JM, *et al.* Glutamic acid decarboxylase 65 and 67 kDa proteins are reduced in autistic parietal and cerebellar cortices. Biol Psychiatry 2002; 52: 805-10.

[50]   Yip J, Soghomonian JJ, Blatt GJ. Decreased GAD67 mRNA levels in cerebellar Purkinje cells in autism: pathophysiological implications. Acta Neuropathol 2007; 113: 559-68.

[51]   Roberts EM, English PB, Grether JK, *et al.* Maternal residence near agricultural pesticide applications and autism spectrum disorders among children in the California Central Valley. Environ Health Perspect 2007; 115: 1482-9.

[52]   Pessah IN, Seegal RF, Lein PJ, *et al.* Immunologic and neurodevelopmental susceptibilities of autism. Neurotoxicology 2008; 29: 532-45.

[53]   Zimanyi I, Pessah IN. Pharmacological characterization of the specific binding of [3H]ryanodine to rat brain microsomal membranes. Brain Res 1991; 561: 181-91.

[54]   Ashwood P, Schauer J, Pessah IN, Van de Water J. Preliminary evidence of the *in vitro* effects of BDE-47 on innate immune responses in children with autism spectrum disorders. J Neuroimmunol 2009; 208: 130-5.

[55]   Wilford BH, Shoeib M, Harner T, Zhu J, Jones KC. Polybrominated diphenyl ethers in indoor dust in Ottawa, Canada: implications for sources and exposure. Environ Sci Technol 2005; 39: 7027-35.

[56]   Meironyte D, Noren K, Bergman A. Analysis of polybrominated diphenyl ethers in Swedish human milk. A time-related trend study, 1972-1997. J Toxicol Environ Health A 1999; 58: 329-41.

[57]   Darnerud PO, Eriksen GS, Johannesson T, Larsen PB, Viluksela M. Polybrominated diphenyl ethers: occurrence, dietary exposure, and toxicology. Environ Health Perspect 2001; 109 Suppl 1: 49-68.

[58]   Viberg H, Mundy W, Eriksson P. Neonatal exposure to decabrominated diphenyl ether (PBDE 209) results in changes in BDNF, CaMKII and GAP-43, biochemical substrates of neuronal survival, growth, and synaptogenesis. Neurotoxicology 2008; 29: 152-9.

[59]   Croen LA, Goines P, Braunschweig D, *et al.* Brain-derived neurotrophic factor and autism: maternal and infant peripheral blood levels in the Early Markers for Autism (EMA) Study. Autism Res 2008; 1: 130-7.

[60]   Hashimoto K, Iwata Y, Nakamura K, *et al.* Reduced serum levels of brain-derived neurotrophic factor in adult male patients with autism. Prog Neuropsychopharmacol Biol Psychiatry 2006; 30: 1529-31.

[61]   Katoh-Semba R, Wakako R, Komori T, *et al.* Age-related changes in BDNF protein levels in human serum: differences between autism cases and normal controls. Int J Dev Neurosci 2007; 25: 367-72.

[62]   Blaylock R. The danger of excessive vaccination during brain development: the case for a link to autism spectrum disorders (ASD). Medical Veritas 2008; 5: 1727-41.

[63]   Blaylock RL, Strunecka A. Immune-glutamatergic dysfunction as a central mechanism of the autism spectrum disorders. Curr Med Chem 2009; 16: 157-70.

[64]   Welch MG, Welch-Horan TB, Anwar M, *et al.* Brain effects of chronic IBD in areas abnormal in autism and treatment by single neuropeptides secretin and oxytocin. J Mol Neurosci 2005; 25: 259-74.

CHAPTER 6

Gastrointestinal Disorders and Autism Spectrum Disorders: A Causal Link or a Secondary Consequence?

Anna Strunecka

[1]*Department of Physiology, Faculty of Science, Charles University in Prague, Prague, Czech Republic*

Abstract: Growing evidence confirms that up to 95% of autistic children suffer with the dysfunctions of the gastrointestinal (GI) system. We discuss the cellular and molecular mechanisms underlying these disturbances. Some researchers, physicians, and health care professionals suggest that beneficial effects of dietary intervention on behavior and cognition of some autistic children indicate a functional relationship between the GI tract (GIT) and the CNS pathology of ASD. A possible genetic cause for the association of autism and GI disease is discussed. GI disorders are not included in diagnostic criteria for ASD. Clinical and practical experiences provide the support for association between inflammatory bowel disease and ASD.

INTRODUCTION

Parents of autistic children have much experience from daily life with chronic diarrhea, bloating, abdominal pain, distension, and abnormal stool consistency. Moreover, children with ASD have often unusual feeding patterns and narrow range of preferred dishes [1]. Although not included in the diagnostic criteria, there have been many reports describing GI symptoms in 9 to 84% or more of children with ASD [2-10]. Clinical studies have confirmed that the most common GI symptoms in patients with ASD are constipation, diarrhea, and abdominal distension. Chronic inflammation of the GIT, food intolerance, and recurrent GI symptoms are often recorded by the general practitioners [11,12].

Several clinicians and many parents admit that a treatment of digestive problems may have positive effects on autistic behavior [4]. GI disorders can present as non-GI problems. For example, Horvath and Perman [4] reported disturbed sleep and nighttime awakening for 52 percent of children with ASD who had GI symptoms in comparison with seven percent of age-matched healthy siblings. Some researchers, physicians, and health care professionals suggest that a beneficial effect of dietary intervention on behavior and cognition of some autistic children indicates a functional relationship between the alimentary tract and the CNS pathology of ASD. Increased gut-blood-brain barrier (BBB) permeability might be involved in this link. Altered behaviors are often linked to abdominal pain or discomfort in children with ASD [13-15].

Recently, a multidisciplinary panel reviewed the medical literature with the aim of generating evidence-based recommendations for diagnostic evaluation and management of GI problems in ASD patient population [16]. On May 29 –30, 2008, the multidisciplinary panel convened in Boston, Massachusetts, to review and discuss GI aspects of ASD. Meeting participants were part of a larger group that was organized to develop recommendations for the evaluation and management of GI disorders for individuals with ASD, as well as for future research directions. Working groups comprised 28 experts in child psychiatry, developmental pediatrics, epidemiology, medical genetics, immunology, nursing, pediatric allergy, pediatric gastroenterology, pediatric pain, pediatric neurology, pediatric nutrition, and psychology. A literature search on Medline was conducted to identify relevant articles by using the key words "gastrointestinal disease" and "autism." A consensus report was released on January 2010 [16]. However, the expert panel reached consensus on 23 statements. The panel admits that:

- GI disorders and associated symptoms are commonly reported in individuals with ASD,

- but key issues such as the prevalence and best treatment of these conditions are incompletely understood.

*Address correspondence to: Anna Strunecka, Department of Physiology, Faculty of Sciences, Charles University in Prague, Vinicna 7, 128 00 Prague 2, Czech Republic; E-mail: strun@natur.cuni.cz

- Because of the absence, in general, of high-quality clinical research data, evidence-based recommendations are not possible at the present time.

- Individuals with ASD deserve the same thoroughness and standard of care in the diagnostic workup and treatment of GI concerns as should occur for patients without ASD.

Children with ASD can benefit from adaptation of general pediatric guidelines for the diagnostic evaluation of abdominal pain, chronic constipation, and gastroesophageal reflux disease [17]. These guidelines help health care providers determine when GI symptoms are self-limited and when evaluation beyond a thorough medical history and physical examination should be considered. Children with ASD who have GI disorders may present with behavioral manifestations. Diagnostic and treatment recommendations for the general pediatric population are useful to consider until the development of evidence-based guidelines specifically for patients with ASD.

Many doctors and scientists have ignored the fact that up to 95 % of autistic children have intestinal problems, such as altered bowel function and abdominal distension [14]. There appeared some studies in a literature search of Medline, which document that there were no reports of inflammatory bowel disease or autism over the study period [18]. Kuddo and Nelson found that the frequency of GI symptoms observed in population-based samples of autistic children indicate that GI problems are not nearly as common in children with autism as reports from pediatric gastroenterology clinics suggest [19]. The recent study of Ibrahim and coworkers examined 124 adult patients with ASD and concluded that no significant associations were found between autism case status and overall incidence of GI symptoms or any other GI symptom category [20]. These authors suggest that a neurobehavioral rather than a primary organic GI etiology may account for the higher incidence of these GI symptoms in children with autism.

The first steps to understanding of association of GI dysfunction with ASD provide recent genetic studies, which demonstrate how disruption of a candidate gene *MET* contributes to ASD risk [21-24]. The MET (mesenchymal epithelial transition factor) receptor tyrosine kinase (RTK) participates not only in development of the cerebral cortex and cerebellum, both of which may be altered in ASD (see Chapter 2), but also contributes to GI and immune function, disruptions of which co-occur in some patients with ASD. GI disorders of patients with ASD have been subjects of numerous discussions and great surrounding controversies. Some authors suggested that GI problems have more commonly been linked to regressive forms of ASD, characterized by loss of previously acquired skills and late onset of behavioral anomalies, not observed in the first year of life [2,25-28]. A link has been postulated between measles-mumps-rubella (MMR) vaccine and a form of autism that is a combination of developmental regression and GI symptoms that occur shortly after immunization.

In contrast, no association between developmental regression and GI symptoms has been repeatedly reported [6,29-32]. No significant difference was found in rates of bowel problems or regression in children who received the MMR vaccine during the 20 years from 1979 [33]; strong evidence against association of autism with MMR exposure was provided by Hornig et al [34]. Although great controversy has plagued those who suggested a link between the MMR vaccine and autism, history alone suggests that more research is needed to determine if there is a unique GI lesion in children with PDD [9]. The debate regarding the potential trigger of GI pathology and regressive autism is on going, exceeding the scope of a literature search on Medline.

The aim of this chapter is to review and integrate the scientific studies with clinical and practical experiences and with the new concept that the central mechanism of ASD is immunoexcitotoxicity [12,35] (see Chapter 4). The ultimate goals are to understand the etiology of ASD and to search for the best treatment of these conditions in children and adults with ASD to improve the quality of their life. Several questions thus arise:

- Has GI pathophysiology a causal role in the etiology of ASD?

- Is GI pathophysiology in ASD contributory to autism symptoms?

- Is there a link between GI disorders and autistic behavior?

- Can therapy of GI disorders lead to amelioration of GI symptoms?

GI ABNORMALITIES IN CHILDREN WITH ASD

The history of association of autism with disturbances of the gut functions is seen in an excellent review by Gilger [9]. He reminds us that several authorities in autism research suggested a possible link to gut dysfunction and neuropsychiatric dysfunction in autistic patients since sixties. However, the first more detailed studies investigating GI anomalies in autistic children were reported by Horvath *et al.* [36] and Wakefield *et al.* [2,37]. Horvath with coworkers evaluated the structure and function of the upper GIT in a group of patients with autism who had GI symptoms. Thirty six children (mean age 5.7 ± 2 years) underwent upper GI endoscopy with biopsies, intestinal and pancreatic enzyme analyses, and bacterial and fungal cultures. The most frequent GI complaints were chronic diarrhea, gaseousness, and abdominal discomfort and distension. Histological examination in these 36 children revealed grade I or II reflux esophagitis in 25 (69.4 %), chronic gastritis in 15 (42 %), and chronic duodenitis in 24 (66.6 %). Low intestinal carbohydrate digestive enzyme activity was reported in 21 children (58.3 %), although there was no abnormality found in pancreatic function. Horvath with coworkers suggested that unrecognized GI disorders, especially reflux esophagitis and disaccharide malabsorption, may contribute to the behavioral problems of the non-verbal autistic patients. High prevalence of histological abnormalities in the esophagus, stomach, small intestine and colon, and intestinal permeability were reported in later study [4,5].

In 1998 Wakefield with 13 coworkers published an article describing an ileal lymphoid-nodular hyperplasia (LNH) and non-specific colitis ("autistic colitis") found in nine of the twelve examined children. Colitis with ileal LNH in children with regressive autism has been repeatedly described in subsequent studies [25,27,37-39].

The GI disorders reported in ASD include: inflammation in both the upper and lower intestinal tract (esophagitis, gastritis, duodenitis, enterocolitis) with or without autoimmunity, LNH, increased intestinal permeability, low activities of disaccharidase enzymes, impairment of detoxification, dysbiosis with bacterial overgrowth and food intolerance in many children with ASD [2-4,6-10]. Chronic inflammation of the GIT, food intolerance, and recurrent GI symptoms are frequently recorded by the general practitioners [11]. Other GI abnormalities that have been described for individuals with ASD include gastroesophageal reflux disease, abdominal bloating, and disaccharidase deficiencies, as well as pathologic findings such as inflammation of the GIT and abnormalities of the enteric nervous system. Frequent patient and/or parent complaints have included chronic diarrhea, bloating, abdominal pain, distension, and abnormal stool consistency.

Recently, Russo and Andrews [41] analyzed the complete medical history records of the Autistic Genetic Resource Exchange (AGRE), a DNA repository and family registry sponsored by Autism Speaks, including contributing family members who have had extensive evaluations by a variety of pediatricians, psychiatrists, and other neurodevelopmental specialists. The diagnosis of autism for all patients was made using the standard ADI-R algorithm. The analysis of 692 children (mean age 9.1 ± 5.1 years) with autism and 187 non autistic siblings (mean age 10.5 ± □6.6 years) shows that autistic children compared to non-autistic sibling controls have significantly higher overall GI disease (43 % vs. 12 %), chronic diarrhea (26 % vs. 13 %), and constipation (33 % vs. 13 %). According to expert panel report [16] the most common GI symptoms and signs reported for persons with ASD are chronic constipation, abdominal pain with or without diarrhea, and encopresis as a consequence of constipation.

Are Children with Autism More Likely to Have a History of GI Disorders than Children without Autism?

Black *et al.* [11] evaluated records from UK General Practice Research Database and found that 9 of 96 (9 %) children with a diagnosis of autism (cases) and 41 of 449 (9 %) children without autism had a history of GI disorders before the index date (the date of first recorded diagnosis of autism in the cases and the same date for controls). The authors thus concluded that no evidence was found that children with autism were more likely than children without autism to have had defined GI disorders at any time before their diagnosis of autism. Their study included only obvious GI disease and symptomatology and recognized that they might miss more subtle symptoms of GI disease. No association between chronic GI symptoms and a history of developmental regression was found in several studies [18, 30, 42]. Another study site was a clinic specializing in ASD in a large pediatric medical center serving a 10 county area in the mid-western USA. In a sample of 137 children, age 24-96 months, classified as having autism or ASD by the Autism Diagnostic Observation Schedule-Generic, 24 percent had a history of at least one chronic GI symptom. The most common symptom was diarrhea, which occurred in 17 percent of patients. However, authors concluded that no association between chronic GI symptoms and a history of developmental regression has been found [6].

On the other hand, in the first study of Wakefield *et al.* [2] 12 children (mean age 6 years, range 3-10, 11 boys) were referred to a pediatric gastroenterology unit with a history of normal development followed by loss of acquired skills, including language, together with diarrhea and abdominal pain. Parents associated an onset of behavioral symptoms with MMR vaccination in eight of the twelve children, with measles infection in one child, and otitis media in another. Symptoms appeared one day to two months after immunization. The hypothesis put forward by the authors is that MMR vaccine causes inflammation or dysfunction of the intestine, increasing the absorption of non-permeable peptides, which in turn can cause serious developmental disorders. The Wakefield's group thus identified associated GI disease and developmental regression in a group of previously normal children, which was generally associated in time with possible environmental triggers. Colitis with ileal LNH in children with regressive autism has been repeatedly described in subsequent studies [25,27]. The GI symptoms had developed coincident with the onset of autistic behavior, according to parents. Wakefield postulated the hypothesis that there exists a subset of children who are vulnerable to developing a particular form of regressive autism following previously normal development, in combination with a novel form of inflammatory bowel disease. Onset may occur over weeks or sometimes months, and is triggered by exposure to a measles-containing vaccine, predominantly the MMR. This exposure leads to long term infection with measles virus within key sites, including the intestine where it causes inflammation. This hypothesis has been confirmed by subsequent studies of Wakefield 's group with 60 and 148 autistic children with the regressive autism [38, 39]. In contrast, several authors found no evidence to support a distinct syndrome of MMR-induced autism [32].

Inflammation of GI System in Patients with ASD

We have explained in previous chapters that chronic inflammation plays a significant role in ASD. The sources of chronic stimulation are less clearly defined. Wakefield with coworkers reported the findings of LNH as the sign of inflammation and concluded that a new variant of inflammatory bowel disease "autistic enterocolitis" is present in the group of children with developmental disorders [2,38,39].

Lymph nodules are encapsulated bodies lying within the submucosa of the intestinal wall. These lymph nodules contain lymphocytes and neutrophils. The fluid absorbed from the intestinal lumen by the action of the absorptive epithelial cells is filtered through the lymph nodes. Here, antibodies are formed [13]. The study of Wakefield *et al.* [38] found LNH in 54 of 58 children with developmental disorders. Scores of frequency and severity of inflammation were significantly greater in affected children, compared with controls. In this trial, ileal LNH was present in 93 % of affected children versus 14.3 % of controls and chronic colitis in 88 % of affected children versus 4.5 % of controls. Active inflammation of the ileum (ileitis) was observed in 8 % and chronic inflammation of the colon (colitis) was seen in 88 % of affected children.

In a comment on the Wakefield paper, Sabra *et al.* [43] reported identical pathology (LNH) in the terminal ileum of two children patients diagnosed with pervasive disorders. Immunohistochemistry confirmed a distinct lymphocytic colitis in ASD in which the epithelium appears particularly affected. This is consistent with increasing evidence for gut epithelial dysfunction in autism [25]. In the later study, Wakefield and coworkers [39] investigated 148 consecutive children with ASD (median age 6 years; range 2-16; 127 male) with GI symptoms by ileocolonoscopy. They found that the prevalence of LNH was significantly greater in ASD children compared with controls: being present in the ileum 129/144 (90 %) vs. 8/27 (30 %), and colon 88/148 (59 %) vs. 7/30 (23 %), whether or not controls had co-existent colonic inflammation. The severity of ileal LNH was significantly greater in ASD children compared with controls.

With respect to the upper GIT, Horvath *et al.* [36] investigated 36 autistic children complaining of abdominal pain, bloating or chronic diarrhea by gastroscopy. The most common histological finding was reflux esophagitis (69.4 %), while 41.7 % had chronic gastritis and 66.7 % had chronic duodenitis in the absence of *Haemophilus pylori* infection. The number of Paneth cells in autistic children was also noted to be significantly elevated compared with neurotypical controls. Torrente *et al.* [26,28] found that 11 of the 25 autistic children had a focally enhanced gastritis, while two had mild diffuse gastritis. Immunohistochemistry results demonstrated the pattern of lymphocyte infiltration was most similar to Crohn's disease, with the exception of a striking predominance of CD8-positive over CD4-positive cells and a marked increase in intraepithelial lymphocytes. Another highly specific finding among autistic children was a dense, subepithelial basement membrane immunoglobulin G deposition, which was absent in the other subgroups.

It is important to note that the Horvath *et al.* [36] and Torrente *et al.* [26,28] studies describe inflammation in the upper GIT of autistic children, whereas Wakefield and coworkers observed inflammation in the ileum and colon.

The results of these different studies taken together suggest that inflammation of GIT may accompany ASD. Wakefield *et al.* [2,38,39] suggested a relationship between severe intestinal inflammation secondary to MMR vaccines and ASD. This idea evokes a strong controversy, which is going beyond the scope of scientific interests. Nevertheless, some studies have linked the live measles virus from MMR vaccine to the inflamed GIT. It has not been excluded that a weakened immune system of some children is unable to produce protective antibodies even against an inactivated live virus. Then the live attenuated virus persists, producing low-grade inflammation – in both the gut and the brain [14]. This does not mean that all children who are vaccinated will develop ASD.

In contrast, some evidence against association of autism with persistent measles virus RNA in the GIT or MMR exposure was provided by the study of Hornig *et al.* [34]. This study evaluated the measles virus RNA in ileal and cecal tissues from 25 children with autism and GI disturbances and 13 children with GI disturbances alone (controls). No differences were found between case and control groups in the presence of measles virus RNA in ileum and cecum. Hyman [14] suggests that this finding does not rule out the possibility that the virus did its harm in a "Hit-Run" fashion. In the "Hit-Run" hypothesis, virus infects the periphery but never enters the CNS. The virus sets up an abnormal immunologic milieu for subsequent autoimmunity. Yet, there is compelling evidence that the measles virus does enter the brain and frequently becomes persistent. Katayama *et al.* demonstrated that measles virus commonly persists in the human brain without causing apparent clinical symptoms, probably due to decreased virus replication [44].

Based on their long-term investigations, Wakefield and coworkers concluded that:

- Ileo-colonic LNH is a characteristic pathological finding in children with ASD and GI symptoms.

- LNH is associated with mucosal inflammation.

- Inflammation is much more severe in autistic children compared to children without autism.

It has been mentioned that several authors found no evidence to support a distinct syndrome of MMR-induced autism or of "autistic enterocolitis." The expert panel [16] concluded that:

- At present, there are inadequate data to establish a causal role for intestinal inflammation.

Many doctors and scientists suggest that autistic children are more susceptible to GI inflammation triggered by certain foods, namely gluten and casein. Immune reactivity to dietary proteins may be associated with GI inflammation in ASD children that may be partly associated with aberrant innate immune response against endotoxin, a product of the gut bacteria [45].

Gastrointestinal Tract and Immune System

It has been generally known that the GIT is the largest immune organ in the body, containing up to 80 % of immune globulins-producing cells in the body. In children with ASD, immunohistochemistry and flow-cytometry studies have consistently shown marked panenteric infiltration of lymphocytes and eosinophils in the gut mucosa [25-28]. Torrente *et al.* suggested an autoimmune component to the inflammatory response – co-localized deposition of IgG and complement C1q on the surface epithelium of the GIT [26,28]. These studies suggest an underlying chronic inflammatory process in some individuals with ASD and co-occurring GI disturbances, characterized by LNH, enterocolitis, and mucosal infiltration by immune cells along the length of the GIT. According to Consensus of expert panel, these findings should be considered preliminary and will require confirmation [16].

Some data provide further evidence of a panenteric mucosal immunopathology in children with regressive autism that is apparently distinct from other inflammatory bowel diseases [27,46,47]. A more recent study found that peripheral blood lymphocytes as well as mucosal CD3 + TNF-α and CD3 + IFN-γ cytokine response were significantly increased in children with ASD as compared to non-inflamed control children. The critical difference between children with Crohn's disease and those with ASD was that in the latter, peripheral and mucosal IL-10 responses were markedly lower. This indicated not only a GI autoimmune reaction in autistic children, but a suppression of the cytokine known to regulate immune termination, IL-10.

Evidence suggests that ASD may be accompanied by aberrant (inflammatory) innate immune responses. This may predispose ASD children to sensitization to common dietary proteins, leading to GI inflammation and aggravation of some behavioral symptoms. Jyonouchi with coworkers [48,49] measured IFN-γ, IL-5, and TNF-α production against gliadin, cow's milk protein, and soy by peripheral blood mononuclear cells (PBMCs) from ASD and control children. PBMCs obtained from ASD children with GI symptoms produced more TNF-α/IL-12 than those obtained from control subjects with CMP. They also produced more TNF-α with gliadin. A high prevalence of elevated TNF-α/IL-12 production by GI (+) ASD PBMCs indicates a role of non-allergic food hypersensitivity (NFH) in GI symptoms observed in children with ASD.

The Role of Gut Microflora in the Pathogenesis of GI Disorders in ASD

The human endogenous intestinal microflora is immensely diverse ecosystem. It has important role in regulating epithelial development and instructing innate immunity. Repeated use of antibiotic therapy may disrupt the complex microbial ecosystem and contribute to more favorable colonization by toxin-producing species [50]. Experience from multiple courses of antibiotic therapy is common in children with ASD. Many parents of children with regressive autism have noted antecedent antibiotic exposure followed by chronic diarrhea. Sandler *et al.* [51] recruited 11 children with regressive-onset autism for an intervention trial using a minimally absorbed oral antibiotic vancomycin. Entry criteria included antecedent broad-spectrum antimicrobial exposure followed by chronic persistent diarrhea. Short-term improvement was noted in 8 of 10 children studied. Unfortunately, these gains had largely waned at follow-up. Sandler concluded that although the protocol used is not suggested as useful therapy, these results indicate that a possible gut flora-brain connection warrants further investigation.

Statement number 19 from Consensus Report postulates that the role of gut microflora in the pathogenesis of GI disorders in individuals with ASD is not well understood [16]. Moreover, clinicians should obtain an abnormal culture result from a duodenal aspirate or abnormal stool culture before starting any treatment designed to alter intestinal flora. The experts noted that empirical antibiotic and antifungal therapy in patients with ASD is not recommended.

Finegold with coworkers suggested that some cases of regressive autism may involve abnormal flora [52]. Fecal flora of children with regressive autism was compared with that of control children, and clostridial counts were higher. The number of clostridial species found in the stools of children with autism was greater than in the stools of control children. Children with autism had nine species of *Clostridium* not found in controls, whereas controls yielded only three species not found in children with autism. In all, there were 25 different clostridial species found. In gastric and duodenal specimens, the most striking finding was total absence of non-spore-forming anaerobes and microaerophilic bacteria from control children and significant numbers of such bacteria from children with autism [52].

Several other studies demonstrate significant alterations in the upper and lower intestinal flora of children with ASD. Autistic children have been shown to have higher counts and more species of clostridia than age- and sex-matched controls [53-55]. Parracho *et al.* [55] studied the fecal flora of patients with ASD and compared them with those of two control groups (healthy siblings and unrelated healthy children). They found that the fecal flora of ASD patients contained a higher incidence of the *Clostridium histolyticum* group (*Clostridium* clusters I and II) of bacteria than that of healthy children. However, the non-autistic sibling group had an intermediate level of the *C. histolyticum* group, which was not significantly different from either of the other subject groups. Members of the *C. histolyticum* group are recognized toxin-producers and may contribute towards gut dysfunction, with their metabolic products also exerting systemic effects. Parracho with coworkers thus suggests that strategies to reduce clostridial population levels harbored by ASD patients or to improve their gut microflora profile through dietary modulation may help to alleviate gut disorders common in such patients.

In a number of studies, reaction to commonly found colon bacterial organism are seen to occur. The presence of β-1,5- glucan in the cell wall of the bacteria and yeast appear to be the most powerful immune component. The most common aerobic bacteria found in healthy individuals is *Escherichia coli* and it accounts for 90-95 % of all the aerobic bacteria. In Bioscreen's autistic study the average amount of *E. coli* was found to be quite low at approximately 56 % compared to the normal. In about 22 % of the autistic children the amount of *E. coli* was actually less than 10 %, which is quite an incredible finding [56]. The second most common aerobe is *Enterococcus,* although it is a lot less common than *E. coli,* at an average of five percent. There were also abnormal elevations in the amount of *Enterococcus* found in the feces of autistic children. This was found to be as high as 40 % in autistic children compared to the average of five percent in healthy individuals.

Many children afflicted with autism have had frequent ear infections as young children and have taken large amounts of antibiotics. There have been anecdotal reports of the onset of autism following broad-spectrum antibiotics, suggesting that disruption of the indigenous flora may lead to colonization by neurotoxin-producing bacteria and the overgrowth of yeast *Candida* [8].

Candida infections are often seen in children with ASD and have often been reported as being involved in pathology of ASD. Some clinicians believe that autistic symptoms are made worse by the overgrowth of *Candida albicans.* The role of candida is however still controversial. If it is present in the gut, it will undoubtedly affect the gut wall and increase permeability. It may also act as a source of strong chronic immunologic reactivity, especially if it penetrates the gut wall. When the yeast multiplies, it releases toxins in the body; and these toxins are known to impair the CNS and the immune system. The "leaky gut" theory of autism implies that treating yeast overgrowth should help the GIT to return toward normal and autistic symptoms to improve. Other possible contributors to candida overgrowth are hormonal treatments, immuno-suppressant drug therapy, exposure to herpes, chicken pox, or other "chronic" viruses or exposure to chemicals that might upset the immune system [57]. Thrush, the white yeast infection of the mouth and tongue, which is common in infants, is another well-known example of candida overgrowth.

> *"I am fairly well convinced that there is a connection and that perhaps 5% to 10% of autistic children— those given many courses of antibiotics, or born with thrush or afflicted with thrush soon after birth—will improve when properly treated for candida. However, there is no consensus among physicians on the candida/autism linkage. Meyer noticed that thrush seemed to be mentioned unusually often in the letters and questionnaires sent to us by parents. Treatment for Candida albicans infrequently results in a cure for autism. However, if the person is suffering from this problem, his/her health and behavior should improve following the therapy." Bernard Rimland [57].*

Increased Intestinal Permeability in Children with Autism

Leaky gut syndrome is an increase in permeability of the intestinal mucosa to luminal macromolecules, antigens and toxins associated with inflammatory degenerative and/or atrophic mucosal damage. Increased permeability has been cited as having a key role in various hypotheses regarding the biology of ASD, including excess opiate activity, diminished peptidase activity, and immune dysfunction. D'Eufemia *et al.* determined the occurrence of gut mucosal damage using the intestinal permeability test in 21 autistic children who had no clinical and laboratory findings consistent with known intestinal disorders [58]. An altered intestinal permeability was found in nine of the 21 (43%) autistic patients, but in none of the 40 controls. Compared to the controls, these nine patients showed a similar mean mannitol recovery, but a significantly higher mean lactulose recovery. These authors thus speculated that an altered intestinal permeability could represent a possible mechanism for the increased passage through the gut mucosa of peptides derived from foods with subsequent behavioral abnormalities.

However, the statement number five of Consensus Report postulates that the evidence for abnormal GI permeability in individuals with ASD is limited. Prospective studies should be performed to determine the role of abnormal permeability in neuropsychiatric manifestations of ASD [16].

Tight junctions represent the major barrier within the paracellular pathway between intestinal epithelial cells. Disruption of tight junctions leads to the increased intestinal permeability and is implicated in the pathogenesis of several acute and chronic pediatric disease entities that are likely to have their origin during infancy [59].

The Issue of Gluten and Casein in ASD

It is worth noting that Kanner in his paper describing autism in 1943 also detailed various somatic features as being present including GI disturbance and feeding problems. The original work of Hans Asperger from 1961 following his description of Asperger syndrome talked about a possible connection between this syndrome and problems with foods containing gluten. In 1979 Panksepp described a neurochemical theory of autism proposing the incomplete breakdown and excessive absorption of dietary food. Over the years, these thoughts have been forgotten or ignored [9,15]. The recent definitions of ASD refer exclusively to irregularities in neuropsychological functioning.

It has been known for years that peptides from gluten and casein affect certain ASD children as to their behavior and overall cognitive function. Diet is widely known to affect both physical and mental health. It is not the topic of this chapter to deal with diets, but medicine is often based on the long-time experience. The collective experience of parents and some clinicians and care-givers supports the view that gluten-free (GF) or casein-free (CF) diet or combination of both (GFCF) has been used in autistic children with success. Parental reports detailing experience of removing gluten and/or casein from the diet of their children with ASD started to emerge in the 1980s. The exact way gluten and casein affect the body is not known. In this part, we will focus on the possible mechanisms explaining how gluten and casein might affect the course of autistic symptoms and we will mention the comparison with coeliac disease. However, The Consensus Report postulates that available research data do not support the use of a CF diet, a GF diet, or combined GFCF diet as a primary treatment for individuals with ASD.

Why is Gluten Harmful for Individuals with ASD?

It seems that the long-term experience with benefits of GFCF diets has contributed to general consensus that gluten and/or casein may be harmful for several individuals with ASD. GFCF diet has grown in popularity; however the mechanism explaining how it works is not clear. Some authors suggest that individuals with ASD have NFH. To evaluate an association between cytokine production with common dietary proteins as a marker of NFH and GI symptoms in children with ASD Jyonouchi *et al.* [49] investigated ASD children with (N = 75) or without GI symptoms (N = 34), children with NFH (N = 15), and control subjects (N = 19). Diarrhea and constipation were the major GI symptoms. They measured production of type 1 T-helper cells (Th1), type 2 T-helper cells (Th2), and regulatory cytokines by PBMC stimulated with whole cow's milk protein, its major components (casein, β-lactoglobulin, and α-lactoalbumin), gliadin, and soy. PBMC obtained from ASD children with GI symptoms produced more TNF-α/ IL-12 than those obtained from control subjects with cow's milk protein, β-lactoglobulin, and α-lactoalbumin. They also produced more TNF-α with gliadin, which was more frequently observed in the group with loose stools. PBMC obtained from ASD children without GI symptoms produced more TNF-α/IL-12 with cow's milk protein than those from control subjects, but not with β-lactoglobulin, α-lactoalbumin, or gliadin. Cytokine production with casein and soy were unremarkable. Based on these results Jyonouchi with coworkers concluded that a high prevalence of elevated TNF-α/IL-12 production with cow's milk protein and its major components indicates a role of NFH in GI symptoms observed in children with ASD.

Coeliac Disease and ASD

The best known gluten sensitive disorder is coeliac disease (CD). This clinically diagnosable disease is characterized by malabsorption and typical small-bowel mucosal atrophy. CD is an immune-mediated enteropathy caused by a permanent sensitivity to gluten in genetically susceptible individuals. Based on a number of studies in Europe and the United States, the prevalence of CD in children between 2.5 and 15 years of age in the general population is 3 to 13 per 1000 children, or approximately 1:300 to 1:80 children. Numerous studies demonstrate that children with CD frequently have GI symptoms such as diarrhea with failure to thrive, abdominal pain, vomiting, constipation and abdominal distension [60]. Currently the only available treatment is lifelong adherence to a GF diet.

Assessment for CD should be performed for any child with an ASD and GI symptoms. Testing at a minimum should include a total IgA level and tissue transglutaminase IgA antibodies [17]. The question thus arises: Have individuals with ASD also had CD? Some authors in the past reported the existence of a linkage of CD with ASD. Pavone *et al.* [61] evaluated 120 patients with CD diagnosed at the Pediatric Clinic of the University at Catania, Italy, in order to identify behavioral problems and autistic features: there were 20 controls for this part of the study. At the same time, CD was assayed in 11 patients with infantile autism and 11 age- and sex-matched controls. No CD case was detected among the group of autistic patients and subsequent antibodies determinations and jejunal biopsies gave normal results. Moreover none of the coeliac patients had a positive DSM-III-R test for infantile autism.

More recent study was performed at the Bologne University in 2008. They examined 150 patients with ASD and found CD in six of them. The authors thus recommend performing assessment for CD for any patient with ASD.

Recently, a significant association between maternal history of CD and ASD was observed in a cohort of 3 325 children with ASD in Denmark [62].

However, the experience with GF diet in patients with ASD show that in many cases the diet has beneficial effects and the symptoms of gluten sensitivity do not appear when gluten is reintroduced after several months of diet. Buie *et al.* [17] recommends that children on GF diet should consider testing for CD when gluten is reintroduced.

While CD is the hereditary life-long disease, there is a hope that symptoms of gluten sensitivity in autistic individuals might be treated. How is this possible? The experience show that gluten acts via different mechanisms in CD and ASD. CD is caused by an abnormal immune reaction to partially digested gliadin, one of the proteins from gluten. In CD gluten induces the secretion of autoantibodies which are targeted against transglutaminase 2. These autoantibodies are produced in the small-intestinal mucosa, where they can be found deposited extracellularly below the epithelial basement membrane and around mucosal blood vessels. In addition, during gluten consumption these autoantibodies can also be detected in patients' serum; disappears from the circulation on a GF diet, but remains for a long time in the small intestinal mucosal deposits [63].

While there are very detailed studies on chemistry and immunology of gliadin protein in CD, nothing is known about the participation of various epitopes of gliadin molecule in the autistic gut. Likewise, nothing is known about the defects of gluten and/or gliadin digestion in autistic intestine. Gliadin is extremely rich on glutamine; the sequences -Pro-Ser-Gln-Gln- and -Gln-Gln-Gln-Pro- were demonstrated to be common for toxic gliadin peptides [64,65]. Nothing is known about intestinal glutamine-glutamate homeostasis and the different functional roles of these closely related amino acids in intestinal mucosa in ASD.

Moreover, biologically active peptides derived from gliadin are tyrosine-containing peptides. The tyrosine-containing groups have the capacity to initiate damaging immunological reactions in patients with CD [66]. Tyrosine is the precursor for melatonin synthesis. The healthy intestine contains hundred times more melatonin than the brain. The synthesis of melatonin is compromised in individuals with ASD. These hypothetical lines only illustrate the complexity of possible interactions of gluten and gliadins in the GI system. These areas warrant further studies.

Opioid-Excess Theory

Another area of interest in connection with GFCF diet is the opioid-excess theory of ASD.

This theory suggests that some, but not all, of the symptoms ASD may be the consequence of the action of peptides of exogenous origin affecting neurotransmission within CNS. Gluten and casein are not broken down properly during digestion and small peptides with opioid activity are released from leaky gut to the blood, cross the BBB and enter the brain to exert an effect on neurotransmission, as well as producing other physiologically-based symptoms. There is a surprisingly long history of research accompanying this theory.

In 1979 Panksepp described a neurochemical theory of autism proposing that incomplete breakdown and excessive absorption of dietary food peptides may exert central opioid-like effects [15]. Panksepp thus put forward the idea that autism is an emotional disturbance arising from an upset in the opiate systems in the brain. This theory was extended by Shattock Sunderland group. Shattock and coworkers [67,68] suggest that the food-derived gut peptides may directly, or via formation of ligands, lead to disruption of normal neuroregulation and brain development. They tested many years whether the elimination of the proteins gluten and casein would improve behavior of children with ASD. They have established The Sunderland Protocol "that seeks to encourage the introduction and utilization of these interventions in a rational and logical way so as to maximize the benefits and minimize the chance of side effects or unnecessarily restrictive diets" [15].

The opiod-excess theory became popular among parents of children with ASD since it offers a seemingly simple explanation for understanding the cause of abnormal behavior of their children. It suggests that gluten and casein are harmful because they are sources of morphine-like compounds. The digestion of gliadin, protein of gluten, leads to peptides called gliadinomorphins. Proteins in bovine milk are also a common source of bioactive peptides. In the opioid-excess theory β-casomorphin from casein has the key role.

The biochemistry of casomorphins has been studied in details by Kaminski *et al.* [69]. He found that *in vitro* the bioactive peptide β-casomorphin 7 is yielded by the successive GI proteolytic digestion of bovine β-casein variants A1 and B, but this was not seen in variant A2. In hydrolysed milk with variant A1 of β-casein, β-casomorphin 7 level is four-fold higher than in A2 milk. Variant A1 is the most frequent in milk of Red, Ayrshire, and Holstein-Friesian cattle breeds. In contrast, a high frequency of A2 is observed in milk of Guernsey and Jersey cattle. Epidemiological evidence from New Zealand claims that consumption of β-casein A1 is associated with higher national mortality rates from ischaemic heart disease. These interesting findings demonstrate that differences in casein polymorphism might also play important, but unknown variable, in evaluation of the requirement of CF diet.

To a significant degree, this bioactive peptide crosses GI mucosa and enters blood in certain individuals. These compounds enter the circulation, cross the BBB, and influence neurological functioning [70]. The level of β-casomorphin 7 is elevated significantly in urine and blood of patients with schizophrenia and autism [71,72].

The presence of opioid peptides in urine has been investigated as support for the opioid-excess theory. Reichelt and Knivsberg [73] reported the findings of opioid peptides derived from food proteins in urine of ASD patients. They suggested that this may be due to a genetically based peptidase deficiency, which manifests by a dietary overload of exorphin precursors, such as by increased gut uptake. (The enzyme dipeptidyl peptidase is required for break down casomorphins into inactive dipeptides.) These authors also show highly significant decreases in urine peptides after introducing a GFCF diet in children who were followed for 1- 4 years [74].

The urine analyses have been extensively provided by Sunderland group. They repeatedly observed the elevated levels of trans-indolyl-3-acryloylglycine (IAG) in the urine of people with autism and Asperger syndrome and reported that their results strongly suggest that urinary titers of IAG may constitute an objective diagnostic indicator for ASD [15, 75-77].

In contrast, some researchers failed to confirm the presence of metabolites of gluten or casein in urine. Dettmer *et al.* [78] developed method to analyze gliadinomorphin and β-casomorphin in urine. The method was used to screen 69 urine samples from children with and without ASD for the occurrence of neuropeptides. The target neuropeptides were not detected above the detection limit in either sample set. Cass *et al.* [79] found no significant differences between the HPLC urinary profiles of 65 boys affected by autism and 158 typically developing controls. In those cases where HPLC showed peaks in the locations at which opioid peptides might be expected to be found, they established that these peaks did not, in fact, represent opioid peptides. These authors concluded that given the lack of evidence for any opioid peptiduria in children with autism, opioid peptides can neither serve as a biomedical marker for autism nor be employed to predict or monitor response to a CF and GF diets. The significance of reports of increased levels of metabolites of casein and gluten in the urine of people with ASD remains unclear [80]. Urinary peptides are not used in conventional practice to prescribe or monitor dietary restriction.

MET: A POSSIBLE GENETIC CAUSE FOR THE ASSOCIATION OF ASD AND GI DISEASE

It seems that genetic research offers the important information, which could contribute to endless discussions whether the association of ASD and GI disorders does exist. In 2009, Campbell *et al.* reported that a functional variant in the promoter of the *MET* gene encoding the MET (mesenchymal epithelial transition factor) receptor tyrosine kinase (RTK) located within a chromosome 7q31 autism candidate gene region, is associated with ASD and that disrupted MET signaling may contribute to the risk for ASD that includes familial GI dysfunction [81]. The MET RTK participates not only in development of the cerebral cortex and cerebellum, both of which may be altered in ASD (see Chapter 2), but also contributes to GI repair and immune function, disruptions of which co-occur in some patients with ASD.

Autism Susceptibility Locus on Chromosome 7q and MET Gene

The International Molecular Genetic Study of Autism Consortium (IMGSAC) found evidence for an autism susceptibility locus (AUTS1) on chromosome 7q. The IMGSAC was the first to publish a genome-wide linkage screen for autism, and the first to identify a linkage peak on the chromosome 7q region [82]. Campbell [23] explains a number of reasons for the difficulty in identifying the chromosome 7q autism risk genes, including genetic and phenotypic heterogeneity – not all cases of ASD are impacted by the same genetic variation and the same genetic variation may not result in ASD. Further difficulties in identifying the autism risk gene on chromosome 7q is the broad linkage peak that includes over 200 mapped genes [24,83].

There are at least two distinct genetic etiologies for ASD: rare, private (*de novo*) single gene mutations that may have a large effect in causing ASD; and inherited, common functional variants of a combination of genes, each having a small to moderate effect in increasing ASD risk. Sousa *et al.* [24] reported association of a potential functional variant of the *MET* gene with autism in two patient samples.

The expression of *MET* transcripts and MET RTK protein was studied by Levitt and his team. These researchers found that both the expression of *MET* transcripts and MET RTK protein were reduced in postmortem brains of individuals with autism compared to age- and gender-matched controls [22]. A number of genetic mechanisms may contribute to the decreased expression of MET RTK in ASD. Another study of Campbell and coworkers [21] found the association of the *MET* promoter rs1858830 C allele in a 204 family Italian cohort and in a 539 family US cohort. (Control subjects have allele G). Association of the same genetic allele was reported in a third cohort of 101 US families with autistic offspring [84]. The *MET* promoter variant rs1858830 allele C results in reduced *MET* gene transcription. The results of Sousa *et al.* [24] identified association of another allele with a similar potential to regulate the expression of the *MET* gene. Sousa and co-workers described association of the rs38845 A allele with autism risk in 335 IMGSAC families and 10 IMGSAC trios. These results failed to replicate in an independent sample of 82 Italian trios. The last study screened two cohorts, an Autistic Disorder cohort from South Carolina and a Pervasive Developmental Disorder (PDD) cohort from Italy, for the presence of the C allele variant in rs1858830 [85]. A significant increase in the C allele variant frequency was found in the South Carolina Autistic Disorder patients as compared to South Carolina Controls. In the Italian cohort, no significant association with PDD was found when comparing the CC or CG genotype to the GG genotype. This study is thus the third independent study to find the rs1858830 C variant in the MET gene promoter to be associated with autism [85].

Disrupted MET Signaling May Contribute to Increased Risk for ASD that Includes Familial GI Dysfunction

In the last recent study Levitt and his team hypothesized that association of the ASD-associated *MET* promoter variant may be enriched in a subset of individuals with co-occurring ASD and GI conditions [81]. Subjects were 918 individuals from 214 Autism Genetics Resource Exchange families with a complete medical history including GI condition report. Genotypes at the ASD-associated *MET* promoter variant rs1858830 C allele were determined. In the entire 214-family sample, the *MET* rs1858830 C allele was associated with both ASD and GI conditions. Stratification by the presence of GI conditions revealed that the *MET* C allele was associated with both ASD and GI conditions in 118 families containing at least 1 child with co-occurring ASD and GI conditions. Contra wise, there was no association of the *MET* polymorphism with ASD in the 96 families lacking a child with co-occurring ASD and GI. These results suggest that disrupted *MET* signaling may contribute to increased risk for ASD that includes familial GI dysfunction [81].

More about MET Receptor Tyrosine Kinase

A finding that reduced *MET* gene expression has been implicated in autism susceptibility draw the attention to the role of the MET RTK in brain development and intestinal repair. However, the *MET* gene has been studied for the last decade in connection with cancer development and progression. The MET RTK was discovered as an activated oncogene. For better understanding of the potential impact of alterations in expression of MET RTK in the etiology of ASD, we offer a short survey of information, which was prepared according to a few comprehensive papers [86-89]. The detailed references can be found in these original papers.

Molecular biology of MET RTK and *MET* gene on chromosome 7q is very complicated and it is very difficult to predict potential effects, which could be implicated in the pathophysiology of ASD. The MET RTK is the prototypic member of a small subfamily of growth factor receptors that, when activated, induce mitogenic, motogenic, and morphogenic cellular responses. MET RTK is a membrane receptor composed of highly glycosylated extracellular α-subunit and a transmembrane β-subunit, which are linked together by a disulfide bridge. MET is expressed in epithelial and endothelial cells, neurons, hepatocytes, hematopoietic cells, and melanocytes. The ligand for MET is hepatocyte growth factor/scatter factor (HGF/SF). HGF is produced by fibroblasts. In embryonic development, MET and HGF are crucial for gastrulation, angiogenesis, myoblast migration, nerve sprouting, and development of placenta.

Activation of MET RTK recruits a number of signaling effectors, including extracellular signal-regulated kinase (ERK), phosphatidylinositol-3-kinase (PI3K), phospholipase Cγ (PLCγ), mitogen-activated kinase (MAPK) and Akt pathways with following activation of multiple signal transduction pathways. Probably the most important unifying factor is that the activation of MET RTK is linked to Ca^{2+} signaling pathways. Sudden changes in the levels of $[Ca^{2+}]_i$ from both external and internal sources results in changes of MET tyrosine phosphorylation. Activation of a pleiotropic MET RTK modulates axonal and dendritic growth, synaptic formation, and neuronal differentiation. While normal HGF/SF-MET signaling is involved in many aspects of embryogenesis, abnormal HGF/SF- MET signaling has been implicated in both tumor development and progression [86].

The MET RTK is known to be overexpressed in many solid tumors and plays a crucial role in tumor invasive growth and metastasis. Activation of the MET RTK promotes cell proliferation, scattering, invasion, survival, and angiogenesis. Deregulation of MET promotes tumor formation, growth, progression, metastasis, and therapeutic resistance [89]. Aberrantly active MET RTK triggers tumor growth, formation of new blood vessels that supply the tumor with nutrients, and metastasis. Numerous agents have been developed that are able to target MET expression and/or function, namely kinase inhibitors that prevent ATP binding to MET, and HGF inhibitors.

The investigation of the MET function in zebrafish embryos shows that MET RTK is required for normal touch-evoked behavior [88]. At later stages of development, embryos stop coiling spontaneously and instead respond to touch on the head or tail. Normal MET function is required for proper tail touch-evoked movements. In chick, mouse, and rat, MET RTK is expressed in a subset of spinal motoneurons and the MET ligand HGF is important for the differentiation of these cells.

Do this Findings Show the Association of ASD and GI Disorders?

The demonstration of the common genetic disturbance of potential pathological changes in the brain, GIT, and immune system could suggest that developmental changes proceed simultaneously during prenatal development, making some individuals more susceptible to manifestation of ASD with GI dysfunctions. Moreover, the MET RTK is a key regulator of immune responsiveness, which may influence both brain development and GI function. The knowledge of the functional pleiotropism of the MET RTK can explain many pathological findings of ASD (see Fig. 1). Interstingly, neuropathological findings in autism indicate altered organization of the cerebral cortex and cerebellum, both of which are disrupted in mice with decreased MET signaling activity [21]. Campbell and co-authors hypothesize that the common, functionally disruptive rs1858830 C allele can, together with other vulnerability genes and epigenetic and environmental factors, precipitate the onset of autism. The existence of epistatic interactions among common genetic variants at several different loci is further supported by the association between the rs1858830 C allele and autism in multiplex families and not in simplex families.

Pat Levitt, Vanderbilt Kennedy Center's director and Professor of Pharmacology, said in the interview:

> *"GI disorders don't cause autism. Autism is a disorder of brain development. However, our study brings together genetic risk for autism and co-occurring GI disorders in a way that provides a biologically plausible explanation for why they are seen so often together."*

GUT-BRAIN INTERACTIONS: PUZZLE OR HOLOGRAM?

While evidenced-based science is calling for evidence that food and disturbances of the GI system could affect our mood, emotions, sensory perception, sleeping pattern, aggressiveness, love, and behavior, the collective human intelligence expresses this ancient human experience in many proverbs, habits, songs, poems, and novels. It therefore seems plausible that some food/diets and diseases of GI system could have impact on some neuropsychological symptoms and behavior of individuals with ASD.

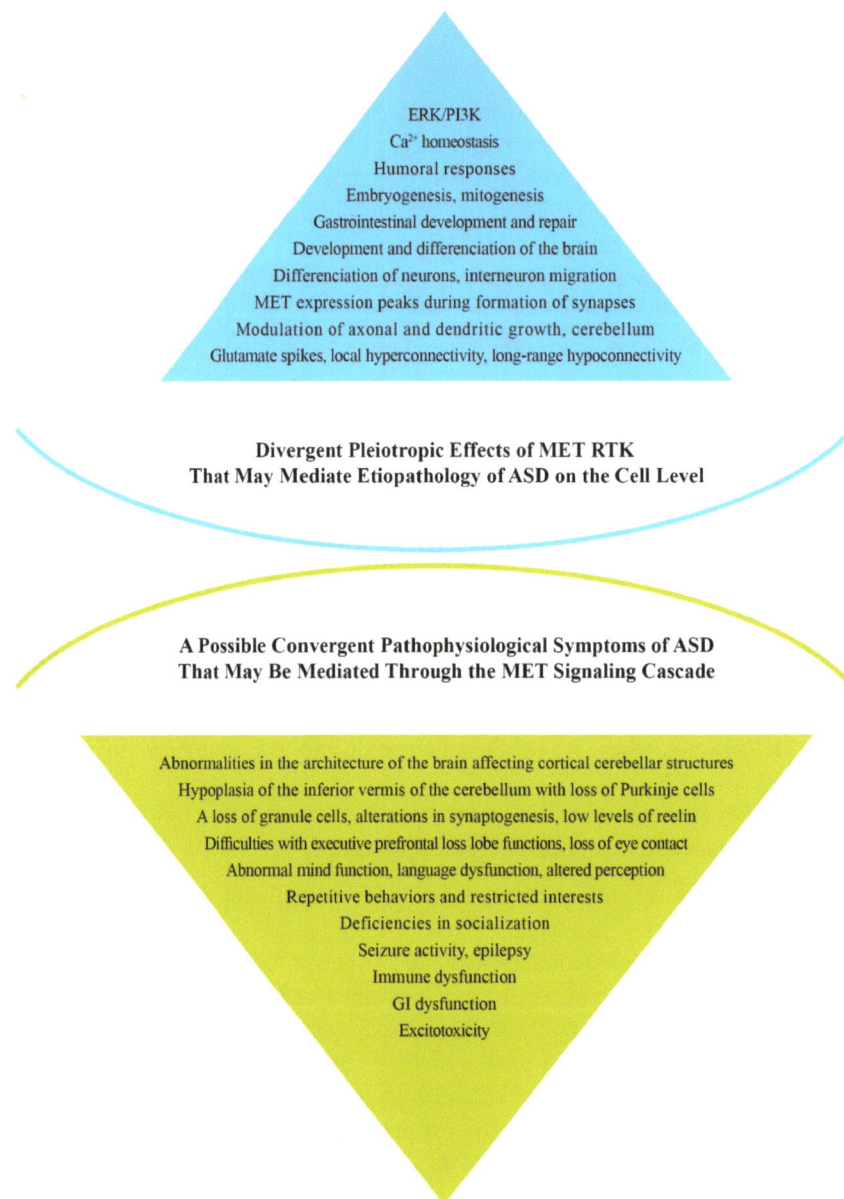

ERK/PI3K
Ca²⁺ homeostasis
Humoral responses
Embryogenesis, mitogenesis
Gastrointestinal development and repair
Development and differenciation of the brain
Differenciation of neurons, interneuron migration
MET expression peaks during formation of synapses
Modulation of axonal and dendritic growth, cerebellum
Glutamate spikes, local hyperconnectivity, long-range hypoconnectivity

**Divergent Pleiotropic Effects of MET RTK
That May Mediate Etiopathology of ASD on the Cell Level**

**A Possible Convergent Pathophysiological Symptoms of ASD
That May Be Mediated Through the MET Signaling Cascade**

Abnormalities in the architecture of the brain affecting cortical cerebellar structures
Hypoplasia of the inferior vermis of the cerebellum with loss of Purkinje cells
A loss of granule cells, alterations in synaptogenesis, low levels of reelin
Difficulties with executive prefrontal loss lobe functions, loss of eye contact
Abnormal mind function, language dysfunction, altered perception
Repetitive behaviors and restricted interests
Deficiencies in socialization
Seizure activity, epilepsy
Immune dysfunction
GI dysfunction
Excitotoxicity

Figure 1: Divergent pleiotropic effects of MET RTK that may mediate etiopathology of ASD on the cell level and possible convergent pathophysiological symptoms of ASD that may be mediated through the MET signaling cascade.

The studies of GI dysfunctions in individuals with ASD show most explicitly that ASD is neither a disease of one gene, neurotransmitter or hormone, nor the disease of one second messenger disturbance. The understanding of enormous increase of autistic behavior during last decade will inevitably require an integrative approach, which brings together not only specialized scientific knowledge, but also knowledge about the homeostatic mechanisms of the whole human being. We simultaneously realize that the living system does not behave as a static jigsaw puzzle. The behavior of a whole cannot be predicted by knowing the separated parts. Rather, the integration of specialized knowledge about molecular and cellular mechanisms leads to understanding of specialized organ systems. The dynamic integration of specialized functions leads to understanding of the human body as a whole.

For example, it has become increasingly evident that a chronic GI inflammatory condition can be markedly affected by interactions between the enteric nervous system (ENS), immune system, and CNS. Both the ENS and the CNS can amplify or modulate the aspects of GI inflammation through secretion of neuropeptides. However, although the available data suggest an important role for neuropeptides in the pathophysiology of intestinal inflammation, there does not yet appear to be a function that can be taken as established for any of these molecules. The complexity of neuroimmune-endocrine systems, conflicting study results, and dual mechanisms of action warrant further research in this field. However, Margolis and Gershon have expressed their belief that clarification of the molecular mechanisms of action of neuropeptides and their effects on immune and inflammatory reactions, will likely yield new treatment options in the future [90]. The effect of potential exorphins on brains of autistic patients has been a subject of many controversial discussions. However, exclusion of gluten and/or casein can help reduce some of the GI disturbance and some of the behaviors at least in some patients with ASD. A beneficial effect of dietary intervention on behavior and cognition of some autistic children could indicate a functional relationship between the GI dysfunction and CNS pathology of ASD (see Table **1**.)

Table 1: The Following Data Have Been Collected in Autism Research Institute (San Diego, USA) from the more than 26,000 Parents who Have Completed the Questionnaires Designed to Collect such Information.

DIET	Got Worse %	No Effect %	Got Better %	No of Cases
Candida diet	3	41	56	941
GFCF diet	3	31	66	2561
CF diet	2	48	50	6360
Removed wheat	2	47	51	3774
Removed sugar	2	47	51	4187

"Worse" Refers Only to Worse Behavior

Enteric Nervous System: Second Brain in the GIT

Textbooks tell us that the digestive system is endowed with its own, local nervous system, ENS, which is embedded in the wall of the digestive tract and extends from esophagus to anus. The magnitude and complexity of the ENS is immense - it contains as many neurons as the spinal cord and has more different types of neuron than the ganglia of any other organ. Enteric neurons are supported by glia rather than by Schwann cells [91]. ENS contains complete reflex circuits that detect the physiological condition of the GIT, integrate information about the state of the GIT, and provide outputs to control gut movement, fluid exchange between the gut and its lumen, and local blood flow [92]. Because of its extent and its degree of autonomy, the ENS has been referred to as a second brain [92]. The ENS can and does function autonomously, but normal digestive function requires communication links between this intrinsic system and the CNS.

Despite considerable progress over the last 15 years in understanding the molecular and cellular mechanisms that control the development of the ENS, several questions remain unanswered [93]. ENS and CNS develop in early embryogenesis separately from two parts of neural crest. Tremendous amount of regulatory factors enters the game during the prenatal development of ENS. For example, recently Welch and co-workers suggest that OT and OTR signaling might be important in early ENS development and function and might play roles in visceral sensory perception and neural modulation of epithelial biology [94]. Later in the prenatal development the ENS and CNS are connected. These links take the form of parasympathetic fibers. Through these cross connections, the gut can provide sensory information to the CNS, and the CNS can affect GI function. The ENS thus has extensive, two-way connections with the CNS, and works in concert with the CNS to control the digestive system in the context of local and whole body physiological demands. The physiology and clinical experiences thus leave us with no doubts about the association of GI system and the brain. These parts of the human body are not separated, but rather affect each other, despite the fact that both are such complicated systems that it is not sure whether the human brain will be able to reveal and understand all the principles and regulatory factors connected with their activities. At present, we only know the inputs (food) and outputs (behavior) of this dual black-box system.

CONCLUSION

GI disorders and associated symptoms are commonly reported in up to 84 % individuals with ASD. Individuals with ASD deserve the same thoroughness and standard of care in the diagnostic workup and treatment of GI concerns as should occur for patients without ASD. Inflammation and changes in gut microflora are the most important symptoms for the potential amelioration of ASD. A beneficial effect of dietary intervention on behavior and cognition of some autistic children could indicate a functional relationship between the GI dysfunction and CNS pathology of ASD.

REFERENCES

[1]   Ahearn WH, Castine T, Nault K, Green G. An assessment of food acceptance in children with autism or pervasive developmental disorder-not otherwise specified. J Autism Dev Disord 2001; 31: 505-11.

[2]   Wakefield AJ, Murch SH, Anthony A, *et al.* Ileal-lymphoid-nodular hyperplasia, non-specific colitis, and pervasive developmental disorder in children. Lancet 1998; 351: 637-41.

[3]   Melmed RD SC, Fabes RA *et al.* Metabolic markers and gastrointestinal symptoms in children with autism and related disorders. J Pediatr Gastroenterol Nutr 2000; 31: S31-2.

[4]   Horvath K, Perman JA. Autistic disorder and gastrointestinal disease. Curr Opin Pediatr 2002; 14: 583-7.

[5]   Horvath K, Perman JA. Autism and gastrointestinal symptoms. Curr Gastroenterol Rep 2002; 4: 251-8.

[6]   Molloy CA, Manning-Courtney P. Prevalence of chronic gastrointestinal symptoms in children with autism and autistic spectrum disorders. Autism 2003; 7: 165-71.

[7]   Valicenti-McDermott M, McVicar K, Rapin I, *et al.* Frequency of gastrointestinal symptoms in children with autistic spectrum disorders and association with family history of autoimmune disease. J Dev Behav Pediatr 2006; 27: S128-36.

[8]   Galiatsatos P, Gologan A, Lamoureux E. Autistic enterocolitis: fact or fiction? Can J Gastroenterol 2009; 23: 95-8.

[9]   Gilger MA, Redel CA. Autism and the gut. Pediatrics 2009; 124: 796-8.

[10]   Wasilewska J, Jarocka-Cyrta E, Kaczmarski M. [Gastrointestinal abnormalities in children with autism]. Pol Merkur Lekarski 2009; 27: 40-3.

[11]   Black C, Kaye JA, Jick H. Relation of childhood gastrointestinal disorders to autism: nested case-control study using data from the UK General Practice Research Database. BMJ 2002; 325: 419-21.

[12]   Blaylock RL, Strunecka A. Immune-glutamatergic dysfunction as a central mechanism of the autism spectrum disorders. Curr Med Chem 2009; 16: 157-70.

[13]   White JF. Intestinal pathophysiology in autism. Exp Biol Med (Maywood) 2003; 228: 639-49.

[14]   Hyman MA. Autism: is it all in the head? Altern Ther Health Med 2008; 14: 12-5.

[15]   Shattock P, Carr K, Todd L, Whitley P. Autism as a metabolic Condition. 1st Edition ed. Sunderland: ESPA Research; 2009.

[16]   Buie T, Campbell DB, Fuchs GJ, 3rd, *et al.* Evaluation, diagnosis, and treatment of gastrointestinal disorders in individuals with ASDs: a consensus report. Pediatrics 2010; 125 Suppl 1: S1-18.

[17]   Buie T, Fuchs GJ, 3rd, Furuta GT, *et al.* Recommendations for evaluation and treatment of common gastrointestinal problems in children with ASDs. Pediatrics 2010; 125 Suppl 1: S19-29.

[18]   Fombonne E. Is there an epidemic of autism? Pediatrics 2001; 107: 411-2.

[19]   Kuddo T, Nelson KB. How common are gastrointestinal disorders in children with autism? Curr Opin Pediatr 2003; 15: 339-43.

[20]   Ibrahim SH, Voigt RG, Katusic SK, Weaver AL, Barbaresi WJ. Incidence of gastrointestinal symptoms in children with autism: a population-based study. Pediatrics 2009; 124: 680-6.

[21]   Campbell DB, Sutcliffe JS, Ebert PJ, *et al.* A genetic variant that disrupts MET transcription is associated with autism. Proc Natl Acad Sci U S A 2006; 103: 16834-9.

[22]   Campbell DB, D'Oronzio R, Garbett K, *et al.* Disruption of cerebral cortex MET signaling in autism spectrum disorder. Ann Neurol 2007; 62: 243-50.

[23]   Campbell DB. When linkage signal for autism MET candidate gene. Eur J Hum Genet 2009; 17: 699-700.

[24]   Sousa I, Clark TG, Toma C, *et al.* MET and autism susceptibility: family and case-control studies. Eur J Hum Genet 2009; 17: 749-58.

[25]   Furlano RI, Anthony A, Day R, *et al.* Colonic CD8 and gamma delta T-cell infiltration with epithelial damage in children with autism. J Pediatr 2001; 138: 366-72.

[26]   Torrente F, Ashwood P, Day R, *et al.* Small intestinal enteropathy with epithelial IgG and complement deposition in children with regressive autism. Mol Psychiatry 2002; 7: 375-82, 34.

[27]   Ashwood P, Anthony A, Pellicer AA, *et al*. Intestinal lymphocyte populations in children with regressive autism: evidence for extensive mucosal immunopathology. J Clin Immunol 2003; 23: 504-17.

[28]   Torrente F, Anthony A, Heuschkel RB, *et al*. Focal-enhanced gastritis in regressive autism with features distinct from Crohn's and Helicobacter pylori gastritis. Am J Gastroenterol 2004; 99: 598-605.

[29]   Fombonne E. Inflammatory bowel disease and autism. Lancet 1998; 351: 955.

[30]   Payne C, Mason B. Autism, inflammatory bowel disease, and MMR vaccine. Lancet 1998; 351: 907; author reply 8-9.

[31]   Fombonne E. Are measles infections or measles immunizations linked to autism? J Autism Dev Disord 1999; 29: 349-50.

[32]   Fombonne E, Chakrabarti S. No evidence for a new variant of measles-mumps-rubella-induced autism. Pediatrics 2001; 108: E58.

[33]   Taylor B, Lingam R, Simmons A, *et al*. Autism and MMR vaccination in North London; no causal relationship. Mol Psychiatry 2002; 7 Suppl 2: S7-8.

[34]   Hornig M, Briese T, Buie T, *et al*. Lack of association between measles virus vaccine and autism with enteropathy: a case-control study. PLoS One 2008; 3: e3140.

[35]   Blaylock R. Chronic microglial activation and excitotoxicity secondary to excessive immune stimulation: possible factors in Gulf War Syndrome and autism. J Am Phys Surg 2004; 9: 46-51.

[36]   Horvath K, Papadimitriou JC, Rabsztyn A, Drachenberg C, Tildon JT. Gastrointestinal abnormalities in children with autistic disorder. J Pediatr 1999; 135: 559-63.

[37]   Wakefield AJ, Montgomery SM. Autism, viral infection and measles-mumps-rubella vaccination. Isr Med Assoc J 1999; 1: 183-7.

[38]   Wakefield AJ, Anthony A, Murch SH, *et al*. Enterocolitis in children with developmental disorders. Am J Gastroenterol 2000; 95: 2285-95.

[39]   Wakefield AJ, Ashwood P, Limb K, Anthony A. The significance of ileo-colonic lymphoid nodular hyperplasia in children with autistic spectrum disorder. Eur J Gastroenterol Hepatol 2005; 17: 827-36.

[40]   Horvath K. Secretin treatment for autism. N Engl J Med 2000; 342: 1216; author reply 8.

[41]   Russo AJ, Andrews K. Is there a relationship between autism and gastrointestinal disease? Autism Insights 2010; 2: 13–5.

[42]   Fombonne E. The epidemiology of autism: a review. Psychol Med 1999; 29: 769-86.

[43]   Sabra A, Bellanti JA, Colon AR. Ileal-lymphoid-nodular hyperplasia, non-specific colitis, and pervasive developmental disorder in children. Lancet 1998; 352: 234-5.

[44]   Katayama Y, Hotta H, Nishimura A, Tatsuno Y, Homma M. Detection of measles virus nucleoprotein mRNA in autopsied brain tissues. J Gen Virol 1995; 76 (Pt 12): 3201-4.

[45]   Jyonouchi H. Food allergy and autism spectrum disorders: is there a link? Curr Allergy Asthma Rep 2009; 9: 194-201.

[46]   Ashwood P, Anthony A, Torrente F, Wakefield AJ. Spontaneous mucosal lymphocyte cytokine profiles in children with autism and gastrointestinal symptoms: mucosal immune activation and reduced counter regulatory interleukin-10. J Clin Immunol 2004; 24: 664-73.

[47]   Ashwood P, Wills S, Van de Water J. The immune response in autism: a new frontier for autism research. J Leukoc Biol 2006; 80: 1-15.

[48]   Jyonouchi H, Sun S, Itokazu N. Innate immunity associated with inflammatory responses and cytokine production against common dietary proteins in patients with autism spectrum disorder. Neuropsychobiology 2002; 46: 76-84.

[49]   Jyonouchi H, Geng L, Ruby A, Reddy C, Zimmerman-Bier B. Evaluation of an association between gastrointestinal symptoms and cytokine production against common dietary proteins in children with autism spectrum disorders. J Pediatr 2005; 146: 605-10.

[50]   Bolte ER. Autism and Clostridium tetani. Med Hypotheses 1998; 51: 133-44.

[51]   Sandler RH, Finegold SM, Bolte ER, *et al*. Short-term benefit from oral vancomycin treatment of regressive-onset autism. J Child Neurol 2000; 15: 429-35.

[52]   Finegold SM, Molitoris D, Song Y, *et al*. Gastrointestinal microflora studies in late-onset autism. Clin Infect Dis 2002; 35: S6-S16.

[53]   Martirosian G. [Anaerobic intestinal microflora in pathogenesis of autism?]. Postepy Hig Med Dosw (Online) 2004; 58: 349-51.

[54]   Song Y, Liu C, Finegold SM. Real-time PCR quantitation of clostridia in feces of autistic children. Appl Environ Microbiol 2004; 70: 6459-65.

[55]   Parracho HM, Bingham MO, Gibson GR, McCartney AL. Differences between the gut microflora of children with autistic spectrum disorders and that of healthy children. J Med Microbiol 2005; 54: 987-91.

[56]   Butt H ET. Cellular malnutrition and intestinal dysbiosis in Autism. Bioscreen Specialist Medical Testing Laboratory. 2003;

[57]   Rimland B. Candida-caused Autism? Autism Research Review International 1988; 2: 3.

[58]   D'Eufemia P, Celli M, Finocchiaro R, *et al.* Abnormal intestinal permeability in children with autism. Acta Paediatr 1996; 85: 1076-9.

[59]   Liu Z, Li N, Neu J. Tight junctions, leaky intestines, and pediatric diseases. Acta Paediatr 2005; 94: 386-93.

[60]   Hill ID, Dirks MH, Liptak GS, *et al.* Guideline for the diagnosis and treatment of celiac disease in children: recommendations of the North American Society for Pediatric Gastroenterology, Hepatology and Nutrition. J Pediatr Gastroenterol Nutr 2005; 40: 1-19.

[61]   Pavone L, Fiumara A, Bottaro G, Mazzone D, Coleman M. Autism and celiac disease: failure to validate the hypothesis that a link might exist. Biol Psychiatry 1997; 42: 72-5.

[62]   Atladottir HO, Pedersen MG, Thorsen P, *et al.* Association of family history of autoimmune diseases and autism spectrum disorders. Pediatrics 2009; 124: 687-94.

[63]   Stenman SM, Lindfors K, Korponay-Szabo IR, *et al.* Secretion of celiac disease autoantibodies after *in vitro* gliadin challenge is dependent on small-bowel mucosal transglutaminase 2-specific IgA deposits. BMC Immunol 2008; 9: 6.

[64]   Wieser H. Relation between gliadin structure and coeliac toxicity. Acta Paediatr Suppl 1996; 412: 3-9.

[65]   Wieser H. Chemistry of gluten proteins. Food Microbiol 2007; 24: 115-9.

[66]   Cornell HJ, Wills-Johnson G. Structure-activity relationships in coeliac-toxic gliadin peptides. Amino Acids 2001; 21: 243-53.

[67]   Shattock P, Whiteley P. Biochemical aspects in autism spectrum disorders: updating the opioid-excess theory and presenting new opportunities for biomedical intervention. Expert Opin Ther Targets 2002; 6: 175-83.

[68]   Shattock P, Hooper M, Waring R. Opioid peptides and dipeptidyl peptidase in autism. Dev Med Child Neurol 2004; 46: 357; author reply -8.

[69]   Kaminski S, Cieslinska A, Kostyra E. Polymorphism of bovine beta-casein and its potential effect on human health. J Appl Genet 2007; 48: 189-98.

[70]   Sun Z, Zhang Z, Wang X, *et al.* Relation of beta-casomorphin to apnea in sudden infant death syndrome. Peptides 2003; 24: 937-43.

[71]   Reichelt KL. [Gluten-free diet in infantile autism]. Tidsskr Nor Laegeforen 1991; 111: 1286-7.

[72]   Reichelt KL, Knivsberg AM, Lind G, Nodlandm M. Probable etiology and possible treatment of childhood autism. Brain Dysfunction 1991; 4: 308-19.

[73]   Reichelt KL, Knivsberg AM. Can the pathophysiology of autism be explained by the nature of the discovered urine peptides? Nutr Neurosci 2003; 6: 19-28.

[74]   Reichelt KL, Knivsberg AM. The possibility and probability of a gut-to-brain connection in autism. Ann Clin Psychiatry 2009; 21: 205-11.

[75]   Anderson RJ, Bendell DJ, Garnett I, *et al.* Identification of indolyl-3-acryloylglycine in the urine of people with autism. J Pharm Pharmacol 2002; 54: 295-8.

[76]   Bull G, Shattock P, Whiteley P, *et al.* Indolyl-3-acryloylglycine (IAG) is a putative diagnostic urinary marker for autism spectrum disorders. Med Sci Monit 2003; 9: CR422-5.

[77]   Carr K, Whiteley P, Shattock P. Development and reproducibility of a novel high-performance liquid-chromatography monolithic column method for the detection and quantification of trans-indolyl-3-acryloylglycine in human urine. Biomed Chromatogr 2009; 23: 1108-15.

[78]   Dettmer K, Hanna D, Whetstone P, Hansen R, Hammock BD. Autism and urinary exogenous neuropeptides: development of an on-line SPE-HPLC-tandem mass spectrometry method to test the opioid excess theory. Anal Bioanal Chem 2007; 388: 1643-51.

[79]   Cass H, Gringras P, March J, *et al.* Absence of urinary opioid peptides in children with autism. Arch Dis Child 2008; 93: 745-50.

[80]   Levy SE, Hyman SL. Complementary and alternative medicine treatments for children with autism spectrum disorders. Child Adolesc Psychiatr Clin N Am 2008; 17: 803-20, ix.

[81]   Campbell DB, Buie TM, Winter H, *et al.* Distinct genetic risk based on association of MET in families with co-occurring autism and gastrointestinal conditions. Pediatrics 2009; 123: 1018-24.

[82]   IMGSAC IMGSoAC. Further characterization of the autism susceptibility locus AUTS1 on chromosome 7q. Hum Mol Genet 2001; 10: 973-82.

[83]   Lamb JA, Barnby G, Bonora E, *et al.* Analysis of IMGSAC autism susceptibility loci: evidence for sex limited and parent of origin specific effects. J Med Genet 2005; 42: 132-7.

[84]   Campbell DB, Li C, Sutcliffe JS, Persico AM, Levitt P. Genetic evidence implicating multiple genes in the MET receptor tyrosine kinase pathway in autism spectrum disorder. Autism Res 2008; 1: 159-68.

[85] Jackson PB, Boccuto L, Skinner C, *et al.* Further evidence that the rs1858830 C variant in the promoter region of the MET gene is associated with autistic disorder. Autism Res 2009; 2: 232-6.

[86] Furge KA, Zhang YW, Vande Woude GF. Met receptor tyrosine kinase: enhanced signaling through adapter proteins. Oncogene 2000; 19: 5582-9.

[87] Gual P, Giordano S, Williams TA, *et al.* Sustained recruitment of phospholipase C-gamma to Gab1 is required for HGF-induced branching tubulogenesis. Oncogene 2000; 19: 1509-18.

[88] Tallafuss A, Eisen JS. The Met receptor tyrosine kinase prevents zebrafish primary motoneurons from expressing an incorrect neurotransmitter. Neural Dev 2008; 3: 18.

[89] Stellrecht CM, Gandhi V. MET receptor tyrosine kinase as a therapeutic anticancer target. Cancer Lett 2009; 280: 1-14.

[90] Margolis KG, Gershon MD. Neuropeptides and inflammatory bowel disease. Curr Opin Gastroenterol 2009; 25: 503-11.

[91] Gershon MD, Ratcliffe EM. Developmental biology of the enteric nervous system: pathogenesis of Hirschsprung's disease and other congenital dysmotilities. Semin Pediatr Surg 2004; 13: 224-35.

[92] Gershon MD. The enteric nervous system: a second brain. Hosp Pract (Minneap) 1999; 34: 31-2, 5-8, 41-2 passim.

[93] Laranjeira C, Pachnis V. Enteric nervous system development: Recent progress and future challenges. Auton Neurosci 2009; 151: 61-9.

[94] Welch MG, Tamir H, Gross KJ, *et al.* Expression and developmental regulation of oxytocin (OT) and oxytocin receptors (OTR) in the enteric nervous system (ENS) and intestinal epithelium. J Comp Neurol 2009; 512: 256-70.

CHAPTER 7

Biochemical Changes in ASD

Anna Strunecka

Department of Physiology, Faculty of Science, Charles University in Prague, Prague, Czech Republic

Abstract: Metabolic dysfunctions have not been extensively studied in ASD despite the fact that chronic biochemical imbalance is often a primary factor in the development of several neurological diseases. Substantial percentages of autistic patients display peripheral markers of mitochondrial energy metabolism dysfunction, such as elevated lactate and alanine levels in blood and serum carnitine deficiency. We assess the reported biochemical changes in the blood and evidence based on the exploration of brain imaging studies. Even though alterations in mitochondrial and cellular energy metabolism are not specific for ASD, they indicate the potential ethiopathological events. Evidence from several laboratories similarly indicates that biomarkers of oxidative stress may be increased in some autistic children. One of the best documented biochemical changes in ASD is a decrease in cellular glutathione (GSH) levels, a major intracellular antioxidant, and an increase in oxidized glutathione (GSSG). Alterations in methionine –homocysteine cycle have been studied in details in ASD. Significant changes in transmethylation and transsulfuration metabolites in plasma from autistic children were reported. The new finding indicates a significant decrease in methylation capacity and redox potential. Metabolic and mitochondrial defects may have toxic effects on brain cells, causing neuronal loss and altered modulation of neurotransmission systems. The observations of biochemical changes thus further support that the antioxidant therapy and supplementation with some vitamins could prevent and restore the energy metabolism of individuals with ASD. This chapter brings evidence of the impact of observed biochemical changes in ASD for potential amelioration of ASD symptoms and for evidence-based therapy.

INTRODUCTION

The biochemical changes described in patients with ASD include impaired glutathione metabolism, impaired methylation, changes in homocysteine cycle, and impaired metallothionein production. Several changes have been connected with mercury burden, such as decreased glutathione levels, altered porphyrin profiles in urine, and alterations in the transsulfuration pathways [1]. The impaired sulfur status of urine organic acids and a low urinary sulfate as a marker of total body sulfur stores is well documented in ASD.

Several biochemical changes are associated with the impaired digestive functions, such as low activities of disaccharidase enzymes, low amino acids, and defective sulfation of ingested phenolic amines (tylenol) [2,3]. The changes in the levels of organic acids in the blood and urine of ASD patients have been frequently observed. These may be connected with impaired energy metabolism and metabolism of neurotransmitters and hormones. Recent studies have suggested a frequent association of ASD and mitochondrial dysfunction. Mitochondrial disorders are heterogeneous in their pathological expression, but all are characterized by impaired energy production and mitochondrial electron transport chain (ETC) dysfunction. Metabolic and mitochondrial defects may have toxic effects on brain cells, causing neuronal loss and altered modulation of neurotransmission systems. Depletion of cellular energy levels increased the vulnerability toward excitotoxins, leading to cell death.

Several metabolic defects have been associated with autistic symptoms with an incidence higher than that found in the general population. Selective metabolic testing should be done in the presence of suggestive clinical findings. In some patients, early diagnosis of the metabolic disorders and proper therapeutic interventions may significantly improve the long-term cognitive and behavioral outcome [4]. Several inborn errors of metabolism, such as phenylketonuria, creatine deficiency syndromes, metabolic purine disorders, and deficiencies in folic acid, have an autistic phenotype [5].

The goals of this chapter are to provide a review of metabolic changes found in patients with ASD and a critical assessment for searching the potential biochemical markers for ASD diagnosis.

*Address correspondence to: Anna Strunecka, Faculty of Sciences, Charles University in Prague, Vinicna 7, 128 00 Prague 2, Czech Republic; E-mail: strun@natur.cuni.cz

MITOCHONDRIAL DYSFUNCTION AND IMPAIRMENT OF ENERGY METABOLISM IN ASD

Mitochondrial disease is not a single entity but rather, a heterogeneous group of disorders characterized by impaired energy production due to phosphorylation dysfunction. These disorders constitute the most common neurometabolic disease of childhood with an estimated minimal risk of developing mitochondrial disease of one in 5000 [6]. In 1998, Lombard published a hypothesis that an etiological possibility for autism may involve mitochondrial dysfunction with concomitant defects in neuronal oxidative phosphorylation within the CNS [7]. This hypothesis was supported by a frequent association of lactic acidosis and carnitine deficiency in autistic patients. Lombard suggested that strategies to augment mitochondrial function may be beneficial in the treatment of autism. Besides a few case reports there are several studies focused on co-occurrence and/or association between ASD and disorders of mitochondrial functions. The idea that impairment of mitochondrial functions could contribute to ASD pathology or even that it can be the cause of ASD has appeared in several papers.

However, the absence of a reliable biomarkers specific for the screening of mitochondrial disease contributes to diagnostic difficulty [6,8]. Initial evaluation of patients with ASD would include metabolic screening of blood and urine [2]. It appears that mitochondrial disease is significantly under-recognized or neglected in patients with ASD.

Evidence of Mitochondrial Dysfunction in Autistic Patients

Of sixty families collected through a single proband to help to better define infantile autism in Spain, Moreno and Borjas [9] found in 1992 that twenty four of patients showed increased lactate and laboratory findings of metabolic acidosis. This finding confirmed the previous report of four autistic patients who had two coexistent syndromes: the behavioral syndrome of autism and the biochemical syndrome of lactic acidosis [10]. Filipek *et al.* [11] found carnitine deficiency in 100 children with autism and suggested that a mild mitochondrial dysfunction may be the origin of this alteration. Concurrently drawn serum pyruvate, lactate, ammonia, and alanine levels were also available in many of these children. Values of free and total carnitine and pyruvate were significantly reduced while ammonia and alanine levels were considerably elevated in autistic subjects. Carnitine acts as a carrier of long chain fatty acid to mitochondria and to β-oxidation, thus an important molecule for energy production and maintenance of mitochondrial function. Mitochondrial acetyl-carnitine also provides acetyl groups (CH3-C=O⁻) for nuclear histone acetylation [12]. Genetic deficiency of the mitochondrial carnitine/acylcarnitine translocase markedly reduced the nuclear histone acetylation, indicating the significance of the carnitine-dependent mitochondrial acetyl group contribution to histone acetylation.

In a population-based study Oliveira *et al.* [13] screened associated medical conditions in a group of 120 children with autism (mean age 13 years) of which 76% were diagnosed with typical autism, 24% with atypical autism. They used hyperlactacidemia and increased lactate/pyruvate ratios as a marker of mitochondrial dysfunction. In another study, plasma lactate levels were measured in 69 patients, and in 14 they found hyperlactacidemia. Five of 11 patients studied were classified with definite mitochondrial respiratory chain disorder. In a subsequent study these authors confirmed the high frequency of hyperlactacidemia and increased lactate/pyruvate ratio in a significant fraction of 210 autistic patients and concluded that mitochondrial dysfunction may be one of the most common medical conditions associated with autism [14]. In a study comparing 15 autistic children with 15 children with epilepsy without autism, Chugani *et al.* [17] also concluded that higher plasma lactate level in the autistic group was consistent with metabolic abnormalities in some of the autistic children.

In a more recent study, Weissman *et al.* [15] reviewed medical records of 25 patients with idiopathic autism accordingly to DSM-IV criteria, later determined to have enzyme- or mutation-defined mitochondrial ETC dysfunction. Levels of blood lactate, plasma alanine, and serum alanine transaminase (ALT) and/or aspartate aminotransferase (AST) were increased at least once in 76%, 36%, and 52% of patients, respectively. The most common ETC disorders were deficiencies of complex I (64%) and complex III (20%). Although all patients' initial diagnosis was idiopathic autism, careful clinical and biochemical assessment identified clinical findings that differentiated them from children with idiopathic autism. These and prior data suggest a disturbance of mitochondrial energy production as an underlying pathophysiological mechanism in a subset of individuals with autism.

Poling *et al.* [16] described a case of six years girl carefully followed since 19 months. Laboratory tests indicated mitochondrial dysfunction, growth failure, and abnormal muscle histopathology without seizures or a defined chromosomal abnormality. This patient exemplifies important questions about mitochondrial function in autism and developmental regression. It is unclear whether mitochondrial dysfunction results from a primary genetic abnormality, atypical development of essential metabolic pathways, or secondary inhibition of oxidative phosphorylation by other factors. The discussion by Poling and co-authors is very instructive for understanding the possible impact of mitochondrial dysfunction in ASD pathogenesis.

If such dysfunction is present at the time of infections and immunizations in young children, the added oxidative stress from immune activation on cellular energy metabolism is likely to be especially critical for the CNS, which is highly dependent on mitochondrial function. Young children who have dysfunctional cellular energy metabolism therefore might be more prone to undergo autistic regression between 18 and 30 months of age if they also have infections or immunizations at the same time. Although patterns of regression can be genetically and prenatally determined, it is possible that underlying mitochondrial dysfunction can either exacerbate or affect the severity of regression [16]. The diagnosis of mitochondrial dysfunction was based on findings of mildly increased AST and serum creatine kinase (CK) levels. A muscle biopsy showed reduced cytochrome c oxidase activity and marked reduction in enzymatic activities for complex I and complex III. Complex IV (cytochrome c oxidase) activity was near the five percent confidence level. Poling with coworkers retrospectively evaluated the laboratory records of 159 patients with autism and 94 patients of a similar age with other neurological disorders. AST was elevated in 38% of patients with autism compared with 15% of controls. The serum CK level also was abnormally elevated in 22 (47%) of 47 patients with autism. Table **1** shows observations from various studies, which could indicate the mitochondrial disturbance in autistic patients.

Table 1: Biochemical Changes in Blood of ASD Patients, Which Could Indicate the Mitochondrial Disturbance.

DIAGNOSIS	NO OF PATIENTS	BIOCHEMICAL CHANGE IN BLOOD	REFERENCE
		lactate	
autism	4	↑	[10]
autism	60	↑ in 24	[9]
autism	15	↑	[17]
autism	69	↑ in 14	[13]
autism	210	↑	[14]
ASD	25	↑	[15]
autism	30	↑	[18]
		pyruvate	
autism	100	↓	[11]
autism	210	↓	[14]
		lactate/pyruvate ratio	
autism	120	↑	[13]
autism	210	↑	[14]
		alanine	
autism	100	↑	[11]
ASD	25	↑	[15]
autism and Asperger´syndrome		↑	[19]
		carnitine	
autism	100	↓	[11]
		creatine	
autism	100	↓	[11]
		creatine kinase	
autism	47	↑↑ in 22 patients	[16]
autism	30	↑	[18]
		ALT and AST	
autism	25	↑ ↑	[15]
autism	159	normal ↑	[16]

Legend: ASD = autism spectrum disorders; ALT = alanine transaminase; AST = aspartate aminotransferase

CK was assessed in plasma of 30 Saudi autistic patients and compared to 30 age-matching control samples [18]. In addition, ATP, ADP, and AMP were measured calorimetrically in the red blood cells of both groups. Lactate concentration in plasma of both groups was monitored as well. The obtained data recorded 72.35% higher activity of CK in autistic patients, which prove the impairment of energy metabolism in these children compared to age and sex matching healthy controls. There was no significant difference in the levels of ATP, ADP, and AMP in both groups neither in the calculated adenylate energy charge (AEC) values. The unchanged AEC value in autistic patients was easily correlated with the induced activity of CK and ADPase as these two enzymes play a critical role in the stabilization of AEC. Lactate was significantly higher in autistic patients compared to control showing about 40% increase. The present study confirmed the impairment of energy metabolism in Saudi autistic patients, which could be correlated to the oxidative stress previously recorded in the same analyzed samples.

Biochemical Markers of Mitochondrial Dysfunction

The Mitochondrial Medicine Society's Committee on Diagnosis published in 2008 the review to provide an overview of currently available and emerging methodologies for the diagnosis of primary mitochondrial disease, primarily focusing on disorders characterized by impairment of oxidative phosphorylation [8]. It is behind the scope of our book to provide algorithms of a systematic overview of these evaluations; we will only comment on the biochemical assays used in the above mentioned studies with ASD patients, which are also in concordance with recommendations of the above mentioned society (see Table **2**).

Table 2: Baseline Screening Tests of Blood and Urine, and Spinal Fluid for Mitochondrial Disease [8].

BLOOD AND URINE	SPINAL FLUID
Blood lactate, pyruvate and lactate/pyruvate ratio	Lactate and pyruvate
Quantitative plasma amino acids	Quantitative amino acids
Creatinine kinase	Routine tests including glucose and
Plasma acylcarnitine analysis	protein measurement
Quantitative urine organic acids	

Lactic acid elevation in blood (typically considered > 2.1 mM) or cerebrospinal fluid (CSF) can be an important, albeit non-specific, marker of mitochondrial disease. Lactate, the product of anaerobic glucose metabolism in the cytoplasm, accumulates when aerobic metabolism in mitochondria is impaired, which causes accumulation of NADH and shift in the oxidized-to-reduced $NAD^+/NADH$ ratio within mitochondria. Elevations of plasma lactate and/or pyruvate levels may be also seen in a wide range of systemic diseases and metabolic disorders as a result of secondary mitochondrial dysfunction.

Pyruvate produced by glycolysis is actively transported across the inner mitochondrial membrane, and into the matrix where it is oxidized. It is necessary to quantify blood *lactate/pyruvate* ratios, which indirectly reflect the $NAD^+/NADH$ cytoplasmic redox state. Reduction of pyruvate ($CH_3–CO–COO^-$) by the enzyme lactate dehydrogenase produces lactate ($CH_3–HCOH–COO^-$); it means that lactate is oxidized to pyruvate. Pyruvate, via lactate dehydrogenase, is in dynamic equilibrium with lactate; this reaction is essential to regenerate NAD^{++} residues (see Fig. **1**). Even when plasma levels of lactate and pyruvate are normal, CSF lactate levels may be elevated in patients with mitochondrial disease who have predominant brain manifestations. Pyruvate is quite unstable in the blood. The enzyme *alanine transaminase (ALT)* is a transaminase enzyme, also called serum glutamic pyruvic transaminase, converts pyruvate to *alanine*. ALT is found in serum and in various tissues, but is most commonly associated with the liver. ALT measurement is obviously used as evidence for the safety of acute administration of drugs such as naltrexone, risperidone, and haloperidol in children.

Alanine is the major gluconeogenic amino acid. Plasma alanine is used to make glucose in the liver in so called alanine cycle. When gluconeogenesis occurs, plasma alanine concentration decreases. Aldred *et al.* [19] found that patients with autism or Asperger syndrome and their siblings and parents all had raised alanine and glutamic acid. Weissman *et al.* [15] found that plasma alanine and serum ALT were increased at least once in 36% and 52% of autistic patients, respectively.

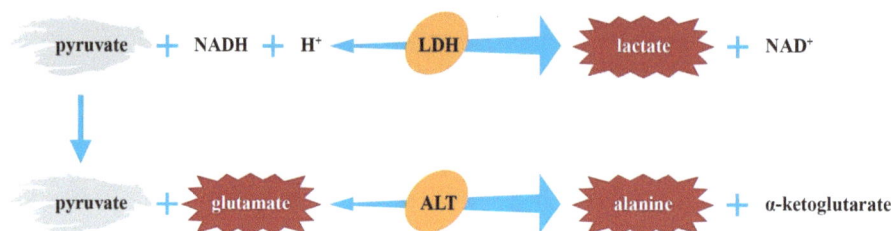

Figure 1: Patients with ASD have raised blood lactate, alanine, and glutamate. LDH - lactate dehydrogenase; ALT - alanine transaminase

Carnitine serves as a mitochondrial shuttle for free fatty acids and a key acceptor of potentially toxic coenzyme A (CoA) esters. It permits restoration of intramitochondrial CoA and removal from the mitochondria of esterified intermediates by enabling their urinary excretion. A frequent association of carnitine deficiency in autistic patients has been reported [7,11]. Moreover, Clark-Taylors [20] hypothesized that dysfunction of fatty acid β-oxidation may be a cause of autism. Quantification of blood total and free carnitine levels, along with acylcarnitine profiling, permits identification of carnitine deficiency as well as fatty acid oxidation defects [8].

Creatine kinase (CK) is an enzyme found mainly in skeletal muscle, the heart, and the brain. The normal function of CK is to catalyze phosphorylation of creatine using ATP, turning creatine into the high-energy molecule phosphocreatine (PCr). PCr is burned as a quick source of energy by muscle cells and acts as an energy buffer, protecting the ATP concentration.

Creatine Kinase in the Brain

The distinct isoenzyme-specific localization of CK isoenzymes found in the brain suggests an important function for CK in brain energetics and points to adaptation of the CK system to the special energy requirements of different neuronal and glial cell types. The presence of muscle-type CK, found exclusively in Purkinje cells, which also express other muscle-specific proteins, is very likely related to the unique calcium metabolism of these neurons [21-23]. Suzuki *et al.* [24] reported that high-functioning adult subjects with autism (n = 12) have abnormal creatine plus PCr concentrations in the hippocampal formation, which may in part account for their aggression. Total creatine (tCr) constitutes one of the most prominent signals in human brain proton magnetic resonance spectroscopy (^{1}H MRS). A significant decrease in the tCr signal indicates a severe disorder of creatine metabolism. Dezortova *et al.* [25] found a very low tCr signal in a 5-year-old boy with severe psychomotor retardation, epilepsy and ASD problems including speech delay, which was approximately three times lower than in his sister. ^{1}H MRS measurements collected from 18 male children with autism and 16 healthy children [26] showed lower levels of PCr and creatine on the left side of thalamus in the autism group compared with controls. Further investigations of this structure are warranted, since it plays an important role in information processing as part of the cortico-thalamo-cortical pathways.

The ratio of each metabolite to creatine is used as a standard reporting value due to fairly uniform CNS creatine levels in most individuals [8]. In contrast, brain CK deficiency syndromes are mainly associated to mental retardation and autism [27]. Thirteen children (7-16 years) with ASD and eight typically developing children were compared on ^{1}H MRS data collected from hippocampus-amygdala and cerebellar regions. The ASD group had significantly lower N-acetyl-aspartate (NAA)/creatine ratios bilaterally in the hippocampus-amygdala but not cerebellum, whereas myo-inositol/creatine ratio was significantly increased in all measured regions. Choline/creatine ratios were also significantly elevated in the left hippocampus-amygdala and cerebellar regions of children with ASD [28].

Carnitine Deficiency and Effects of Carnitine Supplementation

It is not clear if carnitine deficiency is due to some genetic defect or to some nutritional deficiency. It is not clear whether the carnitine deficiency is the primary cause of mitochondrial dysfunction or the secondary response reflecting mitochondrial disorder. Melegh et al [29] reported that treatment of seven children with antibiotic containing pivalic acid for seven days significantly reduced the amounts of total acid-soluble carnitine, free carnitine, and long-chain acyl-carnitines. Pivalic acid is well known as a component of prodrugs by its ability to highly increase drug absorbance in the GIT. Pivalic acid-bound antibiotics are widely used in pediatrics. These antibiotics are available in syrup formulation for small children and are recommended for the treatment of penicillin-resistant bacterial infections. This prodrug is degraded into active antibiotics and pivalic acid by the intestinal epithelial cells; pivalic acid is carnitine-conjugated and excreted into urine. Ito [30] described a case of a one-year-old boy who was administered with several different antibiotics for six months in a row for recurrent otitis media and recurrent upper respiratory tract inflammation. The patient showed tremor of the hands and feet, propagated to generalized convulsion and brought to an emergency outpatient. Hypoglycemia without ketone body production led to an assumption that the fatty acid oxidation is in failure. This case showed about 1/10 of normal blood levels of free carnitine and total carnitine. After a few similar cases were reported, warning for decrease in carnitine has been added to the Drug Package Inserts. It is essential for every clinician to pay attention to possible side-effects of drugs containing pivalic acid [30]. Long-term administration of such antibiotics could induce depletion of carnitine and lead to ketotic hypoglycemia, convulsion, and conscious disturbance.

Ellaway and colleagues conducted a randomized, placebo-controlled, double-blind crossover trial of L-carnitine on 35 girls with Rett syndrome [31]. Eight-week treatment phases were completed for both a placebo and L-carnitine. Medical review showed an improvement on the Hand Apraxia Scale for a higher proportion of girls on L-carnitine. The researchers concluded that while L-carnitine did not lead to major functional changes in ability, the type of changes reported could still have a substantial impact on the girls and their families. Subsequently, an open label trial study was performed in a cohort of 21 Rett syndrome females, with a control group of 62 Rett syndrome females of a similar age, for a 6-month period [32]. Compared with the Rett syndrome controls, treatment with L-carnitine led to significant improvements in sleep efficiency, energy level and communication skills. In addition, before and after comparisons of the treatment group showed improvements in expressive speech. Treatment with L-carnitine seems to be of significant benefit in a subgroup of girls with Rett syndrome.

The use of carnitine in the treatment of ASD has been investigated by randomized double blind placebo controlled study (http://controlled-trials.com/ISRCTN54273114). This recent study is designed to examine mitochondrial dysfunction and how L-carnitine supplementation affects behavior, cognition, muscle strength, and health/physical traits in those with a diagnosed ASD.

The effect of L-carnitine supplementation was intensively studied by Melegh and co-workers. They suggested that L-carnitine supplementation in low-weight newborns may enhance triglyceride utilization [33]. At the second week of life (9 to 14 days of age) the infants were randomly divided into two groups. Five of them received oral L-carnitine supplementation added to pasteurized pooled human milk for seven consecutive days; additional five served as controls. On the seventh day, plasma carnitine and ketone body levels were significantly increased in the supplemented group as compared to controls or to previous values of the same group. Effects of oral L-carnitine supplementation on fat and protein metabolism have been studied in three consequent independent studies [34-36]. Total, free, and esterified carnitine levels were significantly elevated in the plasma at the end of the study period. The increased levels of acyl-carnitines in plasma and urine indicated that the carnitine supplement was taken up by tissues and entered the intermediary metabolism. Plasma triglyceride level was decreased, whereas 3-hydroxybutyrate level was increased at the end of supplementation, indicating an enhanced fat utilization.

ALTERATIONS OF GLUCOSE METABOLISM IN ASD

The findings of impaired mitochondrial metabolism in autistic patients indicate that some disturbance in glucose metabolism could occur both in the brain and peripheral tissues. It has been generally accepted that the essential brain functions depend on a continuous supply of glucose. It is well established that regional cerebral metabolic rates for glucose assessed by [^{18}F]-2-fluoro-2-deoxy-D-glucose (FDG) positron emission tomography (PET) provides a sensitive, *in vivo* metabolic index. However, FDG PET studies conducted with autistic individuals at rest brought very inconsistent findings. There are no published data on Medline regarding reports of changes in blood glucose levels in autistic patients. In this chapter, we could thus only mention a few papers, which evaluated the potential link of ASD and diabetes.

A prospective open-label trial with olanzapine treatment was conducted on 40 male children meeting DSM IV criteria for autism in Kuwait [37]. Children underwent laboratory investigations including urine analysis, serum chemistry, blood glucose, and lipid profiles. Neither significant hepatic enzyme elevation nor any serum chemistry changes were observed before or after treatment.

FDG PET Studies of Autistic Patients

Rumsey *et al.* [38] studied the cerebral metabolic rate for glucose in ten men with well-documented histories of infantile autism. The comparison with a group of 15 age-matched healthy male showed the considerable overlap between the two groups. However, the autistic group demonstrated significantly elevated glucose metabolism in widespread regions of the brain and more subjects demonstrated extreme relative metabolic rates in at least one brain region. No brain region demonstrated a reduced metabolic rate in autistic patients. Similar findings were reported by De Volder and co-workers who measured brain glucose metabolism in 18 autistic children using FDG PET [39]. Global brain glucose utilization was slightly elevated than observed in young adult volunteers, particularly in frontal cortical region. Regional metabolic maps were normal, although there was evidence of heterogeneities. Six children showed a relative hyperfrontality whilst hypofrontality was found in two children. Nevertheless, these heterogeneities were not correlated with clinical symptoms. De Volder *et al.* thus concluded that both the rate and the regional distribution of brain glucose metabolism are normal in autistic children. No significant differences were found between the group of six young autistic men and normal controls in regional cerebral blood flow, oxygen consumption, and glucose consumption [40]. Schifter *et al.* [41] found that seven of 13 autistic patients had normal FDG PET, whilst four of 13 patients had abnormal FDG PET. Sixteen of a total 195 brain areas examined with FDG PET had a hypometabolic abnormality on PET.

Chugani *et al.* [17] investigated the hypotheses that there are increased concentrations of lactate in brain and plasma and reduced brain concentrations of NAA in autistic children. NAA and lactate levels in the frontal lobe, temporal lobe and the cerebellum of nine autistic children were compared to five sibling controls using magnetic resonance spectroscopy (MRS). Preliminary results show lower levels of NAA in the cerebellum in autistic children. These authors found higher lactate level in plasma of 15 autistic children. Lactate was detected in the frontal lobe in one autistic boy, but was not detected any of the other autistic subjects or siblings. They concluded that the findings of altered brain NAA and higher lactate in both plasma and brain suggest evidence of altered energy metabolism in some autistic children.

FDG PET scans of autistic patients and healthy volunteers (n = 17 in each group) were used to examine relative glucose metabolism (rGMR) during performance of a verbal memory task [42]. In the frontal lobe, patients had lower rGMR in medial/cingulate regions but not in lateral regions compared with healthy controls. Patients had higher rGMR in occipital and parietal regions compared with controls, but there were no group differences in temporal lobe regions. Autism patients have dysfunction in some but not all of the key brain regions subserving verbal memory performance.

The FDG PET study of six female patients with Rett syndrome showed a relative decrease in glucose uptake in the lateral occipital areas in relation with the whole brain and a relative increase in the cerebellum. The authors of this study noticed that changes in glucose cerebral metabolism resemble the regional distribution of normal children less than one year of age. Moreover, the changes in frontal areas parallel those in postmortem NMDA receptors density [43].

The investigation of 16 high-functioning adults with a history of infantile autism revealed the more prevalent brain regions with glucose metabolic rate three times higher in comparison with control healthy subjects. These patients had a left > right anterior rectal gyrus asymmetry, as opposed to the normal right > left asymmetry of regional glucose metabolic rate [44].

It is evident that in PET studies conducted with individuals with ASD the findings vary from elevated glucose uptake by those reporting no changes in glucose metabolism or hypometabolic abnormality. To assess the alterations of glucose metabolism in the brain some other variable could be helpful. The most important is the measurement of cerebral oxygen consumption. The mean ratio between oxygen and glucose utilization could indicate the increased glycolysis to lactate. The frequently observed increased lactate levels in serum of patients with ASD can indicate that in connection with disturbance of mitochondria metabolism, the rate of glycolysis is increased in peripheral tissues as well. The biochemical changes in the blood and in the brain of ASD patient could thus indicate hyperglycolysis; the reduction of oxidative metabolism followed by a relative increase of anaerobic glycolysis to maintain energy supply. The study of Frackowiak *et al.* [45] illustrates such view. These researchers studied cerebral oxygen using ^{15}O tracer and glucose using FGD PET in eight patients with biochemically defined mitochondrial myopathies. Four patients had myopathy alone and four had predominantly CNS disease. Patients with major CNS disease showed an uncoupling of glucose and oxygen metabolism when compared with patients without cerebral disease and normal subjects. The mean ratio between oxygen and glucose utilization was 3.8 moles of oxygen per mole of glucose in patients with CNS disease and 6.4 for patients with myopathy alone, compared with 5.6 for controls. Patients with major CNS disease showed a 50% reduction in cerebral oxygen utilization compared with cerebrally unaffected patients and normal subjects.

How Can Metabolism of Glucose Be Affected in the Brain of ASD Patients?

As we mentioned, changes in local brain energy metabolism can now be studied in humans with PET by monitoring alterations in glucose utilization, oxygen consumption, and blood flow during activation of specific areas. Some studies in which these three parameters have been analyzed have yielded unexpected results. Thus, an uncoupling between glucose uptake and oxygen consumption was observed during injury and activation, since the increase in blood flow and in glucose utilization in the activated cortical area was not matched by an equivalent increase in oxygen consumption [45-48]. This observation raises the possibility that, at least during the early stages of activation, the increased energy demand is met by glycolysis rather than by oxidative phosphorylation. We could conclude for example, that some specific areas in the brains of individuals with ASD are permanently activated, or, that in these areas the mitochondrial oxidation is impaired and the energy demand rely on increased rate of glycolysis. Cerebral hyperglycolysis is a pathophysiological response to impairment-induced ionic and neurochemical cascades.

The observation of cerebral hyperglycolysis supports the hypothesis of several researchers that mitochondrial disorders may be the most common metabolic findings in ASD. Mitochondria are vulnerable to a wide array of endogenous and exogenous factors which appear to be linked by excessive free radicals production. It is the main concept of our eBook that the underlying mechanism of most observed biochemical, metabolic, and pathophysiological symptoms of ASD is the excitotoxicity, more accurate the imunoexcitotoxicity. We have previously described the immune-glutamatergic dysfunction as a central mechanism of the ASD [45-52]. Oxidative stress, neuroinflammation, and excitotoxicity are frequently considered distinct but common hallmarks of several neurological disorders. Free radicals are highly reactive and induce oxidative damage to neighboring molecules by extracting electrons. The major intracellular site of free radical generation is the mitochondria, where oxidative phosphorylation occurs in association with ATP production.

Moreover, during the past two decades a close relationship between the energy state of the cell and glutamate neurotoxicity has been documented. The measurement of FDG uptake and the observations of lactate/pyruvate levels are only the first and last markers of very complicated biochemical, metabolic, and neurophysiological processeess that occur in the brain.

Several factors, such as the excessive immune activation systemically inducing a state of chronic brain inflammation, high levels of reactive oxygen species ROS), reactive nitrogen species (RNS), lipid peroxidation products (LPP), inflammatory prostaglandins (PGE), dietary excitotoxins, mercury, fluoride and Al^{3+} can disrupt the glutamate homeostasis and increase extracellular glutamate levels sufficiently to trigger the excitotoxic cascade. Moreover, glia become activated by inflammatory mediators in a wide range of CNS pathologies [53,54].

Excessive activation of the NMDA receptors increase intracellular Ca^{2+} concentrations ($[Ca^{2+}]_i$), triggering a series of cell signaling systems, which can cause an increase in cellular ROS, RNS, and LPP, and activate the PGE reactions. By increasing the activity of inducible nitric oxide synthetase (iNOS), glutamate increases intracellular NO, which in the presence of increased levels of superoxide can generate high levels of peroxynitrite, which is very toxic to mitochondrial energy-producing enzymes. Reducing cellular energy production has been shown to greatly magnify excitotoxicity to a degree where even physiological levels of glutamate can become excitotoxic [52]. There is clear evidence of the role of microglial inflammation in autism pathology [55-57].

Astrocyte end-feet surround intraparenchymal microvessels and represent therefore the first cellular barrier for glucose entering the brain. As such, they are a likely site of prevalent glucose uptake. Magistretti and co-workers studied since 1996 the glutamate-stimulated aerobic glycolysis in astrocytes, and observed the release of lactate from astrocytes [46,48, 58,59]. Concomitant to the stimulation of glucose uptake, glutamate causes a concentration-dependent increase in lactate efflux. These observations suggest that glutamate uptake is coupled to aerobic glycolysis in astrocytes. In addition, since glutamate release occurs following the modality-specific activation of a brain region, the glutamate-evoked uptake of glucose into astrocytes provides a simple mechanism to couple neuronal activity to energy metabolism. Lactate can then contribute to the activity-dependent fueling of the neuronal energy demands associated with synaptic transmission. Analyses of this coupling have been extended *in vivo* and have defined the methods of coupling for inhibitory neurotransmission as well as its spatial extent in relation to the propagation of metabolic signals within the astrocytic syncytium. On the basis of a large body of experimental evidence, Magistretti with coworkers proposed an operational model, "the astrocyte-neuron lactate shuttle" [59]. They suggest that the coupling between synaptic activity and glucose utilization (neurometabolic coupling) is a central physiological principle of brain function that has provided the basis for FDG PET.

In this chapter we show that the measurements of FDG PET indicate the increased uptake of FDG in some cases of ASD patients and in some areas of their brains. Several researchers also reported hyperlactacidemia in the blood of individuals with ASD. The evidence for immunoexcitotoxic processes in the brain of individuals with ASD has been collected in several Blaylock ´s papers [49-52] and in the chapter 4. However, evidence from several laboratories similarly indicates that biomarkers of oxidative stress may be increased in some autistic children [60-64].

Is There a Link Between Type 1 Diabetes Mellitus and ASD?

Freeman with co-workers [65] examined 984 children with type 1 diabetes mellitus attending the Diabetes Clinic at The Hospital for Sick Children, Toronto and nine of them were identified as having ASD. There were seven boys and two girls. The authors concluded that the prevalence of ASD in this group of patients may be greater than that in general population (0.9% vs. 0.34%).

Harjutsalo and Tuomiletho [66] investigated the presence of ASD in 5,178 children diagnosed with type 1 diabetes at age ≤14 years from the Prospective Childhood Diabetes Registry of Finland born between 1980 and 2000. They included autism, Asperger disorder, PDD-NOS, Rett syndrome, and childhood disintegrative disorder in the diagnosis of ASD. Seven cases with type 1 diabetes fulfilled the criteria of ASD, giving a cumulative incidence of 1.35/1,000. The cumulative incidence of ASD did not differ from that in the background population at age 18 years in northern Finland. Thus the observation from Finland does not support the suggestion of a link between type 1 diabetes and ASD.

Iafusco *et al.* [67] presented interesting data from Italy. The prevalence of ASD in the general Italian population is estimated at 0.1%. On the other hand, Italy has a peculiar epidemiology of type 1 diabetes. Sardinia has one of the highest incidences in the world (42.4/100), while peninsular Italy has an overall incidence that is similar to other Mediterranean areas (8.4/100); 11.2/100 in North Italy, and 6.2/100 in South Italy. Only two of 1,373 patients aged <14 years from the Sardinian Registry of type 1 diabetes were diagnosed with ASD (0.146%), a finding similar to the Finnish data. On the contrary, a pattern similar to that observed by Freeman and co-workers [65] has been found in patients with type 1 diabetes aged <14 years from six Italian centers of pediatric diabetology equally distributed in the Italian Peninsula and in Sicily (0.72%).

Peculiarities in the endocrine response to insulin stress in children with early infantile autism were reported in 1975 [68]. Autistic children subjected to insulin-induced hypoglycemia showed slower recovery of blood glucose. The cortisol response was faster and intractable for three hours following the stress. Maher and co-workers accounted this peculiarity to the abnormal behavior seen in ASD.

Whilst the convincing evidence of the association of diabetes with Alzheimer's disease exists [69], the co-morbidity of ASD and diabetes has not been proved. The lack of studies dealing with blood glucose levels in autistic patients probably indicates the lack of manifestation of the links of ASD and diabetes.

IMPAIRED TRANSSULFURATION, METHYLATION, AND OXIDATIVE STRESS IN ASD: THE PUZZLE OF HOMOCYSTEINE CYCLE

It has recently been shown that there is a characteristic metabolic profile in many autistic children involving disturbances in methionine and glutathione metabolism. A number of studies clearly indicate microglial inflammation, the presence of redox imbalance, and chronic oxidative stress in autism [55,70-72]. They also provide the consistent picture of biochemical and metabolic alterations in individuals with ASD in the methionine-homocysteine cycle and in all three independent pathways involved in folate-dependent methionine transmethylation and transsulfuration: in the folate cycle, the methionine cycle, and the transsulfuration pathway leading to glutathione synthesis.

Glutathione is pivotal for the maintenance of intracellular redox homeostasis and defense against oxidative damage in eukaryotic cells. Moreover, it has been demonstrated in rats that the transsulfuration pathway products of glutathione and sulfate are related to mercury excretion rates, and that the heme synthesis pathway products of urinary porphyrins can provide specific profiles that reflect mercury toxicity [73,74]. Geier *et al.* [1] reported that their 28 study participants with a diagnosis of ASD were observed to have significant decreased levels of the transsulfuration metabolites of cysteine, sulfate, and reduced glutathione. Since urinary porphyrin testing is clinically available, relatively inexpensive, and noninvasive, Geier and co-workers suggested it may be used as biomarker of environmental toxicity and susceptibility in ASD.

Biomarkers of Oxidative Stress in ASD

There is evidence that oxygen free radicals play an important role in the pathophysiology of many neuropsychiatric disorders. Plasma biomarkers of oxidative stress have been reported in autistic children by several authors. Children with autism have been shown to exhibit evidence of lipid peroxidation [61,63], reduced antioxidant activity [62,75], and elevated NO levels [60].

Yorbik *et al.* [75] investigated the plasma levels of glutathione peroxidase (GPx) and superoxide dismutase (SOD), and red blood cell (RBC) levels of GPx in 45 autistic children compared with 41 normal controls. Activities RBC SOD, RBC and plasma GPx in autistic children were significantly lower than in controls. Authors concluded that these results indicate that autistic children have low levels of activity of blood antioxidant enzyme systems; if similar abnormalities are present in brain, free radical accumulation could damage brain tissue. The increased RBC NO level was found by Sogut *et al.* [60] in a group of 27 patients with autism. In contrary, they found the increased plasma GPx activity. Increased RBC SOD was reported by Zoroglu *et al.* [62] in 27 autistic patients.

Chauhan *et al.* [61] compared lipid peroxidation status in the plasma of children with autism, and their developmentally normal non-autistic siblings by quantifying the levels of malonyldialdehyde, an end product of fatty acid oxidation. Lipid peroxidation was found to be elevated in autism indicating that oxidative stress is increased in this disorder. Levels of major antioxidant proteins namely, transferrin (iron-binding protein) and ceruloplasmin (copper-binding protein) in the serum, were significantly reduced in autistic children as compared to their developmentally normal non-autistic siblings. A striking correlation was observed between reduced levels of these proteins and loss of previously acquired language skills in children with autism.

Ming *et al.* [63] evaluated children with autism for the presence of two oxidative stress biomarkers. Urinary excretion of 8-hydroxy-2-deoxyguanosine (8-OHdG) and 8-isoprostane-F2α (8-iso-PGF2α) were determined in 33 children with autism and 29 healthy controls. 8-iso-PGF2α levels were significantly higher in children with autism. The isoprostane levels in autistic subjects were variable with a bimodal distribution. The majority of autistic subjects showed a moderate increase in isoprostane levels while a smaller group of autistic children showed dramatic increases in their isoprostane levels. There was a trend for an increase in 8-OHdG levels in children with autism but it did not reach statistical significance. These results suggest that the lipid peroxidation biomarker is increased in this cohort of autistic children, especially in the subgroup of autistic children. These findings were supported by study of Yao *et al.* [64].

Glutathione Redox Imbalance

The glutathione (GSH) redox system is important for reducing oxidative stress. GSH, a radical scavenger, is converted to oxidized glutathione (GSSG) (glutathione disulfide) through GPx, and converted back to GSH by glutathione reductase (GR). Glutathione is an important antioxidant that can detoxify hydrogen peroxide (H_2O_2), preventing the formation of hydroxyl radicals. Measurements of GSH, GSSG and its related enzymatic reactions are thus important for evaluating the redox and antioxidant status of the cell or a system.

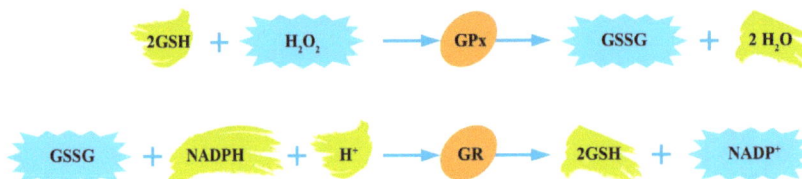

Figure 2: The glutathione redox system. GPx - glutathione peroxidase, GR - glutathione reductase.

James *et al.* [76] evaluated intracellular redox status in lymphoblastoid cells (LCLs) derived from autistic children and unaffected controls to assess relative concentrations of GSH and GSSG in cell extracts and isolated mitochondria as a measure of intracellular redox capacity. Their results indicated that the GSH/GSSG redox ratio was decreased and percent oxidized glutathione increased in both cytosol and mitochondria in the autism LCLs. Exposure to oxidative stress via the sulfhydryl reagent thimerosal resulted in a greater decrease in GSH/GSSG ratio and increase in free radical generation in autism compared to control cells. Acute exposure to physiologic levels of NO decreased mitochondrial membrane potential to a greater extent in the autism LCLs although GSH/GSSG and ATP concentrations were similarly decreased in both cell lines. These results suggest that the autism LCLs exhibit a reduced glutathione reserve capacity in both cytosol and mitochondria that may compromise antioxidant defense and detoxification capacity under pro-oxidant conditions.

Homocysteine Cycle and ASD

In 1969, McCully's proposed his pioneering theory that linked homocysteine and heart disease. He explained this theory in his famous book *Homocysteine Revolution.* Many epidemiological and experimental studies have provided evidence that hyperhomocysteinemia is an important and independent risk factor for a variety of human cardiovascular diseases and a range of neurodegenerative conditions [77]. Elevated homocysteine is associated with neural tube defects [78], seizures [79,80], schizophrenia [81], and neurobehavioral toxicity of chemotherapeutic agents [82]. Zou *et al.* [83] demonstrated that homocysteine promotes proliferation and up-regulates the expression of a marker of microglial activation (CD11b).

Homocysteine is an amino acid, which is not obtained from the diet. Homocysteine is produced from methionine by transmethylation. First, methionine receives an adenosine group from ATP, a reaction catalyzed by S-adenosyl-methionine synthetase, to give S-adenosylmethionine (SAM). SAM then transfers the methyl group to an acceptor molecule. The adenosine is then hydrolyzed to yield homocysteine. Homocysteine circulates through the blood stream or is converted back into methionine via tetrahydrofolate by a metabolic pathway called remethylation. L-methylfolate (the methylated form of folic acid) and methylcobalamin (vitamin B_{12}) are needed for this conversion. Homocysteine can be converted into cysteine by a metabolic pathway called transsulfuration (see Fig. **3**).

Vitamin B_6 is needed for this conversion. Elevated levels of homocysteine result from abnormalities in the function of enzymes involved in homocysteine metabolism or from deficiencies of the vitamin cofactors: folate, cobalamin (B_{12}), and vitamin B_6.

The transsulfuration pathway starts with homocysteine conversion by cystathionine β-synthase, which initiates the synthesis of cysteine, glutathione, taurine, and sulphate. Glutathione is a tripeptide of cysteine, glycine, and glutamate that is synthesized in every cell of the body. It has been mentioned that reduced/oxidized glutathione redox equilibrium regulates a pleiotropic range of functions that includes ROS and RNS scavenger, detoxification, cell membrane integrity and signal transduction, and apoptosis [1]. Under normal physiologic conditions, glutathione reductase activity is sufficient to maintain the high GSH/GSSG ratio. Excessive intracellular oxidative stress could result in GSSG export to the plasma. Thus an increase in plasma GSSG is a strong indicator of intracellular oxidative stress.

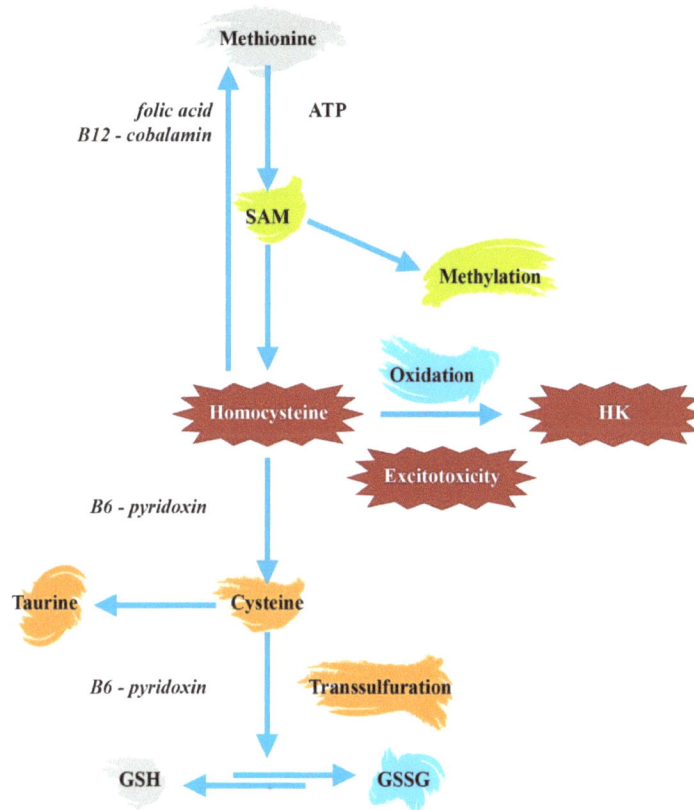

Figure 3: A simplified scheme of the methionine-homocysteine cycle. SAM is adenosylmethionine, GSH is an active reduced form of glutathione, GSSG is inactive oxidized disulfide form.

Recent ASD studies showed abnormal metabolites within these pathways. The plasma levels of transmethylation/transsulfuration metabolites were investigated both in individuals with ASD [1,70,71,84-86] and their mothers [72]. Table **3** shows the levels of some selected metabolites of the homocysteine cycle from studies of James and co-workers and Geier at al. These data show that GSH and GSSG levels are significantly changed in autistic individuals. Of possible pathophysiological relevance, plasma cysteine levels were significantly reduced in individuals with ASD. Cysteine belongs also to conditionally essential amino acids that are dependent on adequate methionine status. Moreover, in James laboratory, plasma concentrations of methionine, S-adenosylmethionine (SAM), S-adenosylhomocysteine (SAH), adenosine, homocysteine, cystathionine, cysteine, GSH and GSSG were repeatedly measured. First in 20 children with autism and in 33 control children [70] and in a subsequent study the group of 80 autistic cases was investigated [71].

Highly significant (P ≤ 0.001) decrease of SAM and SAH was also observed in subjects with ASD. James and co-workers suggested that an increased vulnerability to oxidative stress and a decreased capacity for methylation may contribute to the development and clinical manifestation of autism.

The objective of their subsequent study was to determine whether or not treatment with the metabolic precursors, methylcobalamin and folinic acid, would improve plasma concentrations of transmethylation/transsulfuration metabolites and glutathione redox status in autistic children. In an open-label trial, 40 autistic children were treated with 75 µg/kg methylcobalamin (two times/wk) and 400 µg folinic acid (two times/d) for three months. Metabolites in the transmethylation/transsulfuration pathway were measured before and after treatment. The results indicated the three-months intervention resulted in significant increases in methionine, SAM, SAH, cysteine, and GSH concentrations. The GSSG was decreased and the glutathione redox ratio increased after treatment [87].

Table 3: Changes in Metabolites of Methionine-Homocysteine Cycle in Autistic Cases. Geier *et al.* [1, 84] measured the free plasma cysteine, while James *et al.* [71] probably used the value for total cysteine. However, the both values indicate the decrease in ASD subjects.

µ MOL/L	ASD CHILDREN GEIER *ET AL.* (N)	HEALTHY CONTROLS GEIER *ET AL.* (N)	ASD CHILDREN JAMES *ET AL.* N = 80	HEALTHY CONTROLS JAMES *ET AL.* N = 73	REFERENCE VALUES FOR CHILDREN AND YOUTH
methionine (µmol/L)	17.6 (28)	31.5	20.6 ± 5.2	28.0 ± 6.5	28
GSH (µmol/L)	3.1 ± 0.53 (38)	4.2 ± 0.72 (120)	1.4 ± 0.5	2.2 ± 0.9	4.2 ± 0.74
GSSG (nmol/L)	0.46 ± 0.16 (38)	0.35 ± 0.05 (120)	0.40 ± 0.2	0.24 ± 0.1	???
homocysteine (µmol/L)	5.87 (12)	9.46 (120)	5.7 ± 1.2	6.0 ± 1.3	4.3 - 10
taurine (µmol/L)	48.6 ± 14.0 (38)	97.5 ± 8.8 (27)	-	-	95 ± 9
cysteine (µmol/L)	17.8 ± 9.5 (38)	23.2 ± 4.2 (64)	165 ± 14	207 ± 22	238 ± 22

Legend: GSH = Active reduced form of glutathione; GSSG = inactive oxidized disulfide form.

Abnormal Transmethylation/Transsulfuration Metabolism in Parents of Autistic Children

Based on reports of abnormal methionine and glutathione metabolism in autistic children, James with co-workers examined the same metabolic profile in the parents [72]. Their results indicate that parents share similar metabolic deficits in methylation capacity and glutathione-dependent antioxidant/detoxification capacity observed in many autistic children. The significant difference in metabolic profiles between control and case parents was an unexpected finding.

Case parents were mothers and fathers of children who had been diagnosed with Autistic Disorder based on Diagnostic and Statistical Manual of Mental Disorders-Fourth Edition (DSM-IV) and the Childhood Autistic Rating Scales (CARS). A total of 86 parents participated in the study consisting of 46 mothers and 40 fathers. Of the 86 parents, 72 were mother-father pairs (36 pairs), and ranged in age from 21 to 45 years. Control parents for the metabolic study consisted of 200 mothers with a mean age of 28 (range 17–43 years) who were control participants in a case-control study of maternal risk factors for inherrited heart defects [88]. The mean levels of homocysteine and SAH were significantly elevated in case mothers compared to control mothers. Statistical evaluation of case fathers was not possible because blood samples from control fathers were not available from this cohort. However, elevations in mean homocysteine and/or SAH levels were not statistically different between case mothers and fathers. Relative to control levels, parents of autistic children were found to have significant decreases in both total and free GSH levels and in the GSH/GSSG ratios whereas the levels of the oxidized disulfide GSSG were significantly increased. Methionine and cysteine levels among the case parents were not significantly different from controls.

In summary, abnormal levels of metabolites in both transmethylation and transsulfuration pathways were surprisingly prevalent among case parents [72]. Both parents had significant increases in mean plasma homocysteine and SAH levels and significant decreases in the SAM/SAH ratio, total and free GSH, and the GSH/GSSG ratio. Whether the parental metabolic abnormalities are genetically-based or the result of adult dietary deficiencies and/or chronic pro-oxidant environmental exposures cannot be determined from the present data.

However, it has been well documented that increased homocysteine has been associated with pregnancy complications. Elevated maternal homocysteine can cross the placental membrane and are correlated with fetal homocysteine levels [89,90]. The clinical consequences could be the lack of utilization of homocysteine for methionine transformation connected with the presence of neurological abnormalities. The higher maternal plasma homocysteine at preconception, prior to fetal neurogenesis, was inversely associated with cognitive achievement in the offspring [91]. Also of related interest, elevated maternal homocysteine during the third trimester was found to be a risk factor for schizophrenia [92,93].

Elevation in fetal homocysteine and SAH could theoretically alter fetal methylation patterns and induce inappropriate gene expression during development that could affect predisposition to autism. These studies raise concern that maternal homocysteine levels should be carefully monitored and controlled especially with high risk pregnancies. Hyperhomocysteinemia is a risk factor in neurodegeneration. It has been suggested that apart from disturbances in methylation processes, the mechanisms of this effect may include excitotoxicity mediated by NMDARs.

Excitotoxicity of Metabolites of Homocysteine Cycle

There is evidence from *in vitro* and *in vivo* studies that homocysteine induces neuronal damage and cell loss by both excitotoxicity and different apoptotic processes. Homocysteine plays a role in a shared biochemical cascade involving overstimulation of NMDARs, oxidative stress, activation of caspases, DNA damage, endoplasmic reticulum dysfunction and mitochondrial dysfunction. Elevated homocysteine has been found in parents of autistic children, while the data in autistic children themselves are not consistent. Lower levels in comparison with reference values for age-matched children were reported by Geier and co-workers [1,84] as well as by James and co-workers [70]. The mean concentrations of homocysteine were not statistically different from those in the control children after three months oral supplementation with folinic acid and betaine, or folinic acid, betaine, and vitamin B_{12} [70,87]. However, the later study of James *et al.* [71] did not find a significant change in the homocysteine level in a larger group of investigated autistic children (n = 80). Significantly high levels of homocysteine in plasma of autistic children compared to control children were found in the study of Pasca *et al.* [85]. These authors investigated a group of 12 children with autism and found plasma homocysteine level 9.83 ± 2.75 μmol/L vs. 7.51 ± 0.93 μmol/L in 9 controls. Hyperhomocysteinemia in children with autism was negatively correlated with GPx activity in hemolysate of RBC and suboptimal vitamin B_{12} in plasma. However, in a subsequent study by Pasca *et al.* [86] investigated a group of 39 individuals with ASD and found no statistically different changes in homocysteine levels in comparison with control group. They also mentioned the much higher variability of homocysteine in the autistic children group.

Homocysteine oxidizes to a number of glutamate-analogues such as homocysteic acid and homocysteine sulfinic acid and aspartate-analogues such as cysteine sulfinic acid and cysteic acid with significantly greater excitotoxic effect than homocysteine itself [94, 95]. It has been documented that homocysteic acid may cause seizures and excitotoxic neuronal death [96-100]. Moreover, homocysteic acid may be a glial neurotransmitter, acting through astrocytic NMDA receptors [101]. These mechanisms are believed to be important in the pathogenesis of both excitotoxicity and apoptotic neurotoxicity.

Lipton *et al.* using cortical cultures of mixed neurons and glia derived from embryonic (fetal day 15 or 16) Sprague–Dawley rats has shown that homocysteine causes direct neurotoxicity by activating the NMDA receptors [96]. Neuronal injury in these cultures was assessed by the leakage of lactate dehydrogenase. These authors found that homocysteine is an agonist at the glutamate site of the NMDA receptor and is therefore a potential excitotoxin.

Homocysteine-induced increases in $[Ca^{2+}]_i$ and ROS was observed in cultured murine cortical neurons [102]. These authors also suggested that homocysteine-induced Ca^{2+} influx through NMDA channel activation, which stimulated glutamate excitotoxicity, as evidenced by treatment with antagonists of the NMDA channel and mGluRs, respectively. Oldreive and Doherty exposed cultures of embryonic cerebellar Purkinje neurons to a range of concentrations of homocysteine and determined its effects on their survival [103]. Their experiments revealed that all concentrations of homocysteine studied, from 50 to 500 μM, caused a significant decrease in cerebellar Purkinje cell number. These authors demonstrated that homocysteine is toxic to Purkinje cells *in vitro*, inhibiting both their survival and the outgrowth of neurites. Zieminska and coworkers [97,104] studied the homocysteine induced neurotoxicity in cultured rat cerebellar granule neurones and found involvement of both NMDA receptors and the group I mGluRs.

The baseline concentrations of metabolites in the methionine cycle and the transsulfuration pathway were significantly different between the autistic children and the control age-matched children. Despite the findings of James *et al.* [72], which indicated that parents share similar metabolic deficits in methylation capacity and glutathione-dependent antioxidant/detoxification capacity observed in many autistic children, some important differences in the levels of metabolites between mothers and autistic children were found. While mothers (and probably also the fathers) have significantly increased plasma homocysteine levels, it does not seems that children with ASD have hyperhomocysteinemia. However, oxidized homocysteine metabolites, along with glutamate, inflammatory cytokines, chemokines, and inflammatory prostaglandins, could trigger the autotoxic injury to a widespread area surrounding the immune reactions, thus explaining the autopsy picture seen in the autistic brain [51].

The low methionine concentrations reported in all above mentioned studies would suggest a reduction in methionine synthase activity favoring the increase of plasma homocysteine levels [105,106]. Several observations in autistic children also report the increased frequency of *MTHFR* allele variants, which reduces the synthesis of metabolically active folate. The decreased methionine levels among most autistic children are consistent with low intracellular folate activity; however such change would also favor the increase of homocysteine in plasma and CSF [71,86]. On the other hand, changes in GSH concentrations and low antioxidant enzyme activity [60,62,75] in autistic children provide additional support for oxidative stress and excitotoxicity as a key part of the etiology of ASD. A strong negative correlation between the antioxidant enzyme GPx and homocysteine level in autistic children is in agreement with finding that GPx synthesis is decreased with elevated level of homocysteine [85]. It is evident that homocysteine induced oxidative stress and etiology of ASD is connected with impaired antioxidant activity and excitotoxicity. As it has been reported in previous chapters, autistic children have difficulty resisting infection, suffer from recurrent infections, neuroinflammation, and GI inflammation.

The Role of Taurine in ASD Etiology

An assessment of transsulfuration metabolites among the individuals with ASD in comparison to the neurotypical controls examined in the study conducted at the Autism Treatment Center (Dallas, Texas) revealed that the participants with ASD had significantly decreased levels of plasma taurine [1]. Table **3** shows that individuals with ASD also have lower level of plasma cysteine and plasma methionine.

Taurine is the product of transsulfuration pathways along with glutathione. It is a simple sulfur-containing amino acid (2-aminoethanesulfonic acid) which is not utilized in protein synthesis, but rather is found free. Taurine is a conditionally-essential amino acid. Both plasma taurine and cysteine levels follow an oral methionine load [107]. The deficiency in methionine and/or cysteine may be thus one of the reason for taurine deficiency.

The localization of taurine was investigated in several tissues of the mouse. Terauchi with co-workers used immunohistochemical methods using a polyclonal antibody for taurine derived from rabbits [108]. This method was used since it is a simple procedure and the results are clear and reliable. For purposes of comparison, radioautography with ^{3}H-taurine was performed. Interestingly, immunoreactivity was broadly observed in Purkinje cells of the cerebellum, glia cells of the brain tissue, cardiac muscle cells, matrices of the bone, mucus granules of goblet cells of the intestines, and brown adipose cells of the fetus. In addition, taurine reactivity was observed in cell nuclei.

Taurine is present at high concentrations in the mammalian brain, with several proposed roles in neurotransmission, neuromodulation, control of calcium influx, and cell excitability. Bride and Frederickson proposed taurine's role as a neurotransmitter in the cerebellum [109]. Taurine acts as the most abundant neurotransmitter in the developing neocortex in the mouse fetal brain. During the proliferative stage of neurogenesis between E13 and E17, relatively high levels of glutamate, aspartate, taurine and glycine were detected, consistent with a possible trophic influence of these neurotransmitters during cortical development prior to synaptogenesis [110].

Taurine has been shown to be essential in certain aspects of mammalian development. Earlier experiments with female cats showed that the reproductive performance by the taurine-depleted queens was poor with increased reproductive wastage, whereas those receiving dietary taurine had normal pregnancies and deliveries. Surviving offspring from the taurine-depleted mothers exhibited a number of neurological abnormalities and substantially reduced concentrations of taurine in the body tissues and fluids [111].

In humans, low maternal taurine levels result in low fetal taurine levels. Taurine-deficiency in the mother leads to growth retardation of the offspring, and to impaired perinatal development of the CNS. The adult offspring of taurine-deficient mothers display signs of impaired neurological function, impaired glucose tolerance and vascular dysfunction; they may develop gestational diabetes and transmit the effects to the next generation [112]. This transgenerational effect of taurine-deficiency in the perinatal period fits into the concept of fetal origin of adult disease.

Saransari and Oja reviewed the involvement of taurine in neuron-damaging conditions, including hypoxia, hypoglycemia, ischemia, oxidative stress, and the presence of free radicals, metabolic poisons and an excess of ammonia [113]. They suggested that increase in extracellular levels of taurine in the immature hippocampus in ischemia may serve as an important protective mechanism against excitotoxicity, to which the developing brain is particularly vulnerable, and contribute to the resistance of the immature brain to hypoxia. Saransari and Oja demonstrated that excitotoxic concentrations of glutamate also stimulate taurine release in mouse hippocampal slices. Taurine released simultaneously with an excess of excitatory amino acids in the hippocampus under ischemic and other neuron-damaging conditions may constitute an important protective mechanism against excitotoxicity. The release of taurine may thus prevent excitation from reaching neurotoxic levels.

Neuroprotective actions of taurine were observed against the excitotoxicity of various GluR agonists [114]. El Idrissi with co-workers showed for the first time that taurine neuroprotection is related to the activation of GABA receptors. These authors studied the taurine interactions with the inhibitory (GABA) and excitatory (glutamate) systems in taurine-fed mice. They found that increased expression of glutamate decarboxylase (GAD) was accompanied by increased levels of GABA. They demonstrated *in vitro*, that taurine regulates neuronal calcium homeostasis and Ca^{2+}-dependent processes. The Ca^{2+}-dependent protein kinase C (PKC) was regulated by taurine, whereas the activity of protein kinase A (PKA), a cAMP-dependent kinase, was not affected. Furthermore, as a consequence of calcium regulation, taurine counteracted glutamate-induced mitochondrial damage and cell death. Their *in vitro* data obtained in primary neuronal cultures brought evidence that taurine acts as a low affinity agonist for GABA(A) receptors, protects neurons against kainate excitotoxic insults and modulates calcium homeostasis. Therefore, taurine is potentially capable of treating seizure-associated brain damage [115].

Parenteral injection of kainic acid, a GluR agonist, causes severe and stereotyped behavioral convulsions in mice and is used as a rodent model for human temporal lobe epilepsy. El Ildrissi group found that taurine (43 mg/kg, s.c.) had a significant antiepileptic effect when injected 10 min prior to kainic acid. Acute injection of taurine increased the onset latency and reduced the occurrence of tonic seizures. Taurine also reduced the duration of tonic-clonic convulsions and mortality rate following kainic acid-induced seizures [115].

Taurine prevents or reduces glutamate excitotoxicity through both the enhancement of mitochondrial function and the regulation of intracellular calcium homeostasis. In cultured cerebellar granule cells, glutamate induces a rapid and sustained elevation in $[Ca^{2+}]_i$, causing the collapse of the mitochondrial electrochemical gradient and subsequent cell death. Pretreatment of cultured neurons with taurine prevents or greatly suppresses the elevation of $[Ca^{2+}]_i$ induced by glutamate. Furthermore, taurine was found to inhibit the influx but not the efflux of $^{45}Ca^{2+}$ in cultured neurons [116,117]. Pre-treatment with taurine, did not affect the level of Ca^{2+} uptake with glutamate but rather reduced its duration [118]. Several groups of investigators thus brought evidence that taurine has potent protective function against glutamate-induced excitotoxicity presumably through its function in regulation of $[Ca^{2+}]_i$.

Metabolic actions of taurine include bile acid conjugation, detoxification, and osmoregulation. Taurine is a conjugator of bile acids; it helps increase cholesterol elimination in the bile, helps with fat absorption and elimination of toxins. It is commonly reported by parents of children on the ASD that their fecal matter has sand or a sandy appearance to it. Sandy stool is associated with a bile acid imbalance caused by a taurine deficiency. When there is a lack of taurine a sand-like substance can develop in the GIT and show up in the stool. GI inflammation can contribute to this problem. Taurine helps to produce bile acids in the liver. Taurine as a supplement of 100 to 500mg per day can be helpful for some children (http://drkurtwoeller.blogspot.com/2008/12/sandy-appearance-in-stools-of-autism.html).

Clinically, taurine has been used with varying degrees of success in the treatment of a wide variety of conditions, including: cardiovascular diseases, hypercholesterolemia, epilepsy and other seizure disorders, macular degeneration, Alzheimer's disease, hepatic disorders, alcoholism, and cystic fibrosis.

CONCLUSIONS

Some researchers have proposed that mitochondrial disorders may be the most common metabolic finding in ASD. Mitochondria are vulnerable to a wide array of endogenous and exogenous factors which appear to be linked by excitotoxicity. This understanding further confirms the view that strategies to augment mitochondrial function, either by decreasing production of endogenous toxic metabolites, reducing the effects of environmental toxins, such as mercury, fluoride and aluminum, aspartame and glutamate, or stimulating mitochondrial enzyme activity may be beneficial in the treatment of autism. Whilst the biochemical markers of mitochondrial disease are not specific for ASD, the diagnosis and therapy of mitochondrial disturbance could be beneficial for augmentation of some ASD symptoms. The firmly established evidence of biochemical changes in the methionine-homocysteine cycle with concomitant alterations of transmethylation and transsulfuration events demonstrates the complex metabolic pathways, which are altered in autism. It also brings scientific understanding for the efficacy of folate, vitamin B_6 and vitamin B_{12}, together with taurine supplementation and antioxidant therapy in autistic patients.

The investigations of biochemical changes in parents of autistic children suggest that parents (mostly mothers) may share milder version of characteristics with children who have the full disorder. The trans-generational effect of impaired transmethylation/transsulfuration metabolites and taurine-deficiency in the perinatal period fits into the concept of fetal origin of child/adult disease. These findings have implication for interventions of both the mothers and autistic children to enhance their health and quality of life but simultaneously to ameliorate the metabolic discrepancies.

REFERENCES

[1]   Geier DA, Kern JK, Garver CR, *et al.* A prospective study of transsulfuration biomarkers in autistic disorders. Neurochem Res 2009; 34: 386-93.

[2]   Buie T, Campbell DB, Fuchs GJ, 3rd, *et al.* Evaluation, diagnosis, and treatment of gastrointestinal disorders in individuals with ASD: a consensus report. Pediatrics 2010; 125 Suppl 1: S1-18.

[3]   Buie T, Fuchs GJ, 3rd, Furuta GT, *et al.* Recommendations for evaluation and treatment of common gastrointestinal problems in children with ASD. Pediatrics 2010; 125 Suppl 1: S19-29.

[4]   Manzi B, Loizzo AL, Giana G, Curatolo P. Autism and metabolic diseases. J Child Neurol 2008; 23: 307-14.

[5]   Benvenuto A, Manzi B, Alessandrelli R, Galasso C, Curatolo P. Recent advances in the pathogenesis of syndromic autisms. Int J Pediatr 2009; 2009: 198736.

[6]   Haas RH, Parikh S, Falk MJ, *et al.* Mitochondrial disease: a practical approach for primary care physicians. Pediatrics 2007; 120: 1326-33.

[7]   Lombard J. Autism: a mitochondrial disorder? Med Hypotheses 1998; 50: 497-500.

[8]   Haas RH, Parikh S, Falk MJ, *et al.* The in-depth evaluation of suspected mitochondrial disease. Mol Genet Metab 2008; 94: 16-37.

[9]   Moreno H, Borjas L, Arrieta A, *et al.* [Clinical heterogeneity of the autistic syndrome: a study of 60 families]. Invest Clin 1992; 33: 13-31.

[10]   Coleman M, Blass JP. Autism and lactic acidosis. J Autism Dev Disord 1985; 15: 1-8.

[11]   Filipek PA, Juranek J, Nguyen MT, Cummings C, Gargus JJ. Relative carnitine deficiency in autism. J Autism Dev Disord 2004; 34: 615-23.

[12]   Madiraju P, Pande SV, Prentki M, Madiraju SR. Mitochondrial acetylcarnitine provides acetyl groups for nuclear histone acetylation. Epigenetics 2009; 4: 399-403.

[13]   Oliveira G, Diogo L, Grazina M, *et al.* Mitochondrial dysfunction in autism spectrum disorders: a population-based study. Dev Med Child Neurol 2005; 47: 185-9.

[14]   Correia C, Coutinho AM, Diogo L, *et al.* Brief report: High frequency of biochemical markers for mitochondrial dysfunction in autism: no association with the mitochondrial aspartate/glutamate carrier SLC25A12 gene. J Autism Dev Disord 2006; 36: 1137-40.

[15]   Weissman JR, Kelley RI, Bauman ML, *et al.* Mitochondrial disease in autism spectrum disorder patients: a cohort analysis. PLoS One 2008; 3: e3815.

[16]   Poling JS, Frye RE, Shoffner J, Zimmerman AW. Developmental regression and mitochondrial dysfunction in a child with autism. J Child Neurol 2006; 21: 170-2.

[17]   Chugani DC, Sundram BS, Behen M, Lee ML, Moore GJ. Evidence of altered energy metabolism in autistic children. Prog Neuropsychopharmacol Biol Psychiatry 1999; 23: 635-41.

[18]   Al-Mosalem OA, El-Ansary A, Attas O, Al-Ayadhi L. Metabolic biomarkers related to energy metabolism in Saudi autistic children. Clin Biochem 2009; 42: 949-57.

[19]   Aldred S, Moore KM, Fitzgerald M, Waring RH. Plasma amino acid levels in children with autism and their families. J Autism Dev Disord 2003; 33: 93-7.

[20]   Clark-Taylor T, Clark-Taylor BE. Is autism a disorder of fatty acid metabolism? Possible dysfunction of mitochondrial beta-oxidation by long chain acyl-CoA dehydrogenase. Med Hypotheses 2004; 62: 970-5.

[21]   Hamburg RJ, Friedman DL, Olson EN, *et al.* Muscle creatine kinase isoenzyme expression in adult human brain. J Biol Chem 1990; 265: 6403-9.

[22]   Holtzman D, Tsuji M, Wallimann T, Hemmer W. Functional maturation of creatine kinase in rat brain. Dev Neurosci 1993; 15: 261-70.

[23]   Kaldis P, Hemmer W, Zanolla E, Holtzman D, Wallimann T. 'Hot spots' of creatine kinase localization in brain: cerebellum, hippocampus and choroid plexus. Dev Neurosci 1996; 18: 542-54.

[24]   Suzuki K, Nishimura K, Sugihara G, *et al.* Metabolite alterations in the hippocampus of high-functioning adult subjects with autism. Int J Neuropsychopharmacol 2009; 1-6.

[25]   Dezortova M, Jiru F, Petrasek J, *et al.* 1H MR spectroscopy as a diagnostic tool for cerebral creatine deficiency. Magma 2008; 21: 327-32.

[26]   Hardan AY, Minshew NJ, Melhem NM, *et al.* An MRI and proton spectroscopy study of the thalamus in children with autism. Psychiatry Res 2008; 163: 97-105.

[27]   Arias-Dimas A, Vilaseca MA, Artuch R, Ribes A, Campistol J. [Diagnosis and treatment of brain creatine deficiency syndromes]. Rev Neurol 2006; 43: 302-8.

[28]   Gabis L, Wei H, Azizian A, *et al.* 1H-magnetic resonance spectroscopy markers of cognitive and language ability in clinical subtypes of autism spectrum disorders. J Child Neurol 2008; 23: 766-74.

[29]   Melegh B, Kerner J, Bieber LL. Pivampicillin-promoted excretion of pivaloylcarnitine in humans. Biochem Pharmacol 1987; 36: 3405-9.

[30]   Ito T. Children's toxicology from bench to bed--Liver injury (1): Drug-induced metabolic disturbance--toxicity of 5-FU for pyrimidine metabolic disorders and pivalic acid for carnitine metabolism. J Toxicol Sci 2009; 34 Suppl 2: SP217-22.

[31]   Ellaway C, Williams K, Leonard H, *et al.* Rett syndrome: randomized controlled trial of L-carnitine. J Child Neurol 1999; 14: 162-7.

[32]   Ellaway CJ, Peat J, Williams K, Leonard H, Christodoulou J. Medium-term open label trial of L-carnitine in Rett syndrome. Brain Dev 2001; 23 Suppl 1: S85-9.

[33]   Melegh B, Kerner J, Sandor A, Vinceller M, Kispal G. Oral L-carnitine supplementation in low-birth-weight newborns: a study on neonates requiring combined parenteral and enteral nutrition. Acta Paediatr Hung 1986; 27: 253-8.

[34]   Melegh B, Kerner J, Sandor A, Vinceller M, Kispal G. Effects of oral L-carnitine supplementation in low-birth-weight premature infants maintained on human milk. Biol Neonate 1987; 51: 185-93.

[35]   Melegh B, Szucs L, Kerner J, Sandor A. Changes of plasma free amino acids and renal clearances of carnitines in premature infants during L-carnitine-supplemented human milk feeding. J Pediatr Gastroenterol Nutr 1988; 7: 424-9.

[36]   Melegh B, Kerner J, Szucs L, Porpaczy Z. Feeding preterm infants with L-carnitine supplemented formula. Acta Paediatr Hung 1990; 30: 27-41.

[37]   Fido A, Al-Saad S. Olanzapine in the treatment of behavioral problems associated with autism: an open-label trial in Kuwait. Med Princ Pract 2008; 17: 415-8.

[38]   Rumsey JM, Duara R, Grady C, *et al.* Brain metabolism in autism. Resting cerebral glucose utilization rates as measured with positron emission tomography. Arch Gen Psychiatry 1985; 42: 448-55.

[39]   De Volder A, Bol A, Michel C, Congneau M, Goffinet AM. Brain glucose metabolism in children with the autistic syndrome: positron tomography analysis. Brain Dev 1987; 9: 581-7.

[40]   Herold S, Frackowiak RS, Le Couteur A, Rutter M, Howlin P. Cerebral blood flow and metabolism of oxygen and glucose in young autistic adults. Psychol Med 1988; 18: 823-31.

[41]   Schifter T, Hoffman JM, Hatten HP, Jr., *et al.* Neuroimaging in infantile autism. J Child Neurol 1994; 9: 155-61.

[42]   Hazlett EA, Buchsbaum MS, Hsieh P, *et al.* Regional glucose metabolism within cortical Brodmann areas in healthy individuals and autistic patients. Neuropsychobiology 2004; 49: 115-25.

[43]   Villemagne PM, Naidu S, Villemagne VL, *et al.* Brain glucose metabolism in Rett Syndrome. Pediatr Neurol 2002; 27: 117-22.

[44]   Siegel BV, Jr., Asarnow R, Tanguay P, *et al.* Regional cerebral glucose metabolism and attention in adults with a history of childhood autism. J Neuropsychiatry Clin Neurosci 1992; 4: 406-14.

[45]   Frackowiak RS, Herold S, Petty RK, Morgan-Hughes JA. The cerebral metabolism of glucose and oxygen measured with positron tomography in patients with mitochondrial diseases. Brain 1988; 111 (Pt 5): 1009-24.

[46] Magistretti PJ, Pellerin L. Cellular mechanisms of brain energy metabolism. Relevance to functional brain imaging and to neurodegenerative disorders. Ann N Y Acad Sci 1996; 777: 380-7.

[47] Bergsneider M, Hovda DA, Shalmon E, *et al.* Cerebral hyperglycolysis following severe traumatic brain injury in humans: a positron emission tomography study. J Neurosurg 1997; 86: 241-51.

[48] Magistretti PJ, Pellerin L. Cellular mechanisms of brain energy metabolism and their relevance to functional brain imaging. Philos Trans R Soc Lond B Biol Sci 1999; 354: 1155-63.

[49] Blaylock RL. A possible central mechanism in autism spectrum disorders, part 1. Altern Ther Health Med 2008; 14: 46-53.

[50] Blaylock RL. A possible central mechanism in autism spectrum disorders, part 3: the role of excitotoxin food additives and the synergistic effects of other environmental toxins. Altern Ther Health Med 2009; 15: 56-60.

[51] Blaylock RL. A possible central mechanism in autism spectrum disorders, part 2: immunoexcitotoxicity. Altern Ther Health Med 2009; 15: 60-7.

[52] Blaylock RL, Strunecka A. Immune-glutamatergic dysfunction as a central mechanism of the autism spectrum disorders. Curr Med Chem 2009; 16: 157-70.

[53] Bal-Price A, Brown GC. Inflammatory neurodegeneration mediated by nitric oxide from activated glia-inhibiting neuronal respiration, causing glutamate release and excitotoxicity. J Neurosci 2001; 21: 6480-91.

[54] Brown GC, Bal-Price A. Inflammatory neurodegeneration mediated by nitric oxide, glutamate, and mitochondria. Mol Neurobiol 2003; 27: 325-55.

[55] Pardo CA, Vargas DL, Zimmerman AW. Immunity, neuroglia and neuroinflammation in autism. Int Rev Psychiatry 2005; 17: 485-95.

[56] Vargas DL, Nascimbene C, Krishnan C, Zimmerman AW, Pardo CA. Neuroglial activation and neuroinflammation in the brain of patients with autism. Ann Neurol 2005; 57: 67-81.

[57] Li X, Chauhan A, Sheikh AM, *et al.* Elevated immune response in the brain of autistic patients. J Neuroimmunol 2009; 207: 111-6.

[58] Magistretti PJ, Pellerin L. Metabolic coupling during activation. A cellular view. Adv Exp Med Biol 1997; 413: 161-6.

[59] Magistretti PJ. Role of glutamate in neuron-glia metabolic coupling. Am J Clin Nutr 2009; 90: 875S-80S.

[60] Sogut S, Zoroglu SS, Ozyurt H, *et al.* Changes in nitric oxide levels and antioxidant enzyme activities may have a role in the pathophysiological mechanisms involved in autism. Clin Chim Acta 2003; 331: 111-7.

[61] Chauhan A, Chauhan V, Brown WT, Cohen I. Oxidative stress in autism: increased lipid peroxidation and reduced serum levels of ceruloplasmin and transferrin--the antioxidant proteins. Life Sci 2004; 75: 2539-49.

[62] Zoroglu SS, Armutcu F, Ozen S, *et al.* Increased oxidative stress and altered activities of erythrocyte free radical scavenging enzymes in autism. Eur Arch Psychiatry Clin Neurosci 2004; 254: 143-7.

[63] Ming X, Stein TP, Brimacombe M, *et al.* Increased excretion of a lipid peroxidation biomarker in autism. Prostaglandins Leukot Essent Fatty Acids 2005; 73: 379-84.

[64] Yao Y, Walsh WJ, McGinnis WR, Pratico D. Altered vascular phenotype in autism: correlation with oxidative stress. Arch Neurol 2006; 63: 1161-4.

[65] Freeman SJ, Roberts W, Daneman D. Type 1 diabetes and autism: is there a link? Diabetes Care 2005; 28: 925-6.

[66] Harjutsalo V, Tuomilehto J. Type 1 diabetes and autism: is there a link? Diabetes Care 2006; 29: 484-5; discussion 5.

[67] Iafusco D, Vanelli M, Songini M, *et al.* Type 1 diabetes and autism association seems to be linked to the incidence of diabetes. Diabetes Care 2006; 29: 1985-6.

[68] Maher KR, Harper JF, Macleay A, King MG. Peculiarities in the endocrine response to insulin stress in early infantile autism. J Nerv Ment Dis 1975; 161: 180-4.

[69] Strunecká A, Grof P. The role of glucose in the pathogenesis of Alzheimer's disease revisited: What does it tell us about the therapeutic use of lithium? Central Nervous System Agents in Medicinal Chemistry 2006; 6: 175-92.

[70] James SJ, Cutler P, Melnyk S, *et al.* Metabolic biomarkers of increased oxidative stress and impaired methylation capacity in children with autism. Am J Clin Nutr 2004; 80: 1611-7.

[71] James SJ, Melnyk S, Jernigan S, *et al.* Metabolic endophenotype and related genotypes are associated with oxidative stress in children with autism. Am J Med Genet B Neuropsychiatr Genet 2006; 141B: 947-56.

[72] James SJ, Melnyk S, Jernigan S, *et al.* Abnormal Transmethylation/transsulfuration Metabolism and DNA Hypomethylation Among Parents of Children with Autism. J Autism Dev Disord 2008; 38: 1976.

[73] Ballatori N, Clarkson TW. Biliary secretion of glutathione and of glutathione-metal complexes. Fundam Appl Toxicol 1985; 5: 816-31.

[74] Woods JS. Altered porphyrin metabolism as a biomarker of mercury exposure and toxicity. Can J Physiol Pharmacol 1996; 74: 210-5.

[75] Yorbik O, Sayal A, Akay C, Akbiyik DI, Sohmen T. Investigation of antioxidant enzymes in children with autistic disorder. Prostaglandins Leukot Essent Fatty Acids 2002; 67: 341-3.

[76] James SJ, Rose S, Melnyk S, *et al.* Cellular and mitochondrial glutathione redox imbalance in lymphoblastoid cells derived from children with autism. FASEB J 2009; 23: 2374-83.

[77] Miller AL. The methionine-homocysteine cycle and its effects on cognitive diseases. Altern Med Rev 2003; 8: 7-19.

[78] van der Put NM, van Straaten HW, Trijbels FJ, Blom HJ. Folate, homocysteine and neural tube defects: an overview. Exp Biol Med (Maywood) 2001; 226: 243-70.

[79] Quinn CT, Griener JC, Bottiglieri T, Kamen BA. Methotrexate, homocysteine, and seizures. J Clin Oncol 1998; 16: 393-4.

[80] Folbergrova J, Druga R, Otahal J, *et al.* Seizures induced in immature rats by homocysteic acid and the associated brain damage are prevented by group II metabotropic glutamate receptor agonist (2R,4R)-4-aminopyrrolidine-2,4-dicarboxylate. Exp Neurol 2005; 192: 420-36.

[81] Levine J, Stahl Z, Sela BA, *et al.* Elevated homocysteine levels in young male patients with schizophrenia. Am J Psychiatry 2002; 159: 1790-2.

[82] Quinn CT, Griener JC, Bottiglieri T, *et al.* Elevation of homocysteine and excitatory amino acid neurotransmitters in the CSF of children who receive methotrexate for the treatment of cancer. J Clin Oncol 1997; 15: 2800-6.

[83] Zou CG, Zhao YS, Gao SY, *et al.* Homocysteine promotes proliferation and activation of microglia. Neurobiol Aging 2009;

[84] Geier DA, Geier MR. A clinical and laboratory evaluation of methionine cycle-transsulfuration and androgen pathway markers in children with autistic disorders. Horm Res 2006; 66: 182-8.

[85] Pasca SP, Nemes B, Vlase L, *et al.* High levels of homocysteine and low serum paraoxonase 1 arylesterase activity in children with autism. Life Sci 2006; 78: 2244-8.

[86] Pasca SP, Dronca E, Kaucsar T, *et al.* One Carbon Metabolism Disturbances and the C667T MTHFR Gene Polymorphism in Children with Autism Spectrum Disorders. J Cell Mol Med 2008;

[87] James SJ, Melnyk S, Fuchs G, *et al.* Efficacy of methylcobalamin and folinic acid treatment on glutathione redox status in children with autism. Am J Clin Nutr 2009; 89: 425-30.

[88] Hobbs CA, Cleves MA, Melnyk S, Zhao W, James SJ. Congenital heart defects and abnormal maternal biomarkers of methionine and homocysteine metabolism. Am J Clin Nutr 2005; 81: 147-53.

[89] Guerra-Shinohara EM, Paiva AA, Rondo PH, *et al.* Relationship between total homocysteine and folate levels in pregnant women and their newborn babies according to maternal serum levels of vitamin B12. BJOG 2002; 109: 784-91.

[90] Murphy MM, Scott JM, Arija V, Molloy AM, Fernandez-Ballart JD. Maternal homocysteine before conception and throughout pregnancy predicts fetal homocysteine and birth weight. Clin Chem 2004; 50: 1406-12.

[91] Murphy MM, Fernandez-Ballart JD, Arija V, *et al.* Maternal homocysteine at preconception is negatively correlated with cognitive achievement in children at 4 months and 6 years of age. Conference Proceedings, 6th International Conference on Homocysteine Metabolism. Clinical Chemistry and Laboratory Medicine. 2007; 45: A23.

[92] Bleich S, Frieling H, Hillemacher T. Elevated prenatal homocysteine levels and the risk of schizophrenia. Arch Gen Psychiatry 2007; 64: 980-1.

[93] Brown AS, Bottiglieri T, Schaefer CA, *et al.* Elevated prenatal homocysteine levels as a risk factor for schizophrenia. Arch Gen Psychiatry 2007; 64: 31-9.

[94] Thompson GA, Kilpatrick IC. The neurotransmitter candidature of sulphur-containing excitatory amino acids in the mammalian central nervous system. Pharmacol Ther 1996; 72: 25-36.

[95] Shi Q, Savage JE, Hufeisen SJ, *et al.* L-homocysteine sulfinic acid and other acidic homocysteine derivatives are potent and selective metabotropic glutamate receptor agonists. J Pharmacol Exp Ther 2003; 305: 131-42.

[96] Lipton SA, Kim WK, Choi YB, *et al.* Neurotoxicity associated with dual actions of homocysteine at the N-methyl-D-aspartate receptor. Proc Natl Acad Sci U S A 1997; 94: 5923-8.

[97] Zieminska E, Stafiej A, Lazarewicz JW. Role of group I metabotropic glutamate receptors and NMDA receptors in homocysteine-evoked acute neurodegeneration of cultured cerebellar granule neurones. Neurochem Int 2003; 43: 481-92.

[98] Mares P, Folbergrova J, Kubova H. Excitatory amino acids and epileptic seizures in immature brain. Physiol Res 2004; 53 Suppl 1: S115-24.

[99] Vladychenskaya EA, Tyulina OV, Boldyrev AA. Effect of homocysteine and homocysteic acid on glutamate receptors on rat lymphocytes. Bull Exp Biol Med 2006; 142: 47-50.

[100] Boldyrev AA. Molecular mechanisms of homocysteine toxicity. Biochemistry (Mosc) 2009; 74: 589-98.

[101] Benz B, Grima G, Do KQ. Glutamate-induced homocysteic acid release from astrocytes: possible implication in glia-neuron signaling. Neuroscience 2004; 124: 377-86.

[102] Ho PI, Ortiz D, Rogers E, Shea TB. Multiple aspects of homocysteine neurotoxicity: glutamate excitotoxicity, kinase hyperactivation and DNA damage. J Neurosci Res 2002; 70: 694-702.

[103] Oldreive CE, Doherty GH. Neurotoxic effects of homocysteine on cerebellar Purkinje neurons in vitro. Neurosci Lett 2007; 413: 52-7.

[104] Zieminska E, Lazarewicz JW. Excitotoxic neuronal injury in chronic homocysteine neurotoxicity studied in vitro: the role of NMDA and group I metabotropic glutamate receptors. Acta Neurobiol Exp (Wars) 2006; 66: 301-9.

[105] Finkelstein JD. The metabolism of homocysteine: pathways and regulation. Eur J Pediatr 1998; 157 Suppl 2: S40-4.

[106] Finkelstein JD. Metabolic regulatory properties of S-adenosylmethionine and S-adenosylhomocysteine. Clin Chem Lab Med 2007; 45: 1694-9.

[107] Obeid OA, Johnston K, Emery PW. Plasma taurine and cysteine levels following an oral methionine load: relationship with coronary heart disease. Eur J Clin Nutr 2004; 58: 105-9.

[108] Terauchi A, Nakazaw A, Johkura K, Yan L, Usuda N. Immunohistochemical localization of taurine in various tissues of the mouse. Amino Acids 1998; 15: 151-60.

[109] McBride WJ, Frederickson RC. Taurine as a possible inhibitory transmitter in the cerebellum. Fed Proc 1980; 39: 2701-5.

[110] Benitez-Diaz P, Miranda-Contreras L, Mendoza-Briceno RV, Pena-Contreras Z, Palacios-Pru E. Prenatal and postnatal contents of amino acid neurotransmitters in mouse parietal cortex. Dev Neurosci 2003; 25: 366-74.

[111] Sturman JA, Gargano AD, Messing JM, Imaki H. Feline maternal taurine deficiency: effect on mother and offspring. J Nutr 1986; 116: 655-67.

[112] Aerts L, Van Assche FA. Taurine and taurine-deficiency in the perinatal period. J Perinat Med 2002; 30: 281-6.

[113] Saransaari P, Oja SS. Taurine and neural cell damage. Amino Acids 2000; 19: 509-26.

[114] El Idrissi A, Trenkner E. Taurine as a modulator of excitatory and inhibitory neurotransmission. Neurochem Res 2004; 29: 189-97.

[115] El Idrissi A, Messing J, Scalia J, Trenkner E. Prevention of epileptic seizures by taurine. Adv Exp Med Biol 2003; 526: 515-25.

[116] Chen WQ, Jin H, Nguyen M, et al. Role of taurine in regulation of intracellular calcium level and neuroprotective function in cultured neurons. J Neurosci Res 2001; 66: 612-9.

[117] Wu JY, Wu H, Jin Y, et al. Mechanism of neuroprotective function of taurine. Adv Exp Med Biol 2009; 643: 169-79.

[118] El Idrissi A, Trenkner E. Taurine regulates mitochondrial calcium homeostasis. Adv Exp Med Biol 2003; 526: 527-36.

CHAPTER 8

Searching the Role of Mercury in Autism Spectrum Disorders

Anna Strunecka[1,*] and Russell L. Blaylock[2]

[1]*Department of Physiology, Faculty of Science, Charles University in Prague, Prague, Czech Republic and* [2]*Institute for Theoretical Neuroscience, LLC, and Visiting Professor of Biology, Belhaven University, Ridgeland, MS 39157, USA*

Abstract: Mercury is a ubiquitous environmental toxin that causes a wide range of adverse health effects in humans. The population is now exposed to mercury mostly from seafood consumption, dental amalgam, vaccines, and certain pharmaceuticals. Exposure to mercury can cause neurological, immune, sensory, motor, and behavioral dysfunctions similar to traits defining or associated with autism spectrum disorders (ASD). Because of an observed increase in autism in the last decades, which parallels cumulative mercury exposure, it was proposed that autism may be, in part, caused by mercury. The autism-mercury hypothesis has generated much interest and controversy. Thimerosal, a preservative added to many vaccines, has become one of the potential culprits in the pathogenesis of ASD. This chapter shows that the overwhelming preponderance of the evidence favors acceptance of the hypothesis that mercury exposure is capable of causing some ASD if exposure occurs at critical developmental periods. Special attention is paid to forms of mercury of current public health concern, which include vapors of metallic mercury from dental amalgam, methylmercury in edible tissues of fish and whales, and ethylmercury from thimerosal added to certain vaccines. Mercury in all of its forms is toxic to the fetus and children, and efforts should be made to reduce exposure to pregnant women and children as well as the general population. Moreover, emerging evidence supports the theory that some ASD may result from a combination of genetic/biochemical susceptibility and synergistic action of mercury with other excitotoxic contaminants from the environment.

INTRODUCTION

It is generally accepted that mercury and mercurial compounds are among environmentally ubiquitous substances most toxic for humans and animals. Industrial activities can raise the natural exposure to toxic levels directly or through the use or misuse of the liquid metals or synthesized mercurial compounds. Mercury causes a wide range of adverse health effects [1-3]. Exposure to mercury can result in mood disorders, personality disturbances, intellectual deficits, and epilepsy. It has a strong impact on sensory, neurological and psychomotor functions, cognition and behavior. Exposure to mercury compounds during gestation can results in severe injury to the infant brain and severe developmental and behavioral disorders, such as seizures, autism, childhood schizophrenia, early onset emotional disturbances and attention deficit disorders (ADD) [4]. Mercury has been on the list of toxic chemicals, which are known to cause neurodevelopmental disabilities such as intellectual retardation, dyslexia, attention deficit hyperactivity disorder (ADHD), autism, learning disabilities, and propensity to violence. The review of American Academy of Pediatrics (AAP) has provided pediatricians with comprehensive information on mercury toxicity and prevention of mercury exposure [5]. The authors explain that the developing fetus and young children are thought to be disproportionately affected by mercury exposure, because many aspects of development, particularly brain maturation, can be disturbed by the presence of mercury. The AAP recommends parents reduce methylmercury exposure to their children by limiting the amount of fish with high mercury content consumed during pregnancy and lactation and the amounts eaten by children [5]. Minimizing mercury exposure is, therefore, essential to optimal child health.

The statement from the document, A Research-Oriented Framework for Risk Assessment and Prevention of Children's Exposure to Environmental Toxicants [6], is valid today:

"However, serious impediments to prevention exist: too few chemicals are tested for toxicity to early brain development, knowledge of infants' and children's special vulnerabilities and unique exposures is

---

*Address correspondence to: **Anna Strunecka,** Faculty of Science, Charles University in Prague, Vinicna 7, 128 00 Prague 2, Czech Republic; E-mail: strun@natur.cuni.cz

scant, and paradigms for environmental risk assessment have only begun to address the hazards confronting infants and children. "

Because of an observed increase in autism in the last decades, it was proposed that autism may be in part caused by mercury. The hypothesis that autism represents an unrecognized mercurial syndrome suggested by Bernard *et al.* [7] has generated much interest, numerous studies and meta-analysis, discussions and much controversy. Bernard and co-workers suggested that mercury in vaccines – particularly thimerosal, as the source of organic ethylmercury - may be an important factor in the pathogenesis of autism [7,8]. Thimerosal, a preservative added to many vaccines, has become a major source of organic ethylmercury in children who, within their first two years, may have received a quantity of mercury that exceeds safety guidelines. By 1999, expanding recommendations for infant vaccination meant that US children who received a complete series of vaccines that contained thimerosal potentially received up to 187.5 µg of ethylmercury during the first 6 months of life [9]. The thimerosal controversy continues on many levels of experts, doctors and scientists, many parents, public health officials, governments, vaccine makers and drug companies [10-13]. The campaign against thimerosal exceeded the laboratory and clinical investigations; it has had several legal, social and political consequences.

In this review we pay special attention to forms of mercury of current public health concern, which include vapors of metallic mercury from dental amalgam, methylmercury in edible tissues of fish and whales, and ethylmercury in the form of a preservative thimerosal added to certain vaccines. In most cases, children are exposed to all of these sources. Health risks from mercury have been the subject of several large epidemiological investigations and continue to be the subject of intense debate.

While the official conclusion is that there are no reliable data indicating that mercury is a primary cause of ASD, the overwhelming preponderance of the evidence from the peer-reviewed scientific and medical literature favors acceptance that medicinal, dental and environmental mercury exposures are capable of causing the set of symptoms commonly used to diagnose autism, particularly in children who are biochemically and/or genomically susceptible to mercury toxicity. However, one cannot rule out the possibility that the individual gene profile and/or gene-environment interactions may play a role in modulating the response to acquired risk by modifying the individual's susceptibility [12,14]. Emerging evidence supports the theory that some ASD may result from a combination of genetic/biochemical susceptibility, specifically a reduced ability to excrete mercury, and exposure to mercury at critical developmental periods.

SOURCES AND BIOAVAILABILITY OF MERCURY

Mercury is toxic in all of its forms. Three forms of mercury - elemental, inorganic, and organic exist, and each has its own profile of toxicity. Exposure to mercury typically occurs by inhalation or ingestion. In an uncontaminated environment the general population is often exposed to elemental mercury vapors from dental amalgam. Potential toxicity from exposure to mercury vapor from dental amalgam fillings is the subject of current public health debates by researchers, dentists, and health officials in many countries [2,15-18]. Today, mercury vapor emitted from amalgam tooth fillings are considered to be the most widespread human exposures to mercury [1,19].

Additionally, industry emissions with resulting ambient air pollution through the use of liquid metals, are an important source of inhaled mercury. Bacteria in lakes, streams, and ocean sediments can convert elemental mercury to organic mercury compounds (e.g. methylmercury), which may then accumulate in animal tissues and thus enter the food chain. Large, long-lived, predatory ocean animals, such as tuna, swordfish, sharks, dolphins and whales, may have increased methylmercury content because of exposure of the seas to industrial sources of mercury. Major sources of human exposure to methylmercury are thus primarily from eating high levels of seafood. Health risks from methylmercury in edible tissues of fish have been the subject of several large epidemiological investigations. The ingestion of wheat and barley seed treated with an alkyl mercury fungicide for sowing, by a largely illiterate population in Iraq, led to a major outbreak of poisoning with a high fatality rate.

Ethylmercury, in the form of thimerosal, has been used as an effective preservative for killed vaccines and other biological agents for medical therapy. It is the most recent form of mercury that has become a public health concern. Ethylmercury is the most discussed organic compounds of mercury as a potential culprit of ASD.

Ubiquitous sources of inorganic mercury include many pharmaceuticals, which contain mercurials (mercuric salts). Mercurials may be found as preservatives in cosmetics, toothpastes, dental fillings, lens solutions, vaccines, and allergy test and immunotherapy solutions; in antiseptics, disinfectants, and many other products.

Dental Amalgam

Dental amalgam, a composite metal that is about 50% pure elemental mercury, has been used to fill decayed teeth since the 1820s [5]. The other half of amalgam filling is a powdered alloy of silver, tin, and copper. Mercury is used to bind the alloy particles together into a strong, durable, and solid filling. In 1991, the WHO confirmed that mercury contained in dental amalgam is the greatest source of mercury vapors in non-industrialized settings, exposing the concerned population to mercury levels significantly exceeding those set for food and for air (http://www.who.int/ipcs/publications/cicad/en/cicad50.pdf).

An average-sized amalgam filling contains 750-1000 mg of mercury [20,21] and has an average serviceable life span in the human mouth of 7-9 years. Clinical studies in subjects with amalgam fillings demonstrated that quite substantial amounts of mercury vapor are released into intra-oral air from dental amalgam, being six fold higher in subjects who chewed gum for 10 minutes in comparison with pre-chewing levels. The intra-oral mercury vapor concentration remained elevated during 30 minutes of continuous gum chewing; and after cessation of chewing, the mouth mercury vapor concentration declined slowly to pre-chewing levels over a period of 90 minutes. Mercury vapor concentrations are highest immediately after placement and removal of dental amalgam, but decline thereafter [22].

Brushing the teeth with commercial toothpaste will also stimulate the release of mercury vapor from amalgam surfaces. Re-calculations of almost all the available daily dose data showed a mean daily dose value of about 1.3 µg Hg/day (range, 0.3-2.2 µg Hg/day). The mean swallowed amount of mercury from intra-oral mercury vapor was calculated as being in the order of 10 µg Hg/day (range, 2.4-17 µgHg/day), resulting in an estimated absorption of about 1 µg Hg/day from the gastrointestinal tract (GIT) [23-25]. In a study of 21 subjects with amalgams, Halbach found the daily dose 4.8 µg Hg, which the author points out is well below the acceptable daily intake of 40 µg per day for the general population [26]. More recently, Halbach *et al.* [27] estimated an average daily dose of mercury from amalgam in 82 patients to be up to 3 µg for an average number of fillings and at 7.4 for a high amalgam load. WHO Policy Paper gives the range of daily intake of mercury from amalgam fillings from 1.2-27 µg Hg [3]. The biological half-life of inorganic mercury in the whole human body is approximately 40 days.

Canadian researchers demonstrated that when radioactive ^{203}Hg is mixed with dental Hg/silver fillings (amalgam) and placed in teeth of adult sheep, the isotope will appear in various organs and tissues within 29 days. Evidence of mercury uptake, as determined by whole-body scanning and measurement of the isotope in specific tissues, revealed three uptake sites: lung, gastrointestinal, and jaw tissue absorption. Once absorbed, high concentrations of dental amalgam mercury rapidly localize in kidneys and liver [20]. In another study Vimy with co-workers established a time-course distribution for amalgam mercury in body tissues of adult and fetal sheep [21]. Under general anesthesia, five pregnant ewes had twelve occlusal amalgam fillings containing radioactive ^{203}Hg placed in teeth at 112 days gestation. Blood, amniotic fluid, feces, and urine specimens were collected at 1- to 3-day intervals for 16 days. From days 16-140 after amalgam placement (16-41 days for fetal lambs), tissue specimens were analyzed for radioactivity, and total mercury concentrations were calculated. Their results demonstrated that mercury from dental amalgam appeared in maternal and fetal blood and amniotic fluid within 2 days after placement of amalgam tooth restorations. All tissues examined displayed mercury accumulation. Highest concentrations of mercury from amalgam in the adult occurred in kidney and liver, whereas in the fetus the highest amalgam mercury concentrations appeared in liver and pituitary gland. The placenta progressively concentrated mercury as gestation advanced to term, and the mother's milk concentrated the mercury from the amalgam as well.

There is limited clinical information about the potential effects of dental amalgam fillings on pregnant women and their developing fetuses, and on children under the age of 6, including breastfed infants. Additionally, infants may have incurred additional mercury exposure through breast milk if they were born to mothers with amalgam fillings. However, the estimated amount of mercury in breast milk attributable to dental amalgam is low and falls well below general levels for oral intake that the Environmental Protection Agency (EPA) considers safe. The Food and Drug

Administration (FDA) concludes that the existing data support a finding that infants are not at risk for adverse health effects from the breast milk of women exposed to mercury vapor from dental amalgam. The estimated daily dose of mercury vapor in children under age 6 with dental amalgams is also expected to be at or below levels that the EPA and the Centers for Disease Control and Prevention (CDC) consider safe. Pregnant or nursing mothers and parents with young children should talk with their dentists if they have concerns about dental amalgam. However, the FDA has concluded that the existing data support a finding that infants are not at risk for adverse health effects from the breast milk of women exposed to mercury vapors from dental amalgam.

Globally, dental mercury use is in decline as a result of regulation and cultural preferences for "white" composite amalgam materials. While the FDA concluded that clinical studies have not established a causal link between dental amalgam and adverse health effects in adults and children, the European Union is implementing a strategy of reducing the supply and demand for mercury. Norway, Sweden, and Denmark totally banned mercury fillings in 2008. Mutter and co-workers expressed their conclusion as scientists who are involved in preparing a German federal guideline regarding dental amalgam [28]. It is useful to consider the points of their assessment:

a) Dental amalgam is the main source of human total mercury body burden, because individuals with amalgam have 2-12 times more mercury in their body tissues compared to individuals without amalgam.

b) There is no significant correlation between mercury levels in blood, urine, or hair and in body tissues, and none of the parameters correlate with severity of symptoms.

c) The half-life of mercury deposits in brain and bone tissues could last from several years to decades, and thus mercury accumulates over time of exposure.

d) Mercury, in particular mercury vapor, is known to be the most toxic non-radioactive element, and is toxic even in very low doses.

e) Some studies, which conclude that amalgam fillings are safe for human beings, have important methodological flaws. Therefore, they have no value for assessing the safety of amalgam.

f) Finally, mercury levels in amniotic fluid and breast milk correlate significantly with the number of maternal dental amalgam fillings.

Methylmercury and Seafood Consumption

Mercury accumulation in fish is a well known example of biomagnification (bioaccumulation) – the increase in concentration of a substance that occurs in the food chain [29]. An important source of methylmercury exposure is fish and whale consumption. Risk assessment for methylmercury from fish consumption is mainly based on human data coming from large scale epidemiological studies in various fish eating communities around the world. Longitudinal prospective studies in island populations were established in the 1970s and 1980s to evaluate the effects of moderate methylmercury exposure from frequent fish consumption during pregnancy in the Faroe Islands and in the Seychelle Islands [30-34].

The Faroe Islands are located southeast of Iceland in the Norwegian Sea. They are inhabited by a homogeneous and isolated population of people who consume small amounts of fish (1–3 meals of cod per week) and have episodic feasts of pilot whale. The fish have very low mercury concentrations, but pilot whale meat has a mean content of methylmercury of 1.9 ppm (parts per million). The Seychelles are equatorial islands in the Indian Ocean inhabited by a stable, cohesive, and homogeneous population of people who eat fish frequently (12 fish meals per week). The fish have relatively low methylmercury concentrations (mean, 0.3 ppm). The long-term intakes of both total and methylmercury from eating pilot whale (*Globicephalus meleanus*) in the Faroe Islands far exceed the Provisional Tolerable Weekly Intakes (PTWI) recommended by WHO. For the general population, the PTWI's are 300 and 200 µg/person/week for total and methylmercury, respectively. The calculated intake of methylmercury in this study approaches the lower value (1200 µg /person/week) of the recognized critical level of methylmercury intoxication in the general population.

It was concluded that the general Faroe Island population should significantly restrict the consumption of pilot whale foods [30]. It had been recommended that pregnant women probably should not eat pilot whale foods at all, as the critical levels for methylmercury intoxication of pregnant women and fetuses are lower by a factor of 2-5 than for the general population. Frequent ingestion of whale meat dinners during pregnancy and, to a much lesser degree, frequent consumption of fish, and increased parity or age were associated with high mercury concentrations in cord blood and hair. Grandjean with co-workers [31] reported that umbilical cord blood from 1,023 consecutive births in the Faroe Islands showed a median blood-mercury concentration of 121 nmol/L (24.2 μg/L); 250 of those samples (25.1%) had blood-mercury concentrations that exceeded 200 nmol/L (40 μg/L). Maternal hair mercury concentrations showed a median of 22.5 nmol/g (4.5 μg /g), and 130 samples (12.7%) contained concentrations that exceeded 50 nmol/g (10 μg/g). Later, Weihe *et al.* [35] found that mercury concentrations in umbilical cord blood showed a maximum of 351 μg /L. The large variation in mercury exposure is associated with differences in the frequency of whale dinners. Following an official recommendation in the Faroe Islands that women should abstain from eating mercury-contaminated pilot whale meat, a survey was carried out to obtain information on dietary habits and hair samples for mercury analysis. Weihe with co-workers reported that a letter was sent to all 1180 women aged 26-30 years who resided within the Faroes, and the women were contacted again one year later [36]. A total of 415 women responded to the first letter; the second letter resulted in 145 repeat hair samples and 125 new responses. Questionnaire results showed that Faroese women, on average, consumed whale meat for dinner only once every second month, but the frequency and meal size depended on the availability of whale in the community. In comparison with previously published data on hair-mercury concentrations in pregnant Faroese women, this study found substantially lower exposures as well as a further decrease temporally associated with the issue of a stricter dietary advisory.

The pattern of methylmercury consumption in these two island populations is different: with the Faroe Islands pattern being more episodic and the Seychelles pattern being more constant. Moreover, pilot whales consumed in the Faroe Islands also contain polychlorinated biphenyls (PCB), which are known to have a long-term impact on neurodevelopment and intellectual functions in infants and young children [37,38]. Prenatal PCB exposure was associated with greater impulsivity, poorer concentration, and poorer verbal, pictorial, and auditory working memory in children, which were tested in their homes at age 11 years [39].

The Seychelles Islands study has explored the potential for protection against methylmercury toxicity by nutrients present in fish, particularly ω-3 fatty acids and selenium [40]. Docosahexaenoic acid (DHA) is a necessary structural component of the brain and eye that may benefit early brain development. Some observational studies have also found that umbilical cord tissue or blood levels of DHA were associated with better development in infants [41]. As the most rapid uptake of DHA into the brain occurs in late pregnancy, it is possible that prenatal exposure is even more important. Moreover, the consumption of oils rich in long-chain *n-3* polyunsaturated fatty acids (n-3 LC-PUFA) during pregnancy reduces the risk for early premature birth.

It was concluded that the DHA from sea foods raised IQ, and that in children with the highest Hg intake, their IQ was lowered, but was still higher than children with low DHA intake. In other words, the mercury blunted the beneficial effects of the ω-3 oil. DHA has also been shown to lower methylmercury accumulation in neurons and glia [42]. On the basis of the limited data available, a European expert panel recently recommended an average daily intake of at least 200 mg/day of DHA during pregnancy [43]. Most pregnant women do not consume this amount. Also, it is known that selenium significantly detoxifies mercury in the brain, even though it may raise brain mercury levels [44].

Thimerosal in Vaccines

Thimerosal is sodium ethylmercury thiosalicylate, an organic compound of ethylmercury. Thimerosal contains 49.6 % mercury by weight and is metabolized to ethylmercury and thiosalicylate. Thimerosal (having the common US trade names of Merthiolate and Thimerosal and, in the UK, Thiomersal) has been used as an additive to biologics and vaccines since the 1930s because it is very effective in killing bacteria used in several vaccines and in preventing bacterial contamination, particularly in multidose containers. For decades thimerosal has been marketed as an antimicrobial agent in a range of products, including topical antiseptic solutions and antiseptic ointments for treating cuts, nasal sprays, eye solutions, vaginal spermicides, diaper rash treatments, and perhaps most importantly as a preservative in vaccines and other injectable biological products, including immune globulin preparations [45].

Before fall 1999, there were 25 or 50 μg of mercury in each 0.5-mL dose of most vaccines. Starting in the late 1980s/early 1990s, the cumulative dose of mercury in children received from thimerosal-containing childhood vaccines almost tripled. Looking at cumulative exposure over the first 6 months of life, an infant 6 months old who received all recommended vaccine doses on schedule could be exposed to up to 187.5 μg of mercury [9]. Redwood and co-workers found that at birth an infant received 12.5 μg of mercury, 62.5 μg at 2 months, 50 μg at 4 months, 62.5 μg at 6 months and 50 μg at 18 months, for a total mercury burden of 237.5 μg of ethylmercury during the first 18 months of life, which exceed the EPA safety guidelines for an adult [46].

On July 7, 1999, the US Public Health Service (USPHS) and the AAP issued a joint statement that urged

"all government agencies to work rapidly toward reducing children's exposure to mercury from all sources."

However, the statement suggests that there are no data or evidence of any harm caused by the level of exposure that some children may have encountered in following the existing immunization schedule in the US. Further, it states that infants and children who have received thimerosal-containing vaccines do not need to be tested for mercury exposure. Nevertheless, because any potential risk is of concern, the USPHS, the AAP, and vaccine manufacturers agree that

"thimerosal-containing vaccines should be removed as soon as possible".

Similar conclusions were reached in 1999 in a meeting attended by European regulatory agencies, European vaccine manufacturers, and FDA, which examined the use of thimerosal-containing vaccines produced or sold in European countries. However, USPHS and AAP continue to recommend that

"all children should be immunized against the diseases indicated in the recommended immunization schedule. Given that the risks of not vaccinating children far outweigh the unknown and much smaller risk, if any, of exposure to thimerosal-containing vaccines over the first 6 months of life, clinicians and parents are encouraged to immunize all infants even if the choice of individual vaccine products is limited for any reason."

Thimerosal was widely used in multidose vaccine vials in the US and Europe until 2001 and continues to be used in many countries throughout the world. The process of public policy formulation in the case of thimerosal in vaccines was described in details by Freed *et al.* [47]. Some excerpted parts from this document could illustrate the difficulties in searching for some consensus of the participating institutions in the process of removal of thiomersal from vaccines, with the aim to protect the health of children:

"An early assessment of the health risks from all forms of mercury by the WHO found that insufficient information was available to perform risk calculations for human exposure to ethylmercury compounds, the type of mercury contained in thimerosal. However, the WHO did note that the limited data available suggested that ethylmercury was probably less hazardous than methylmercury, because it is metabolized faster in the body...To quickly bring representatives of several organizations involved in immunization policymaking together to discuss the issue, a meeting was organized by Dr Halsey and Dr Cooper for June 30, 1999, at the AAP offices in Washington, DC. Invitees included representatives of the CDC, FDA, EPA, AAP, vaccine manufacturers, and toxicologic consultants."

Some participants believed strongly that the potential threat to health from thimerosal was significant; others believed that there was no clear evidence that thimerosal was harmful, particularly when compared with the clear health risks of delaying childhood vaccines.

Thimerosal was re-introduced into the US routinely recommended childhood vaccine schedule in 2002 with mandatory influenza vaccines. The multi-dose vials contain in each 0.5-mL dose 50 μg of thimerosal. Moreover, the 2002 recommendation has been continually expanded to the point that, in 2008, the US CDC recommended that all pregnant women should receive an influenza vaccine during the "flu" season and that all infants should receive two doses of influenza vaccine in the first year of life, with one influenza vaccine administered on a yearly basis thereafter until 19 years. In addition, the CDC recommends that children under nine who only received one influenza vaccine initially should be given two doses of influenza vaccine, at least two weeks apart.

At the same time, public concern about real or perceived adverse events associated with vaccines has increased. This heightened level of concern often results in an increase in the number of people refusing vaccines.

MERCURY TOXICITY

Mercury in any form is toxic. The difference lies in how it is absorbed, the clinical signs and symptoms, and the response to treatment modalities. Mercury poisoning can result from vapor inhalation, ingestion, injection, or absorption through the skin. Brain, GIT, and renal systems are the most commonly affected organ systems in mercury exposure. Mercury poisoning is usually misdiagnosed because of the insidious onset, nonspecific signs and symptoms, and lack of knowledge within the medical profession. Risk assessment for mercury is mainly based on human data coming from the massive episodes of poisoning in Japan and Iraq.

Toxicity of Elemental Mercury

Elemental mercury easily vaporizes at room temperature and is well absorbed (80%) through inhalation. Its lipid-soluble property allows for easy passage through the alveoli into the bloodstream and red blood cells. Once inhaled, elemental mercury is mostly converted to an inorganic divalent (Hg^{2+}) or mercuric form by catalase in red blood cells. This inorganic form has similar properties to inorganic mercury (e.g., poor lipid solubility, limited permeability to the blood-brain barrier (BBB), and excretion in feces). Small amounts of nonoxidized elemental mercury continue to persist and account for CNS toxicity [48]. Elemental mercury as a vapor has the ability to penetrate the CNS, where it is ionized and trapped, attributing to its significant toxic effects. Elemental mercury is not well absorbed by the GIT and, therefore, when ingested (e.g., from thermometers), is only mildly toxic. Excretion of Hg^{2+}, as with organic mercury, is mostly through feces. Renal excretion of mercury is considered insufficient and contributes to its chronic exposure and accumulation within the brain, causing CNS effects. Dental professionals who are in contact with amalgam must follow specific guidelines to avoid exposure to toxic amounts of aerosolized elemental mercury. Patients with dental amalgam fillings have slightly elevated levels in their urine, but these findings have not correlated with any systemic disease [49].

Acute exposure caused by inhaled elemental mercury can lead to pulmonary symptoms. Initial signs and symptoms, such as fever, chills, shortness of breath, metallic taste, and pleuritic chest pain, may be confused with metal fume fever. Other possible symptoms could include stomatitis, lethargy, confusion, and vomiting. Chronic and intense acute exposure causes cutaneous and neurological symptoms. The classic triad found in chronic toxicity includes tremors, gingivitis, and erethism (i.e., a constellation of neuropsychiatric findings that includes insomnia, shyness, memory loss, emotional instability, depression, anorexia, vasomotor disturbance, uncontrolled perspiration, and blushing). Additional findings may include headache, visual disturbance (e.g., tunnel vision), peripheral neuropathy, salivation, insomnia, and ataxia [48].

Toxicity of Organic Mercury Compounds

The toxicity of organic mercury compounds is dependent on the specific compound, route of exposure, dose, and age of the person at exposure. In general, organic mercury compounds are lipid soluble, and 90 to 95 percent is absorbed from the GIT. Once absorbed, the alkyl compounds are converted to their inorganic forms and possess similar toxic properties to inorganic mercury. Alkyl organic mercury has high lipid solubility and is distributed uniformly throughout the body, accumulating in the brain, kidney, liver, hair, and skin. Organic mercurials also cross the BBB and placenta and penetrate erythrocytes, contributing to neurological symptoms, teratogenic effects, and high blood to plasma ratios, respectively.

Methylmercury and ethylmercury cross cell membranes as complexes with small molecular weight thiol compounds, entering the cell, in part, as a cysteine complex on the large neutral amino acid carriers and exiting the cell, in part, as a complex with reduced glutathione on endogenous carriers [19]. Methylmercury has a high affinity for sulfhydryl groups, which is attributed to its effect on enzyme dysfunction. One enzyme that is inhibited is choline acetyl transferase, which is involved in the final step of acetylcholine production. This inhibition may lead to acetylcholine deficiency, contributing to the signs and symptoms of motor dysfunction.

The mean half-life for methylmercury in blood is approximately 40-65 days [48] with a range of 20–70 days for adults. The primary mechanism for excreting mercury from the body involves binding to glutathione and then being excreted in the bile. Ninety percent of methylmercury is excreted through bile in feces [5, 50]. Infants are especially vulnerable to mercury poisoning because they are poor excretors of mercury due to low production of glutathione.

Risk assessment for alkylmercury compounds – methylmercury and ethylmercury, in humans is mainly based on human data coming from the massive episodes of poisoning in Japan and Iraq, and from the epidemiological studies with fish-eating populations [51]. Some authors declare that the primary source of human exposure to "toxic" mercury is from fish. The data obtained so far from epidemiological studies on fish-eating populations are not consistent. A recommended reference dose of 0.1 µg methylmercury per kilogram of wet weight per day has been established by the US EPA based on a study on Iraqi children. However, these accidental exposures were not typical of lower chronic exposure levels associated with seafood consumption.

Minamata Disease

The first well-documented outbreak of acute methylmercury poisoning by consumption of contaminated fish occurred in Minamata, Japan, in 1953 when a Chisso Corp. plant discharged large quantities of a mercury catalyst into the bay [52]. This highly toxic chemical accumulated in shellfish and other fish, which when eaten by the local inhabitants and animals resulted in similar accumulation. There were thousands of victims poisoned by organic mercury, identified by professor Kitamura as methylmercury [53]. Cat, dog, and pig deaths continued over more than 30 years. A second outbreak of Minamata disease occured in Niigata Prefecture in 1965. It was also caused by organic mercury released by a Showa Denko K.K. plant. In total, nearly 2,000 people have died from the disease, and babies have suffered birth defects. Recently, a re-estimation stated that it is expected to involve about 20,000 of some 30,000 people who are seeking recognition as Minamata disease suffers. The clinical picture was officially recognized and called Minamata disease in 1956. Symptoms of acute exposure to methylmercury in adults (Minamata disease) may include impairments in all primary sensory modalities (blurred vision, with bilateral and symmetric constriction of the visual fields, bilateral deficits in hearing, with speech discrimination more impaired than pure tone, deficits in smell and taste, and distal paresthesias. Residents living around the sea were exposed to low-dose methylmercury through fish consumption for about 20 years (at least from 1950 to 1968). Ekino *et al.* reported that these patients with chronic methylmercury poisoning continue to complain of distal paresthesias of the extremities and the lips even 30 years after cessation of exposure to methylmercury [54]. Based on findings in these patients, the symptoms and lesions in methylmercury poisoning were reappraised. The persisting somatosensory disorders after discontinuation of exposure to methylmercury were induced by diffuse damage to the somatosensory cortex, but not by damage to the peripheral nervous system, as previously believed [54]. Yorifugi *et al.* found that the subjects in the Minamata area manifested neurologic signs very frequently [55]. They examined associations between long-term exposure to methylmercury and the following neurologic signs measured on clinical examination: paresthesia of whole body, paresthesia of extremities, paresthesia around the mouth, ataxia, dysarthria, tremors, and pathologic reflexes. The highest prevalence odds ratio was observed for paresthesia around the mouth. There were at least 30 cases of profound brain injury in infants born to mothers who ingested contaminated fish during pregnancy. Many infants with "cerebral palsy" in villages where adult cases occurred were established as having congenital Minamata disease. Developing brains were affected by methylmercury through transplacental exposure and even by breastfeeding [56]. Eto *et al.* [57] found that neuropathological lesions found in chronic human Minamata disease tend to be localized in the calcarine cortex of occipital lobes, the pre- and postcentral lobuli, and the temporal gyri. The mechanism for the selective vulnerability is still not clear.

Ethylmercury Poisoning in Humans

Beginning in 1955, the Iraqi Ministry of Agriculture supplied farmers with seeds dusted with a mercury-containing fungicide. Ethylmercury p-toluene sulfonanilide (Granosan) has been used in Iraq and farmers had been given frequent warnings against using the treated seed for food. Nevertheless, some farmers and their families consumed the seed in the preparation of home-made bread and became the victims of mercury poisoning [58]. It affected a large number of farmers and their families. In 1956 many cases of mercury poisoning were observed in the North of Iraq, and more than 100 cases were admitted to Mosul Hospital with 14 deaths. In 1960, many farmers from the central part of Iraq were affected and 221 patients were admitted to one hospital in Baghdad. Many systems were affected, including the kidneys, the GIT, the skin, the heart, and the muscles, but involvement of the nervous system was the most constant with disturbance of speech, cerebellar ataxia, and spasticity. Mental abnormalities were occasionally observed – some patients had headache, insomnia, confusions or excitement [58].

In these communities, exposed pregnant women, who themselves had no or minimal symptoms, had babies with devastating neurological handicaps, including delayed attainment of developmental milestones, blindness, deafness and cerebral palsy. Whereas adult mercury exposure causes localized damage to discrete areas of the brain, exposure in fetal life causes diffuse and widespread neurological damage [59,60]. Analytical data indicate that the predominant route of exposure for the infant was through breast milk in which approximately 60% of total mercury was determined to be organic mercury. Abnormal neurological signs in these infants became more obvious with time: hyperreflexia was observed in 8 of 22 infants at the first examination, and in 17 of 22 at the second examination. Delayed motor development became evident at the second and third examinations. The frequency of pathological reflexes and delayed motor developmental milestones was so high as to be considered significant even in the absence of a controlled study.

Forty-one patients in the Peoples Republic of China were poisoned by ethylmercury chloride, caused by the ingestion of rice that had been treated with the chemical. Five months after the onset of the intoxication, the patients were still in poor condition [61]. Later, a significant series of patients in Russia was observed to suffer from serious toxic outcomes following ingestion of ethylmercury and occupational exposure to ethylmercury [62]. Early signs of exposure included general weakness, pains, tachycardia, and headache. There were also complaints of nausea, liquid stool, disordered sleep, decreased memory, and pain in the extremities. Four case reports were presented of patients in Romania who ate the meat of a hog inadvertently fed seed treated with fungicides containing ethylmercury chloride. The clinical, electrophysiological, and toxicological, and in two of the patients, pathological data, showed that this organic mercury compound had very high toxicity, not only for the brain, but also for the spinal motoneurones, peripheral nerves, skeletal muscles, and myocardium [63].

An excellent comprehensive review on health risks and toxicity of thimerosal (Merthiolate) with specific historical considerations regarding thimerosal safety and effectiveness was published by Geier *et al.* [64]. Biologists and biochemists have used thimerosal as a laboratory tool in their cell experiments. Interestingly, Salle and Lazarus determined in 1935 that thimerosal was 35.3 times more toxic for embryonic cells than for the bacterial cells that thimerosal was supposed to kill. Engley found that thimerosal was significantly toxic to human tissue culture cells at a concentration of 10 ppb [65]. These recent investigations also show that thimerosal affects the viability of cells in concentrations as low as 1 nM-10 μM [66].

Ethylmercury induces cell death due to microtubule depolymerization and inhibition of tubulin synthesis and/or β-tubulin degradation [67]. A pulse of thimerosal induces an instantaneous, complete and long-lasting microtubule interphasic network disassembly in mouse primary oocytes, correlated with the irreversible inhibition of meiosis. Thimerosal exerts a much more dramatic effect on resumption of meiosis than any other pharmacological manipulation [68].

Mercury Pharmacokinetics and Toxicity in Animals

Since there are still some debates regarding biotransformation of ethylmercury from thimerosal in humans, we should compare some laboratory studies. Quanstrom with co-workers [69] performed determination of methylmercury, ethylmercury, and inorganic mercury in mouse tissues, following administration of thimerosal, by species-specific isotope dilution. Mice were treated with thimerosal, 10 mg/L in drinking water ad libitum for 1, 2.5, 6, or 14 days. Using complicated isotope dilution technique, the authors found that the ethylmercury component of thimerosal was rapidly taken up in the organs of the mice (kidney, liver, and mesenterial lymph nodes), and concentrations of ethylmercury as well as Hg^{2+} increased over the 14 days of thimerosal treatment. This shows that ethylmercury in mice to a large degree is degraded to Hg^{2+}.

Burbacher *et al.* [70] evaluated infant monkeys following oral administration of methylmercury or injected doses of thimerosal, comparable to the dosing schedule (weight- and age-adjusted) that US children received during the 1990s. Researchers determined the total blood mercury levels 2, 4, and 7 days after each exposure. Total and inorganic brain mercury levels were assessed 2, 4, 7, or 28 days after the last exposure. They found that the maximum mercury content in the brains of the thimerosal-treated infant monkeys averaged about 40-50 ppb. They calculated that the half-life for organic mercury in the brain of the infant monkeys was about 14 days. By contrast, they determined that maximum mercury content in the brains of methylmercury treated infant monkeys averaged about 80-120 ppb. In this study, they calculated that the half-life for organic mercury in the brain of the infant monkeys examined was about 58 days. Based on these results, it was demonstrated that infant low-dose ethylmercury mercury exposure was able to induce significant levels of mercury in the monkey brain, and that the mercury from ethylmercury was present in the brain for several weeks postdosing. A higher percentage of the total mercury in the brain was in the form of inorganic mercury for the thimerosal-exposed monkeys (34% vs. 7%). However, brain concentrations of total mercury were significantly lower by approximately 3-fold for the thimerosal-exposed monkeys when compared with the methylmercury infants, whereas the average brain-to-blood concentration ratio was slightly higher for the thimerosal-exposed monkeys.

Harry *et al.* [71] examined the distribution of mercury to the brain following an injection of methylmercury or ethylmercury in immature mice. Postnatal day 16 CD1 mice received methylmercury chloride by intramuscular injection. At 24 h and 7 days post-injection, total mercury concentrations were determined in blood, kidney, brain, and muscle. The amount of mercury found in tissues after injection of methylmercury was approximately one third of that obtained from oral administration. Approximately 0.6 % of the injected methylmercury was found in the brain and 1.1 % in the kidney after 24 h. The highest mercury concentration was in the kidney; its concentration in the brain was the same as in the blood. For ethylmercury approximately 0.06 % of the injected mercury was found in the brain and 0.6 % in the kidney after 24 h.

HEALTH RISKS OF MERCURY BURDEN

WHO has identified the adverse effects of mercury pollution as a serious global environmental and human health problem. Mercury has long been known to affect neurodevelopment in both humans and experimental animals. Despite extensive literature and research, the threshold dose for organic mercury neurotoxic effects is still unclear. In assessment of the population exposure from various sources of mercury, mercury concentrations are determined in blood, urine (as creatinine), and hair. The comparison of mercury content in blood, urine, and hair of subjects from various geographic areas without mercury burden (non-fish-eating) is given in Table **1**.

Table 1: The Comparison of Some Selected Values of Total Mercury Concentrations in Blood, Urine, and Hair of Adults and Children from Various Geographic Areas Without Mercury Burden.

	BLOOD MG.L^{-1} (PPB)	URINE MG.G^{-1} OF CREATININE	HAIR MG.G^{-1} (PPM)	REFERENCE
HBM lower limits	5	5		[73]
US EPA limit	< 5.8	< 10-20	1	[5,74]
NOAEL WHO for fetus toxicity			50 10	
ADULTS				
USA (women)	1.2		1.4*	[74]
USA (men)	1.16		0.27	[75]
CZECH REPUBLIC	0.89	1.1	0.33	[76,77]
SWEDEN	0.1		0.06	[78]
SWEDEN (pregnant women)	0.27-2.1		0.04-0.83	[79]
CHILDREN				
USA	0.3		0.4*	[74]
USA 6-10 years		0.7-0.9	0.3-0.4	[80]
GERMANY 8-10 years		0.22	0.18	[81]
CZECH REPUBLIC 6-10 years	0.45 0.35	0.3 0.16	0.13 0.18	[76,77]
SPAIN – preschool and newborns		2	0.94 1.68	[82]

HBM - Recommendations of the Commission on Human Biological Monitoring of the German Federal Environmental Agency;

NOAEL -Hg levels exceeding the no observed adverse effects level

*data express the 90th percentile. It means that 90% of investigated subjects had values below it, and 10% above it.

Hair as a Biomarker of Mercury Burden

Due to its ability to avidly accumulate methylmercury from blood, scalp hair has been widely used as a biological monitor for human exposure. The EPA reference dose for methylmercury was established using data from populations with greater exposures than those typical of the US. Few data are available on potential adverse health effects at lower levels. Surkam *et al.* [72] examined relationships between hair mercury levels and neuropsychological outcomes in a population of US children. This study included data from 355 children ages 6-10 enrolled in the New England Children's Amalgam Trial. Data on total hair mercury levels, sociodemographic information and neuropsychological function were collected. They evaluated associations between hair mercury and neuropsychological test scores with linear regression methods and used generalized additive models to determine the shape of associations that departed from linearity. Models controlled for relevant covariates, including the potential beneficial effects of consuming fish. These authors reported that they observed no significant linear relationships between hair mercury level and any test score. The association was positive for hair mercury levels below 0.5 µg/g and negative for levels between 0.5 and 1.0 µg /g. Overall, test scores of children with hair mercury levels 1.0 µg /g appeared to be lower than those of children with levels < 1.0 µg /g, but few children had levels in this upper range and these differences did not reach statistical significance. Hair mercury levels below 1.0 µg /g in US school-age children were not adversely related to neuropsychological function. Mean mercury levels in mothers' hair were 4.3 ppm (0.2–39.1 ppm) in the Faroe Islands and 6.8 ppm (range: 0.5–27 ppm) in the Seychelles.

The question arises concerning hair as an important route of elimination of methylmercury from the body. Taking original publications and reviews on the physiology of hair (including growth by weigh and density) and on the deposition parameters for methylmercury in the body (including the hair to blood concentration ratio of methylmercury), one can calculate the rate of elimination of methylmercury in hair. The result indicates that hair accounts for only a small fraction, less than 10%, of the total elimination of methylmercury from the body. This relationship is expected to be maintained at every level when the dominant form of mercury is the methyl form [83]. In that study, motor retardation was seen in children whose mothers had hair mercury levels in the range of 10 to 20 ppm. Administration of thimerosal containing vaccines to infants was reported to result in a substantial increase (446%) in hair Hg levels [45].

Adverse Health Effects of Mercury from Dental Amalgam

Mercury exposure from dental amalgams has provoked concerns about subclinical or unusual neurologic effects ranging from subjective complaints, such as chronic fatigue, to demyelinating neuropathies, including multiple sclerosis. Although amalgam fillings have been suspected of causing clinical toxicity since they were introduced, studies have been hampered by insensitive analytic techniques and idiosyncratic outcome measures. Although dental amalgams are a source of mercury exposure and are associated with slightly higher urinary mercury, an expert panel for the National Institutes of Health (NIH) has concluded that existing evidence indicates dental amalgams do not pose a health risk and should not be replaced merely to decrease mercury exposure.

Mercury poisoning from dental amalgam through a direct nose-brain transport was described in humans by Stortebecker [84]. In contrast, Maas with coworkers [85] concluded that their results do not support the hypothesis of a significant flow of mercury from dental amalgam fillings to the cranial cavity by a direct oro-nasal route. These authors measured mercury concentration in the olfactory bulb and the trigeminal ganglion in 55 deceased persons and have emphasized that content of mercury in olfactory bulb, the trigeminal ganglion, and the pituitary gland did not correlate with the number of dental amalgam fillings. However, in the olfactory bulb, the mercury concentration was significantly higher (mean 17.4 µg/kg wet weight) than in the occipital lobe cortex (9.2 µg /kg w. w.). Mercury was also found in the pituitary gland (30.0 µg/kg w. w.). The highest mercury concentrations (93.1 µg/kg w. w.) were detected in the kidney cortex. In contrast, a statistically significant correlation was found between the number of dental amalgam fillings and the mercury concentration in the kidney cortex [85].

The olfactory bulb contains powerful detoxification systems to neutralize a wide range of toxins, including mercury [86]. Those with intact olfactory bulb detoxification would have lower mercury levels in the olfactory tracts and entorhinal areas of the brain. It has been conclusively shown that mercury in the form of vapors readily enters the olfactory nerves, travels to the olfactory tract and enters the entorhinal area and then hippocampus, amygdala and prefrontal cortex [87,88]. Studies have also shown that the greatest toxicity (by release of frontal glutamate) was with the lower dose of mercury, not the highest.

In contrast, several authors and institutions have emphasized that exposure to mercury from amalgam dental fillings is at low levels and that it is insufficient to adversely affect human health [16]. FDA also considers dental amalgam fillings safe for adults and children ages 6 and above [49]. FDA based the statement on the randomized clinical trial of DeRouen *et al.* [89]. In this study, a total of 507 children in Lisbon, Portugal, aged 8 to 10 years were over 7 years of follow-up investigated for one group receiving amalgam restorations for posterior lesions (n = 253) and the other group receiving resin composite restorations instead of amalgam (n = 254). During the seven-year trial period, children had a mean of 18.7 tooth surfaces restored in the amalgam group and 21.3 restored in the composite group. There were no statistically significant differences in measures of memory, attention, visuomotor function, or nerve conduction velocities for the amalgam and composite groups over all seven years of follow-up. The authors concluded that these findings suggest that amalgam should remain a viable dental restorative option for children.

Bellinger with co-workers conducted a randomized controlled trial involving 534 6- to 10-year-old urban and rural children who were assessed yearly for five years using a battery of tests of intelligence, achievement, language, memory, learning, visual-spatial skills, verbal fluency, fine motor function, problem solving, attention, and executive function [16]. These authors found that exposure to elemental mercury in amalgam at the levels experienced by the children who participated in the trial did not result in significant effects on neuropsychological function within the five-year follow-up period. Clarkson demonstrated that there could be a long delay between mercury exposure and symptoms—as much as five months following exposure [90].

Methylmercury and Ethylmercury

The mean half-life for methylmercury in blood is 40 to 50 days for adults. The primary mechanism for excreting mercury from the body involves binding to glutathione and then being excreted in the bile. Ninety percent of methylmercury is excreted through bile in feces [5,50]. Infants are especially vulnerable to mercury poisoning because they are poor excretors of mercury.

Ethylmercury, although it may have similar toxicity to methylmercury, has been less studied. The proponents of thimerosal safety have argued that ethylmercury does not enter the brain and that it is rapidly eliminated from the body:

> *"Methylmercury is a small molecule that can get into the brain and takes almost TWO MONTHS to break down... High concentrations of methylmercury can be found in tuna, swordfish and shark from contaminated waters. Now, let's contrast that with Ethylmercury, which is/was the type of mercury used in vaccine preservatives. Ethylmercury (thimerosal is an example) is rapidly eliminated from the body within a WEEK. Compared to methylmercury, ethylmercury is a much larger molecule that cannot enter the brain"[91].*

Animal experiments as well as *post mortem* investigations of persons after ethylmercury poisoning clearly document that the ethylmercury as well as the ethylmercury component of thimerosal is rapidly taken up in the organs (blood, kidney, liver, and brain), and to a large degree is degraded to inorganic mercury. It remains in the brains of infant monkey for several weeks [70]. Additionally, it was previously reported that, as a result of the significant Hg^{2+} fraction of mercury observed in the brain following injection of thimerosal, a longer biological retention was observed for the mercury deposited in the brain from ethylmercury than for the mercury from methylmercury [12]. The comparison of methylmercury and ethylmercury toxicity was done in studies with laboratory animals [64,70,71]. Knowledge of the toxicokinetics and developmental toxicity of thimerosal is needed to afford a meaningful assessment of the developmental effects of thimerosal-containing vaccines.

Major obstacles for estimation of a threshold dose for both methylmercury and ethylmercury include the limited knowledge of cellular and molecular processes underlying the pathological and developmental neurological changes and the delayed appearance of the neurodevelopmental effects following prenatal exposure. In this respect, a strategy which aims at identifying sensitive molecular targets of methyl- vs ethylmercury at cellular or molecular relevant levels may prove particularly useful to risk assessment. According to the information from laboratory investigations we could conclude that for mercury and mercury compounds there is no value for NOAEL and that every minute amount of mercury can potentially evoke an adverse health effect.

The Pros and Cons of Thimerosal

Several studies thus appeared in the last decade to show vaccine safety and benefits of vaccination and warning that vaccine refusal not only increases the individual risk of disease but also increases the risk for the whole community [92-97]. For example, Heron *et al.* [94] concluded that they could find no convincing evidence that early exposure to thimerosal had any deleterious effect on neurologic or psychological outcome in their prospective cohort study in the UK. The study was monitoring >14,000 children who are from the geographic area formerly known as Avon, UK, and were delivered in 1991-1992. The authors concluded that contrary to expectation, it was common for the unadjusted results to suggest a beneficial effect of thimerosal exposure. For example, exposure at 3 months was inversely associated with hyperactivity and conduct problems at 47 months; motor development at 6 months and at 30 months; difficulties with sounds at 81 months; and speech therapy, special needs, and "statementing" at 91 months.

Andrews and co-workers [93] presented a retrospective cohort study using 109 863 children who were born from 1988 to 1997 and were registered in general practices in the UK that contributed to a research database [93]. This study was funded by the WHO grant and it was designed to investigate whether there is a relationship between the amount of thimerosal that an infant receives via diphtheria-tetanus-whole-cell pertussis (DTP) or diphtheria-tetanus (DT) vaccination at a young age and subsequent neurodevelopmental disorders. The disorders investigated were general developmental disorders, language or speech delay, tics, ADD, autism, unspecified developmental delays, behavior problems, encopresis, and enuresis. Exposure was defined according to the number of DTP/DT doses received by 3 and 4 months of age and also the cumulative age-specific DTP/DT exposure by 6 months. Each DTP/DT dose of vaccine contained 50 µg of thimerosal (25 µg of ethylmercury). Hazard ratios for the disorders were calculated per dose of DTP/DT vaccine or per unit of cumulative DTP/DT exposure. The authors concluded that there was no evidence of a rise in blood concentrations above "safe values" and showed that mercury in ethylmercury is eliminated rapidly via the stools. Accordingly to authors, this provided additional evidence that three doses of DTP given at monthly intervals do not provide evidence that thimerosal exposure via DTP/DT vaccines causes neurodevelopmental disorders.

A remarkable study on the safety of thimerosal from vaccine was published in New England Journal of Medicine in 2007. Thompson *et al.* [98] enrolled 1047 children between the ages of 7 and 10 years and administered standardized tests assessing 42 neuropsychological outcomes. The authors did not assess ASD and children with ASD or ADHD were not involved in this study. The median cumulative exposure to mercury from thimerosal from birth to 7 months was 112.5 µg; less than 11% of children were exposed to thimerosal prenatally through maternal vaccination or receipt of immune globulins. Surprisingly, the authors found that among boys, higher prenatal mercury exposure was associated with significantly better performance on the Stanford–Binet copying test and poorer performance on the WISC-III digit-span test of backward recall. It means that, according to authors, higher prenatal mercury exposure was associated with better performance on one measure of language and poorer performance on one measure of attention and executive functioning. Among girls, there were no significant associations. Increasing exposure to mercury during the neonatal period (birth to 28 days) was related to significantly poorer performance on the GFTA-2 measure of speech articulation. Among boys there was a significant positive association with performance IQ, and among girls there was a significant negative association with verbal IQ. Among boys, higher exposure to mercury from birth to 7 months was associated with significantly better performance on letter and word identification, and a higher likelihood of motor and phonic tics, as reported by the children's evaluators. In conclusion, this study did not support a causal association between early exposure to mercury from thimerosal-containing vaccines and deficits in neuropsychological functioning at the age of 7 to 10 years and demonstrated the benefits of ethylmercury on development of children' intelligence.

Researchers from the US and Argentina conducted a study to assess blood levels and elimination of ethylmercury after immunization of infants with thimerosal-containing vaccines [92,95]. The study was conducted in Buenos Aires, where thimerosal continues to be used as a preservative in infant vaccines. Healthy infants from three age cohorts (newborns, two-month-olds, and six-month-olds) were enrolled in the project. There were 72 children in each age cohort. All children received age-appropriate vaccines as routinely administered in Argentina. Samples of blood, urine, and stool were obtained on study children both before vaccination and at one randomly assigned time interval after vaccination, with time intervals after vaccination ranging from 12 hours to 30 days. Mercury pharmacokinetics were estimated based on a one-compartment model that averaged all accounted for baseline mercury levels, ethylmercury dosage, and timing of vaccination. They estimated the blood half-life of mercury after administration of thimerosal to be 3.7 days, which did not vary significantly by age group. These authors concluded that administration of vaccines containing thimerosal does not seem to raise blood concentrations of mercury above safe values in infants. Ethylmercury seems to be eliminated from blood rapidly via the stools after parenteral administration of thimerosal in vaccines.

However, this should be accepted with caution. As discussed later in this chapter, small exposures to mercury increases the release of glutamate within the brain and this can increase cerebral activity, which may explain the tics and excitability see by some observers. This excess brain glutamate stimulation may improve some functions and harm others. For example, it may improve memory but interfere with attention. Behavioral problems, such as depression and anxiety have been linked to abnormalities in prefrontal glutamate levels. Most of these studies were done on very young children. A clearer discernment of cerebral dysfunction may become evident as higher order brain function comes into play later in life.

Unfortunately, Pichichero's sample size was small and the study may not have included infants with increased susceptibility. However, the newborn presents several physiological degrees of immaturity in the excretory system (kidneys and bile formation) and CNS. These features are inversely accentuated by gestational age and birth weight. Under such circumstances, unbound circulating ethylmercury in a newborn (and immature) may not be eliminated as fast as in older baby and thus will be more likely to cross the more vulnerable BBB. The newborn BBB increases in effectiveness with age; therefore, the free ethylmercury can more easily penetrate the immature CNS [99]. Stajich *et al.* [100] showed that preterm infants do not metabolize mercury efficiently. These authors measured total mercury levels before and after the administration of hepatitis B vaccine in 15 preterm and 5 term infants. Comparison of pre- and post-vaccination mercury levels showed a significant increase in both preterm and term infants after vaccination. Additionally, post-vaccination mercury levels were significantly higher in preterm infants as compared with term infants. Dorea *et al.* [101] studied the exposure of newborns in Brazil to ethylmercury from hepatitis B vaccines; hospital records (21,685) were summarized for the years 2001 to 2005 regarding date of birth, vaccination date, and birth weight. Over the 5 years, there was an increase in vaccinations from 7.4 % in 2001 to 87.8% in 2005. Nearly 94.6% of infants are now being vaccinated within the first 24 hours. Range of mercury exposure spread from 4.2 to 21.1 µg mercury/kg body weight. It is evident that further study of ethylmercury pharmacodynamics in infants is warranted.

From these studies it seems that thimerosal-containing vaccines have no record of overt clinical neurological consequences due to ethylmercury and that sufficient evidence about thimerosal safety exists. It is interesting to note that the crucial study, which demonstrates a significant positive association of ethylmercury from thimerosal in prenatal and early development with performance IQ in boys and no causal association with deficits in neuropsychological disorders [98], was supported by the CDC and most of pharmacological companies producing vaccines.

Dr. Thompson reports being a former employee of Merck; Dr. Marcy, receiving consulting fees from Merck, Sanofi Pasteur, GlaxoSmithKline, and MedImmune; Dr. Jackson, receiving grant support from Wyeth, Sanofi Pasteur, GlaxoSmithKline, and Novartis, lecture fees from Sanofi Pasteur, and consulting fees from Wyeth and Abbott and serving as a consultant to the FDA Vaccines and Related Biological Products Advisory Committee; Dr. Lieu, serving as a consultant to the CDC Advisory Committee on Immunization Practices; Dr. Black, receiving consulting fees from MedImmune, GlaxoSmithKline, Novartis, and Merck and grant support from MedImmune, GlaxoSmithKline, Aventis, Merck, and Novartis; and Dr. Davis receiving consulting fees from Merck and grant support from Merck and GlaxoSmith-Kline.

Can scientists agree that "the immunization system in the US remains healthy and intact"[47]? Does it really correspond to all knowledge and evidence collected by numerous researchers during many decades?

Thimerosal induces in genetically susceptible mice a systemic autoimmune syndrome [102]. Hornig and co-workers hypothesized that autoimmune propensity influences outcomes in mice following thimerosal challenges that mimic routine childhood immunizations [103]. Autoimmune disease-sensitive SJL/J mice showed growth delay; reduced locomotion; exaggerated response to novelty; and densely packed hyperchromic hippocampal neurons with altered GluRs and transporters. Strains resistant to autoimmunity, C57BL/6J and BALB/cJ, were not susceptible. These findings implicate genetic influences and provide a model for investigating thimerosal-related neurotoxicity.

The well-established negative effects of mercury on the immune system led to a study examining whether natural immunomodulator glucan can overcome the immunosuppressive effects of mercury [104]. Thimerosal was administered in a dose of 2-8 mg/L of drinking water to mice. After 2 weeks, all mice exhibited profound suppression of both cellular (phagocytosis, natural killer cell activity, and mitogen-induced proliferation) and humoral (antibody formation and secretion of interleukin-6 (IL-6), IL-12, and interferon-γ) responses. The mice were then fed with a diet containing a standard dose of glucan. This study showed that simultaneous treatment with mercury and glucan resulted in significantly lower immunotoxic effects of mercury, which suggests that glucans can be successfully used as a natural remedy of low-level exposure to mercury [104].

Thimerosal, as a known inhibitor of intracellular Ca^{2+} receptors, affects the onset of neurogenesis in mouse neocortex [105]. It can be expected that thimerosal's effects on calcium homeostasis could affect the activity of neuronal cells on a cellular and molecular level, with impacts on animal behavior. Mercuric chloride (HgCl2) is a highly toxic compound that inhibits glutamate uptake in astrocytes, resulting in excessive extracellular glutamate accumulation, leading to excitotoxicity and neuronal cell death. Mutkus *et al.* [106] demonstrated that thimerosal accumulation in the brain also contributes to dysregulation of glutamate homeostasis. In another study it has been shown that thimerosal in 1 nM-10 µM concentrations in a human neuroblastoma cell line could alter nerve growth factor (NGF)-induced signaling and that NGF provides protection against thimerosal cytotoxicity. Following 24-h exposure to thimerosal, the EC50 for cell death was 38.7 nM, after 48-h exposure the EC50 for cell death was 4.35 nM [107]. It is evident, that thimerosal exerts its biological activity on the cell level in extremely low concentrations.

THE EFFECTS OF MERCURY EXPOSURE ON CHILDHOOD DEVELOPMENT

Dental Amalgam Fillings

While FDA has concluded, as mentioned above, that infants are not at risk for adverse health effects from the breast milk of women exposed to mercury vapors from dental amalgam, several European countries have guidelines suggesting that women should not receive mercury-containing dental amalgam fillings during pregnancy. On the other hand, some recent studies have found no evidence for harm from dental amalgam for repair of dental caries. However, in a large cohort analysis of British children born in 1991-92 (n = 7,375) maternal dental care during pregnancy, including amalgam fillings, was not associated with birth outcomes such as preterm delivery and low birth weight, or language development at age 15 months [108].

A population-based, case-control study was designed to investigate whether placement of mercury-containing fillings in 1993-2000 during pregnancy increased the low-birth-weight risk. Cases and controls were sampled from enrollees of a dental insurance plan with live singleton births in Washington State; 1,117 women with low-birth-weight infants (less than 2,500 g) were compared with a random sample of 4,468 women with infants weighing 2,500 g or more [109]. This study found that having had a dental amalgam filling was not a risk factor for low birth weight, even among women who had up to 11 amalgam fillings placed during the course of pregnancy.

Drasch *et al.* [110] investigated the total mercury concentrations in the liver, the kidney cortex, and the cerebral cortex of 108 children aged 1 day-5 years, and of 46 fetuses. The mercury concentration in the kidney and liver of fetuses and mercury concentration in the kidney and the cerebral cortex of older infants (11-50 weeks of life) correlated significantly with the number of dental amalgam fillings of the mother. These authors thus recommended that future discussion on the pros and cons of dental amalgam should also include fetal exposure.

Palkovicova and co-workers studied maternal amalgam dental fillings as the source of mercury exposure in developing fetus and newborn [17]. Cord blood mercury concentrations were used as the biomarker of prenatal exposure to mercury. The median values of mercury concentrations were 0.63 µg /L (range 0.14-2.9 µg /L) and 0.80 µg /L (range 0.15-2.54 µg /L) for maternal and cord blood, respectively. None of the cord blood mercury concentrations reached the level considered to be hazardous for neurodevelopmental effects in children exposed to mercury *in utero* (e.g., EPA reference dose for Hg of 5.8 µg /L in cord blood). However, a strong positive correlation between maternal and cord blood mercury levels was found. Levels of mercury in the cord blood were significantly associated with the number of maternal amalgam fillings (which ranged from 0 to 20, mean 5.6) and with the number of years since the last filling; these associations remained significant after adjustment for maternal age and education. Palkovicova *et al.* recommend that dental amalgam fillings in girls and women of reproductive age should be used with caution, to avoid increased prenatal Hg exposure.

Methylmercury and Ethylmercury

The clinical findings in the victims of the Japanese and Iraqi outbreaks revealed the particular sensitivity of the fetus to the toxic effects from methylmercury and ethylmercury exposure [59,60,111]. In the Minamata Bay disaster and the Iraq epidemic, mothers who were asymptomatic or showed mild toxic effects gave birth to severely affected infants. Typically, infants appeared normal at birth, but psychomotor retardation, blindness, deafness, and seizures developed throughout time, though the kidneys and immune system may also be affected. Signs of toxicity from acute exposure progress from paresthesias and ataxia to generalized weakness, visual and hearing impairment, and tremor and muscle spasticity to coma and death. The Iraq study involved higher exposures and less sensitive measures of neurodevelopmental outcomes, compared with the prospective epidemiologic studies in the Seychelles and in the Faroe Islands.

Methylmercury is a known teratogen in the fetal brain. In the developing brain, methylmercury is toxic to the cerebral and cerebellar cortex, causing focal necrosis of neurons and destruction of glial cells; it interferes with neuronal migration and the organization of brain nuclei and layering of the cortical neurons. Methylmercury also crosses the placenta. Cord blood mercury levels are equal to or higher than maternal levels. Methylmercury appears in human milk following ingestion.

Studies in fish eating populations have identified adverse neurological and developmental outcomes, but these findings have not been consistent. Human neurodevelopmental consequences of exposure to methylmercury from eating fish remain a question of public health concern. The Seychelles study enrolled 711 mother - child pairs at birth and monitored mercury levels in mothers' hair and in children's hair at 6, 19, and 66 months of age as well as standardized measures of global neurobehavioral function of children at these times [112]. The mean maternal hair total mercury level was almost the same as the mean child hair total mercury level at age 66 months (6.8 ppm and 6.5 ppm), no adverse outcomes at 66 months were associated with either prenatal or postnatal mercury exposure.

In contrast, results from the Faroe Islands study suggested that exposure *in utero* to mercury at lower levels is associated with subtle adverse effects on the developing brain. Memory, attention, and language tests were inversely associated with higher methylmercury exposures in children up to 7 years of age, even after controlling for PCB exposures. Motor function and visual spatial ability were less clearly associated with methylmercury exposure. Adverse effects on development or IQ have not been found in the Seychelles Child Development Study (SCDS) at up to 66 months of age, although exposures were in the same range as the Faroe Islands study [33, 112-114].

A prospective, longitudinal main study with more covariates and expanded endpoints was begun on a new cohort of 779 children. No association with neurodevelopment was seen at 6 1/2, 19, or 29 months of age, but there was an inverse relationship at 29 months in boys only between mercury level and activity as judged by the examiner. In a related study, 32 brains were obtained at autopsy from Seychellois infants. These were examined histologically and analyzed for mercury. No clear histological abnormalities were found. Mercury levels ranged from a background of about 50 ppb up to 300 ppb, and correlated well between brain regions. For 27 of the infants used in the brain analysis study, maternal hair after delivery was available and maternal hair mercury correlated well with brain mercury [34,115].

Results of the Faroe Islands study were used by the National Academy of Sciences (NAS) to establish a reference dose for mercury of 0.1 mg/kg/day. One question that is raised by the difference in findings between the Seychelles and Faroe Islands studies is whether bolus doses of methylmercury administered during sensitive time periods are more likely to cause neurodevelopmental damage than the same doses given cumulatively throughout a time period of several months.

Daniels *et al.* [116] evaluated the association between maternal fish intake during pregnancy and offspring's early development of language and communication skills in a cohort of 7421 British children born in 1991-1992. Fish intake by the mother and child was measured by questionnaire. The child's cognitive development was assessed using adaptations of the MacArthur Communicative Development Inventory at 15 months of age and the Denver Developmental Screening Test at 18 months of age. Mercury was measured in umbilical cord tissue for a subset of 1054 children. In this study, the total mercury concentrations were low and were not associated with neurodevelopment. Fish intake by the mother during pregnancy and by the infant postnatally, was associated with higher mean developmental scores. Daniels with coworkers therefore concluded that when fish is not contaminated, moderate fish intake during pregnancy and infancy may benefit development.

THE POSSIBLE ROLE OF MERCURY IN ASD

Autism children are not unique in being exposed to a number of sources of both metals, but there may be a difference in the metabolism and/or vulnerability seen in the ASD child not seen in the normally developing child. This difference has not been worked out, but we have some hints from experimental studies. For example, certain strains of mice appear to be more susceptible to mercury induced autoimmunity than others [117]. ASD has been linked genetically to autoimmune susceptibility. Low-dose mercury exposure has been shown to accelerate and increase mortality in spontaneous lupus in mice [118].

The Role of Mercury in Developmental Brain Damage

There are also suggestions concerning mercury clearance and metallothionein (MT) physiology. Besides the observed lower levels of MT soon after birth, saturation of available MT with mercury could interfere with the normal anti-inflammatory function of this molecule [119]. MT-1 and MT-2, both of which are normally increased during mercury exposure, have been shown to play a major protective role against excitotoxicity [120].

Other differences in susceptibility to ethylmercury toxicity could include variations in GluRs physiology, alterations in antioxidant enzymes, interaction with other environmental neurotoxins, and differences in glutathione levels, all of which have been shown in ASD cases. Thus there are a number of reasons for differences in vulnerability among children exposed to the various forms of mercury, including ethylmercury.

Of considerable importance, and often ignored in many studies, is the total body burden of mercury and the distribution of the various forms of mercury. We have seen that children are exposed to a number of sources of mercury, including maternal dental amalgam fillings during pregnancy, atmospheric mercury, vaccines containing ethylmercury, methylmercury from contaminated seafood and exposure to mercury-containing pesticides and fungicides.

The dealkylation of ethylmercury and methymercury has been shown to occur within the brain itself and not peripherally, and appears to occur mostly in astrocytes and possibly microglia [121]. The dealkylation process occurs slowly and there is some evidence that once organic mercury is converted to inorganic mercury (Hg^{2+}) it is redistributed in the brain. Shapiro and Chan have shown that oxidative stress increases demethylation [122]. This would indicate that in the ASD brain, due to the high levels of oxidative stress, one would see even greater conversion of ethylmercury and methylmercury to Hg^{2+} than would occur under less inflammatory/oxidative stressed conditions.

Burbacher and co-workers have shown that a larger percentage of ethylmercury (34%) than methylmercury (7%) is converted in the brain into inorganic mercury [70]. They found that the total mercury in the thimerosal-exposed monkeys was 2.6 to 4.6 fold higher in the brain than in the blood, so a considerable amount of mercury was being transferred to the brain. Interestingly, measures of ethylmercury exposed monkeys demonstrated a considerably higher brain inorganic mercury level than was seen in the methylmercury-exposed monkeys; 21 to 86% vs 6 to 10%, respectively. Proponents of thimerosal safety suggest that ethylmercury from vaccines is safer than from methylmercury found in seafood because of the shorter blood half-life of ethylmercury—6.9 vs. 19.1 days. They also note that organic forms of mercury cleared much faster with thimerosal than did methylmercury exposures. What is ignored by these reports is that the proportion of Hg^{2+} in the brain was significantly higher in the thimerosal-exposed animals as opposed to the methylmercury-exposed animals—71% vs. 10%, respectively. In essence the absolute level of Hg^{2+} in the brain of the thimerosal-exposed animals was approximately twice that of the methylmercury-exposed animals.

Charleston and Vahter found that despite earlier suggestions that Hg^{2+} conversions were neuroprotective, conclusive evidence indicated that it was neurodestructive [121,123]. Testing 6 months after exposure to the organic mercury, brain Hg^{2+} levels were higher, indicating a slow conversion of organic mercury to Hg^{2+}, which again has been shown to be more neurotoxic than organic forms. Of critical importance is the prolonged half-life of Hg^{2+} in the brain, which has been estimated to be from 227 days to 540 days. With repeated exposure to various sources of organic mercury, brain levels of Hg^{2+} would continue to rise. Burbacher's study found that dealkylation of ethylmercury was more extensive than that of methylmercury, which would make the former more of a danger to the developing brain.

Effect of Mercury on Neurons, Microglia and Astrocytes

Early on, there was a controversy as to which cell type in the brain accumulates the highest concentration of mercury. Shanker and co-workers determined that the astrocyte was the main site of mercury accumulation [124]. In a subsequent study, using mono- and co-cultures of astrocytes, neurons and mixed cultures, they found that in neuronal mono-cultures, neurons accumulated methylmercury faster and to a greater extent than did astrocytes in mono-culture, especially at higher exposures [125]. When placed in co-cultures containing astrocytes, neuronal mercury levels dropped by one third. They also found that astrocytes were damaged at a lower concentration of methylmercury than were neurons. This is in keeping with Charleston and co-workers observation that there was a significant drop out of astrocytes at 6 months post-exposure in the brains of monkeys exposed chronically to methylmercury [126].

Several workers have found that exposure to mercury dramatically activates glia in the brain, both microglia and astrocytes [127,128]. While most mercury is distributed to the cortical and subcortical grey matter, studies by Warfvinge and co-workers have shown that it also accumulates along white matter fiber tracts [129]. They found greater accumulation of mercury in pyramidal cells of the hippocampus, in the amygdaloid complex and claustrum, as well as along fiber tracts. As previously reported they found significant accumulation of mercury in microglia, both in the grey matter and fiber tracts.

Methylmercury is also known to affect the development of microvasculature and the BBB in the developing brain. In one study, exposure of chick cerebellar embryos to methylmercury was shown to cause significant changes in the cerebellar blood vessels, including immature morphology, poor differentiation of endothelial barriers, and high permeability of the vessels to exogenous proteins [130].

Studies of the effect of mercury on astrocytes have disclosed some interesting mechanisms of neurotoxicity. For example, Shanker and co-workers have shown that methylmercury inhibits cysteine uptake by astrocytes by inhibiting its main transporters, AG(-)and ASC cystine transport system [131]. Entry of cysteine and cystine into the astrocyte is essential for formation of glutathione. Low glutathione greatly increases the vulnerability of neurons to a number of stresses and astrocytes are the major source for neuronal GSH. Thimerosal has been shown to lower neuronal GSH levels and increase neurotoxicity [132]. Likewise, elevation of GSH is significantly protective against thimerosal neurotoxicity.

While mercury is known to interact and inhibit a number of enzymes, of particular importance is the inhibition of glutamine synthetase, an enzyme that regulates extraneuronal glutamate by converting glutamate into glutamine within the astrocyte. Ionic form of mercury has been shown to inhibit this critical enzyme [133]. Dysfunction of this enzyme can lead to excitotoxicity.

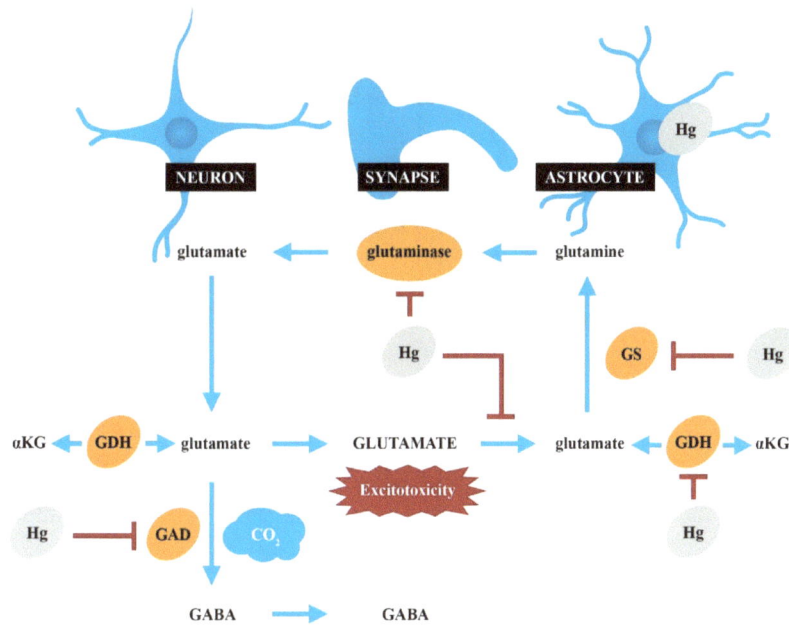

Figure 1: Schematic representation of adverse effects of mercury on glutamate/glutamine neurotransmitter cycling between astrocytes and neurons. Under physiological conditions, this cycle accomplishes removal of glutamate from the synapse via uptake into astrocytes, which convert glutamate to glutamine via glutamine synthetase (GS) and export glutamine to neurons. Mercury inhibits the transport of glutamate into astrocytes; accumulates in astrocytes and inhibits GS. It also inhibits glutaminase and glutamate decarboxylase (GAD). Dysfunction of these enzymes can lead to excitotoxicity. Mercury (Hg) means all forms – Hg^{2+}, methylmercury, and ethylmercury.

Mercury and Excitotoxicity

A number of studies have shown architectonic abnormalities in fetal brain development following maternal mercury exposure [134-136]. This can present as abnormalities in neuronal and glial proliferation, neuronal migration, and the final cytoarchitectonic composition of the brain, especially in the cerebellum. The question remains as to the exact mechanism responsible. There is compelling evidence that immunoexcitotoxicity is playing an essential if not central role in the neurotoxicity of mercury [13,137].

Several studies have suggested a significant role for glutamate excitotoxicity in mercury neurotoxicity. For example, Brookes demonstrated that Hg^{2+} (≤ 1 μM) markedly inhibited glutamate clearance, lowered glutamine content of astrocytes and initiated excitotoxicity [138]. It was concluded that mercury had little effect on neurotoxicity at these low levels in the absence of excitotoxicity.

In most of the studies, interference with glutamate uptake in astrocytes and microglia seem to be at play. Glutamate release by astrocytes, either through reverse transport or impairment of glutamate transport proteins is most often cited as a cause for extraneuronal glutamate elevation. Juarez and co-workers demonstrated mercury-induced glutamate release in freely moving rats [139]. Using sampling from a microdialysis probe implanted in the frontal cortex of Wistar rats, they were able to demonstrate a 9.8-fold increase in glutamate using a 10 μM dose of methylmercury. Interestingly, at the 100 μM dose there was only a 2.4-fold increase in frontal lobe glutamate, probably because of damage to release mechanism at this higher dose. Nanomolar concentrations of mercury have been shown to reduce glutamate transport efficiency by 50%.

It is also known that methylmercury can damage DNA in rat cortex. One study found that by injecting MK-801, a non-competitive NMDA receptor antagonist 1 hour before methylmercury injection, they could significantly attenuate the DNA damage [140]. Miyamoto and co-workers also demonstrated that NMDA receptors were playing a major role in methylmercury toxicity [141]. In the study, they administered oral methylmercury to rats on postnatal day 2 (P2), P16, and P60 (10 mg/kg) for 7 days. Giving MK-801 NMDA receptor antagonist at the same time markedly reduced the neurodegeneration seen when methylmercury was given alone. They also observed a marked reduction in brain nitrotyrosine residues, which are associated with elevations in peroxynitrite. Peroxynitrite is elevated in the course of excitotoxicity and plays a major role in mitochondrial suppression.

One of the most obvious neurotoxic effects of mercury is the generation of abundant amounts of ROS/RNS and LPP and the fact that antioxidants provide considerable protection against mercury-induced neurotoxicity [142]. Yet, a more complicated process may be in operation, since blocking NMDA receptors has been shown to significantly attenuate mercury neurotoxicity and reduces ROS/RNS generation as well [141]. Other studies have shown that free radicals can dramatically increase the toxicity of methylmercury, so that previously nontoxic concentrations become fully toxic, a synergism we also see with excitotoxins [143-145]. Therefore, we see a self-enhancing effect from mercury-induced excitotoxicity, with the excitotoxic process generating a storm of free radicals and these radicals enhancing sensitivity to the toxic effects of the mercury.

Peroxynitrite appears to be central to both mercury neurotoxicity and the excitotoxic process itself. This powerful radical is formed when superoxide, a rather weak radial, combines with nitric oxide (NO). Superoxide has a higher affinity for NO than it does for superoxide. Peroxynitrite has been shown to be a powerful inhibitor of mitochondrial function [146,147].

Mitochondrial dysfunction is known to dramatically enhance excitotoxicity to such an extent that even physiological levels of glutamate can become excitotoxic [148]. The mitochondria appear to be a natural sink for organic mercury. It is also known that both inflammatory cytokines and excess glutamate can inhibit mitochondrial migration within the dendrite along its normal path to the synapse. In combination, this indicates that immunoexcitotoxicity can significantly reduce energy generation within synapses and dendrites. In addition to energy production, mitochondria play a critical role in intracellular Ca^{2+} buffering, along with the smooth endoplasmic reticulum. Dysfunction of either or both of these systems can result in excessive calcium signaling, leading to free radical generation, lipid peroxidation, and activation of cellular death signals. Mercury, by enhancing calcium signaling, further exacerbates these problems, leading to abnormal neurogenesis, and neurodegeneration as well [149,150].

A number of studies have shown that mercury can significantly activate microglia [126,127,151]. While some studies found no connection between microglial activation and mercury neurotoxicity, or even neuroprotection, most evidence indicates that microglial activation over a prolonged period is neurodestructive [152,153]. Neither of the negative studies, both conducted by the same laboratory, is very convincing in eliminating microglia as a major player in mercury neurotoxicity. In one study, they rely on the neuroprotective effect of IL-6 in hypothesizing a similar neuroprotective role in mercury toxicity. While short-term elevation in IL-6 can be neuroprotective, prolonged activation, especially in conjunction with other pro-inflammatory cytokines and glutamate, have been shown to be neurodestructive. The other study assumes that because microglia are not found adjacent to the apoptotic neurons that they must not be associated with toxicity. Neurotoxicity at a distance from microglia is known and is secondary to the release of neurodestructive factors, such as glutamate, quinolinic acid (QUIN), aspartate, pro-inflammatory cytokines, and chemokines.

Because both glutamate and mercury can result in microglial priming, subsequent episodes of immune stimulation, either from vaccines, natural infections, exposure to immune-triggering environmental chemicals or heavy metals, can initiate a more intense and prolonged immunoexcitotoxic response. It should also be appreciated that while astrocytes are major repositories of glutamate, microglia when activated, can secrete large amounts of excitotoxic glutamate, QUIN, and aspartate. This is especially so under conditions of mitochondrial dysfunction, magnesium deficiency and hypoxia/ischemia.

With astrocytes being the major sink for mercury, eventual death of these glial cells can result in a massive release of extraneuronal glutamate, leading to acceleration of excitotoxicity. A loss of astrocytes, as shown by Charleston et al [126] and discussed above, occurs in cases of chronic exposure to methylmercury and would also be expected with ethylmercury as well. In addition, a significant loss of astrocytes would also mean a loss of glutamate buffering and other essential functions of astrocytes in the CNS and would have long-term consequences for adult neurological function.

Extraneuronal glutamate regulation, and thus brain protection, occurs by four basic mechanism: the excitatory amino acid transporters (EAAT1-5), the Xc^- cystine/glutamate antiporter, conversion of glutamate into glutamine by glutamine synthetase, and conversion of glutamate into metabolic products that enter the Kreb's cycle or generate GABA by glutamate dehydrogenase (GDH) and glutamic acid decarboxylase (GAD) (see Chapter 3). Inhibition of EAATs and the other enzymes is via oxidation, since all are redox sensitive. Antioxidants can reverse EAAT inhibition [143, 154]. It is also thought that because the EAAT transporters contain sulfhydryl groups, mercury can inhibit their function by binding to these groups. In addition, mercury inhibits protein kinase C, a vital enzyme in transporter function [155]. One of the protective effects of estrogen is its ability to enhance glutamate transport into the astrocyte [156].

Brookes demonstrated that inhibition of glutamate transport was reversible and the rate and degree of reversal depended on the concentration of the mercury [138]. Very low doses were rapidly and completely reversible and higher doses required longer to reverse. Kim and Choi demonstrated that the form of mercury determined the sensitivity to glutamate uptake inhibition [157]. They found that mercuric chloride (inorganic mercury) at 5 μM cause a 50% inhibition of glutamate uptake in cultured mouse astrocytes, whereas this level of inhibition was reached only at 10 μM using methylmercury chloride. Of other metals tested, $CuCl_2$, $FeCl_2$, $PbCl_2$ and $ZnCl_2$, only lead reduced glutamate uptake, but was less effective than mercury as an inhibitor.

Reverse transport of glutamate from astrocytes is also known to occur with mercury exposure. Mutkus and co-workers found, using Chinese hamster ovary cells transfected with glutamate transporters GLAST (EAAT1) and GLT-1 (EAAT2), a differential effect of methylmercury on these transporters. Methylmercury exposure led to dramatic reductions in glutamate uptake by GLAST and a significant increase activity of GLT-1 activity. In another study, inorganic mercury was shown to inhibit function of both transporters and appeared to do so by direct binding of mercury to transporter thiol groups [158].

There exist a programmed rise and fall in brain glutamate levels during neurodevelopment that plays an essential role in the brain's development. Abnormalities in this glutamate schedule can result in significant alterations in ultimate brain architecture. Kugler and Schleyer found that GLAST was expressed at higher levels earlier in brain development than GLT-1 in the rat hippocampus and that both glutamate transporters and glutamate dehydrogenase were increased after birth and attained adult levels between P20 and P30 [159]. This not only suggests a differential function for these glutamate transporters, but also indicates that this period of alteration in glutamate transporters is critical to successful brain development. Mercury has been shown to also suppress GDH function as well as glutamate transporter function [160].

It has been shown that Purkinje cells are quite dependent on GLAST and EAAT4 for its resistance against excitotoxicity induced by hypoxia/ischemia [161]. This could also explain the dramatic loss of Purkinje cells in autism, since mercury toxicity alone usually spares the Purkinje cells and targets cerebellar granule cells. That is, we are seeing a more global immunoexcitotoxic effect than just mercury toxicity alone, with widespread ROS/RNS, LPP damage, excitotoxicity, pro-inflammatory cytokine injury and enhancement of excitotoxicity, androgen amplification of damage and chronic microglial activation.

CONCLUSIONS

A review of medical literature, WHO, and US government data suggest that mercury is well recognized neurotoxin. Evidence also shows that mercury from the mother reaches the fetus. Prenatal and infantile exposure to mercury can result in neurological damage, mental retardation, incoordination, muscle weakness, seizures, and inability to speak. It is clear that while genetic factors are important to the pathogenesis of ASDs, mercury exposure can induce immune, sensory, neurological, motor, and behavioral dysfunctions similar to traits defining or associated with ASD. Patients with ASD have biochemical evidence of mitochondrial dysfunctions and decreased function in their glutathione pathways; they also excreted significant amounts of mercury post chelation challenge. Thimerosal has become one of the potential culprits in the pathogenesis of ASD. It can be expected that thimerosal's effects on Ca^{2+} homeostasis could affect the activity of neuronal cells on a cellular and molecular level, with impacts on behavior. Thimerosal accumulation in the brain might contribute to dysregulation of glutamate homeostasis. Moreover, thimerosal induces in genetically susceptible mice a systemic autoimmune syndrome. It is evident, that thimerosal exerts its toxicological effects on the cell level in extremely low concentrations.

Our review presents evidence that mercury in very low concentrations triggers both the inflammatory cascade and the excitotoxic cascade. The mechanism of immunoexcitotoxicity explains most of the features of the ASD, including the behavioral difficulties, language problems, repetitive behaviors, intellectual delay, and episodic dyscontrol of anger [137]. It also explains why ASD has not disappeared despite the reduction in ethylmercury exposure from most childhood vaccines, since excessive immune activation is the initiating and sustaining event in ASD. It is to be appreciated that mercury, even in submicromolar concentrations, can initiate excitotoxicity. The vaccination program should be evaluated to reduce the excessive stimulation of the immature immune system.

Another important factor with regard to the effects of mercury on the etiopathogenesis of ASD is **synergistic toxicity** – mercury's enhanced effect when other poisons are present. As a multifaceted disorder, ASD requires a multifaceted approach, one that should include the protection against excitotoxicity/microglial activation by mercury. Evidence for mercury contribution in the ethiopathogenesis of ASD should be considered in the search of therapeutic approaches and prevention.

REFERENCES

[1]   Echeverria D, Aposhian HV, Woods JS, *et al.* Neurobehavioral effects from exposure to dental amalgam Hg(o): new distinctions between recent exposure and Hg body burden. FASEB J 1998; 12: 971-80.
[2]   Clarkson TW, Magos L, Myers GJ. The toxicology of mercury--current exposures and clinical manifestations. N Engl J Med 2003; 349: 1731-7.
[3]   WHO. Mercury in Health Care: Policy Paper. 2005.
[4]   Nelson BK. Evidence for behavioral teratogenicity in humans. J Appl Toxicol 1991; 11: 33-7.
[5]   Goldman LR, Shannon MW, Health CoE. AAP Technical Report: Mercury in the environment: implications for pediatricians. Pediatrics 2001; 108: 197-205.
[6]   A Research-Oriented Framework for Risk Assessment and Prevention of Children's Exposure to Environmental Toxicants. Environ Health Perspect 1999; 107: 510.
[7]   Bernard S, Enayati A, Redwood L, Roger H, Binstock T. Autism: a novel form of mercury poisoning. Med Hypotheses 2001; 56: 462-71.
[8]   Bernard S, Enayati A, Roger H, Binstock T, Redwood L. The role of mercury in the pathogenesis of autism. Mol Psychiatry 2002; 7 Suppl 2: S42-3.
[9]   Clements CJ, Ball LK, Ball R, Pratt D. Thiomersal in vaccines. Lancet 2000; 355: 1279-80.
[10]   Geier MR, Geier DA. Mercury in vaccines and potential conflicts of interest. Lancet 2004; 364: 1217; author reply -8.
[11]   Sugarman SD. Cases in vaccine court--legal battles over vaccines and autism. N Engl J Med 2007; 357: 1275-7.
[12]   Geier DA, King PG, Sykes LK, Geier MR. A comprehensive review of mercury provoked autism. Indian J Med Res 2008; 128: 383-411.
[13]   Blaylock RL. A possible central mechanism in autism spectrum disorders, part 2: immunoexcitotoxicity. Altern Ther Health Med 2009; 15: 60-7.
[14]   Aschner M, Ceccatelli S. Are Neuropathological Conditions Relevant to Ethylmercury Exposure? Neurotox Res 2010; 18: 59-68.
[15]   Counter SA, Buchanan LH. Mercury exposure in children: a review. Toxicol Appl Pharmacol 2004; 198: 209-30.
[16]   Bellinger DC, Trachtenberg F, Zhang A, *et al.* Dental amalgam and psychosocial status: the New England Children's Amalgam Trial. J Dent Res 2008; 87: 470-4.
[17]   Palkovicova L, Ursinyova M, Masanova V, Yu Z, Hertz-Picciotto I. Maternal amalgam dental fillings as the source of mercury exposure in developing fetus and newborn. J Expo Sci Environ Epidemiol 2008; 18: 326-31.
[18]   Geier DA, Kern JK, Geier MR. A prospective blinded evaluation of urinary porphyrins verses the clinical severity of autism spectrum disorders. J Toxicol Environ Health A 2009; 72: 1585-91.
[19]   Clarkson TW, Vyas JB, Ballatori N. Mechanisms of mercury disposition in the body. Am J Ind Med 2007; 50: 757-64.
[20]   Hahn LJ, Kloiber R, Vimy MJ, Takahashi Y, Lorscheider FL. Dental "silver" tooth fillings: a source of mercury exposure revealed by whole-body image scan and tissue analysis. FASEB J 1989; 3: 2641-6.
[21]   Vimy MJ, Takahashi Y, Lorscheider FL. Maternal-fetal distribution of mercury (203Hg) released from dental amalgam fillings. Am J Physiol 1990; 258: R939-45.

[22]   Pleva J. Dental mercury--a public health hazard. Rev Environ Health 1994; 10: 1-27.

[23]   Olsson S, Bergman M. Daily dose calculations from measurements of intra-oral mercury vapor. J Dent Res 1992; 71: 414-23.

[24]   Olsson S, Berglund A, Bergman M. Release of elements due to electrochemical corrosion of dental amalgam. J Dent Res 1994; 73: 33-43.

[25]   Lorscheider FL, Vimy MJ, Summers AO, Zwiers H. The dental amalgam mercury controversy--inorganic mercury and the CNS; genetic linkage of mercury and antibiotic resistances in intestinal bacteria. Toxicology 1995; 97: 19-22.

[26]   Halbach S. Amalgam tooth fillings and man's mercury burden. Hum Exp Toxicol 1994; 13: 496-501.

[27]   Halbach S. [Amalgam: a risk assessment using a review of the latest literature through 2005]. Gesundheitswesen 2006; 68: e1-6; discussion e-15.

[28]   Mutter J, Naumann J, Guethlin C. Comments on the article "the toxicology of mercury and its chemical compounds" by Clarkson and Magos (2006). Crit Rev Toxicol 2007; 37: 537-49; discussion 51-2.

[29]   Clarkson T, Cox C, Davidson PW, Myers GJ. Mercury in fish. Science 1998; 279: 459, 61.

[30]   Andersen A, Julshamn K, Ringdal O, Morkore J. Trace elements intake in the Faroe Islands. II. Intake of mercury and other elements by consumption of pilot whales (Globicephalus meleanus). Sci Total Environ 1987; 65: 63-8.

[31]   Grandjean P, Weihe P, Jorgensen PJ, *et al.* Impact of maternal seafood diet on fetal exposure to mercury, selenium, and lead. Arch Environ Health 1992; 47: 185-95.

[32]   Davidson PW, Myers GJ, Cox C, *et al.* Longitudinal neurodevelopmental study of Seychellois children following in utero exposure to methylmercury from maternal fish ingestion: outcomes at 19 and 29 months. Neurotoxicology 1995; 16: 677-88.

[33]   Myers GJ, Davidson PW, Shamlaye CF, *et al.* Effects of prenatal methylmercury exposure from a high fish diet on developmental milestones in the Seychelles Child Development Study. Neurotoxicology 1997; 18: 819-29.

[34]   Myers GJ, Thurston SW, Pearson AT, *et al.* Postnatal exposure to methyl mercury from fish consumption: a review and new data from the Seychelles Child Development Study. Neurotoxicology 2009; 30: 338-49.

[35]   Weihe P, Grandjean P, Debes F, White R. Health implications for Faroe islanders of heavy metals and PCBs from pilot whales. Sci Total Environ 1996; 186: 141-8.

[36]   Weihe P, Grandjean P, Jorgensen PJ. Application of hair-mercury analysis to determine the impact of a seafood advisory. Environ Res 2005; 97: 200-7.

[37]   Jacobson JL, Jacobson SW. Intellectual impairment in children exposed to polychlorinated biphenyls in utero. N Engl J Med 1996; 335: 783-9.

[38]   Jacobson JL, Jacobson SW. Evidence for PCBs as neurodevelopmental toxicants in humans. Neurotoxicology 1997; 18: 415-24.

[39]   Jacobson JL, Jacobson SW. Prenatal exposure to polychlorinated biphenyls and attention at school age. J Pediatr 2003; 143: 780-8.

[40]   Rice DC. Overview of modifiers of methylmercury neurotoxicity: chemicals, nutrients, and the social environment. Neurotoxicology 2008; 29: 761-6.

[41]   Cetin I, Koletzko B. Long-chain omega-3 fatty acid supply in pregnancy and lactation. Curr Opin Clin Nutr Metab Care 2008; 11: 297-302.

[42]   Kaur P, Schulz K, Aschner M, Syversen T. Role of docosahexaenoic acid in modulating methylmercury-induced neurotoxicity. Toxicol Sci 2007; 100: 423-32.

[43]   Koletzko B, Lien E, Agostoni C, *et al.* The roles of long-chain polyunsaturated fatty acids in pregnancy, lactation and infancy: review of current knowledge and consensus recommendations. J Perinat Med 2008; 36: 5-14.

[44]   Skerfving S. Interaction between selenium and methylmercury. Environ Health Perspect 1978; 25: 57-65.

[45]   Geier DA, Geier MR. A case series of children with apparent mercury toxic encephalopathies manifesting with clinical symptoms of regressive autistic disorders. J Toxicol Environ Health A 2007; 70: 837-51.

[46]   Redwood L, Bernard S, Brown D. Predicted mercury concentrations in hair from infant immunizations: cause for concern. Neurotoxicology 2001; 22: 691-7.

[47]   Freed GL, Andreae MC, Cowan AE, Katz SL. The process of public policy formulation: the case of thimerosal in vaccines. Pediatrics 2002; 109: 1153-9.

[48]   Diner BM BB. Toxicity, Mercury. e-medicine from Web MD 2009; http://emedicine.medscape.com/article/819872-overview:

[49]   FDA. Issues Final Regulation on Dental Amalgam. FDA News Release 2009; http://www.fda.gov/NewsEvents/Newsroom/PressAnnouncements/ucm173992.htm:

[50]   Adams JB, Romdalvik J, Ramanujam VM, Legator MS. Mercury, lead, and zinc in baby teeth of children with autism versus controls. J Toxicol Environ Health A 2007; 70: 1046-51.

[51] Castoldi AF, Coccini T, Ceccatelli S, Manzo L. Neurotoxicity and molecular effects of methylmercury. Brain Res Bull 2001; 55: 197-203.

[52] Tsuda T, Yorifuji T, Takao S, Miyai M, Babazono A. Minamata disease: catastrophic poisoning due to a failed public health response. J Public Health Policy 2009; 30: 54-67.

[53] Harada M. Congenital Minamata disease: intrauterine methylmercury poisoning. Teratology 1978; 18: 285-8.

[54] Ekino S, Susa M, Ninomiya T, Imamura K, Kitamura T. Minamata disease revisited: an update on the acute and chronic manifestations of methyl mercury poisoning. J Neurol Sci 2007; 262: 131-44.

[55] Yorifuji T, Tsuda T, Takao S, Harada M. Long-term exposure to methylmercury and neurologic signs in Minamata and neighboring communities. Epidemiology 2008; 19: 3-9.

[56] Kondo K. Congenital Minamata disease: warnings from Japan's experience. J Child Neurol 2000; 15: 458-64.

[57] Eto K, Yasutake A, Kuwana T, *et al.* Methylmercury poisoning in common marmosets--a study of selective vulnerability within the cerebral cortex. Toxicol Pathol 2001; 29: 565-73.

[58] Jalili MA, Abbasi AH. Poisoning by ethyl mercury toluene sulphonanilide. Br J Ind Med 1961; 18: 303-8.

[59] Amin-Zaki L, Elhassani SB, Majeed MA, *et al.* Methylmercury poisoning in mothers and their suckling infants. Dev Toxicol Environ Sci 1980; 8: 75-8.

[60] Amin-Zaki L, Majeed MA, Greenwood MR, *et al.* Methylmercury poisoning in the Iraqi suckling infant: a longitudinal study over five years. J Appl Toxicol 1981; 1: 210-4.

[61] Zhang J. Clinical observations in ethyl mercury chloride poisoning. Am J Ind Med 1984; 5: 251-8.

[62] Nizov AA, Shestakov NM. [Clinical picture of subacute granosan poisoning]. Sov Med 1971; 34: 150-2.

[63] Cinca I, Dumitrescu I, Onaca P, Serbanescu A, Nestorescu B. Accidental ethyl mercury poisoning with nervous system, skeletal muscle, and myocardium injury. J Neurol Neurosurg Psychiatry 1980; 43: 143-9.

[64] Geier DA, Sykes LK, Geier MR. A review of Thimerosal (Merthiolate) and its ethylmercury breakdown product: specific historical considerations regarding safety and effectiveness. J Toxicol Environ Health B Crit Rev 2007; 10: 575-96.

[65] Engley F. Evaluation of mercurial compounds as antiseptics. Ann. NY Acad. Sci. 1950; 53: 197-206.

[66] Minami T, Miyata E, Sakamoto Y, *et al.* Expression of metallothionein mRNAs on mouse cerebellum microglia cells by thimerosal and its metabolites. Toxicology 2009; 261: 25-32.

[67] Yole M, Wickstrom M, Blakley B. Cell death and cytotoxic effects in YAC-1 lymphoma cells following exposure to various forms of mercury. Toxicology 2007; 231: 40-57.

[68] Alexandre H, Delsinne V, Goval JJ. The thiol reagent, thimerosal, irreversibly inhibits meiosis reinitiation in mouse oocyte when applied during a very early and narrow temporal window: a pharmacological analysis. Mol Reprod Dev 2003; 65: 454-61.

[69] Qvarnstrom J, Lambertsson L, Havarinasab S, Hultman P, Frech W. Determination of methylmercury, ethylmercury, and inorganic mercury in mouse tissues, following administration of thimerosal, by species-specific isotope dilution GC-inductively coupled plasma-MS. Anal Chem 2003; 75: 4120-4.

[70] Burbacher TM, Shen DD, Liberato N, *et al.* Comparison of blood and brain mercury levels in infant monkeys exposed to methylmercury or vaccines containing thimerosal. Environ Health Perspect 2005; 113: 1015-21.

[71] Harry GJ, Harris MW, Burka LT. Mercury concentrations in brain and kidney following ethylmercury, methylmercury and Thimerosal administration to neonatal mice. Toxicol Lett 2004; 154: 183-9.

[72] Surkan PJ, Wypij D, Trachtenberg F, *et al.* Neuropsychological function in school-age children with low mercury exposures. Environ Res 2009; 109: 728-33.

[73] Ewers U, Krause C, Schulz C, Wilhelm M. Reference values and human biological monitoring values for environmental toxins. Report on the work and recommendations of the Commission on Human Biological Monitoring of the German Federal Environmental Agency. Int Arch Occup Environ Health 1999; 72: 255-60.

[74] CDC. From the Centers for Disease Control and Prevention. Blood and hair mercury levels in young children and women of childbearing age--United States, 1999. JAMA 2001; 285: 1436-7.

[75] Bautista LE, Stein JH, Morgan BJ, *et al.* Association of blood and hair mercury with blood pressure and vascular reactivity. WMJ 2009; 108: 250-2.

[76] Wranova K, Cejchanova M, Spevakova V, *et al.* Mercury and methylmercury in hair of selected groups of Czech population. Cent Eur J Public Health 2009; 17: 36-40.

[77] Puklova V, Krskova A, Cerna M, *et al.* The mercury burden of the Czech population: An integrated approach. Int J Hyg Environ Health 2010; 213: 243-51.

[78] Lindberg A, Bjornberg KA, Vahter M, Berglund M. Exposure to methylmercury in non-fish-eating people in Sweden. Environ Res 2004; 96: 28-33.

[79] Gerhardsson L, Lundh T. Metal concentrations in blood and hair in pregnant females in southern Sweden. J Environ Health 2010; 72: 37-41.

[80] Dunn JE, Trachtenberg FL, Barregard L, Bellinger D, McKinlay S. Scalp hair and urine mercury content of children in the Northeast United States: the New England Children's Amalgam Trial. Environ Res 2008; 107: 79-88.

[81] Pesch A, Wilhelm M, Rostek U, *et al.* Mercury concentrations in urine, scalp hair, and saliva in children from Germany. J Expo Anal Environ Epidemiol 2002; 12: 252-8.

[82] Diez S, Delgado S, Aguilera I, *et al.* Prenatal and early childhood exposure to mercury and methylmercury in Spain, a high-fish-consumer country. Arch Environ Contam Toxicol 2009; 56: 615-22.

[83] Magos L, Clarkson TW. Overview of the clinical toxicity of mercury. Ann Clin Biochem 2006; 43: 257-68.

[84] Stortebecker P. Mercury poisoning from dental amalgam through a direct nose-brain transport. Lancet 1989; 1: 1207.

[85] Maas C, Bruck W, Haffner HT, Schweinsberg F. [Study on the significance of mercury accumulation in the brain from dental amalgam fillings through direct mouth-nose-brain transport]. Zentralbl Hyg Umweltmed 1996; 198: 275-91.

[86] Watelet JB, Strolin-Benedetti M, Whomsley R. Defence mechanisms of olfactory neuro-epithelium: mucosa regeneration, metabolising enzymes and transporters. B-Ent 2009; 5 Suppl 13: 21-37.

[87] Tjalve H, Henriksson J. Uptake of metals in the brain via olfactory pathways. Neurotoxicology 1999; 20: 181-95.

[88] Yasutake A, Sawada M, Shimada A, Satoh M, Tohyama C. Mercury accumulation and its distribution to metallothionein in mouse brain after sub-chronic pulse exposure to mercury vapor. Arch Toxicol 2004; 78: 489-95.

[89] DeRouen TA, Martin MD, Leroux BG, *et al.* Neurobehavioral effects of dental amalgam in children: a randomized clinical trial. Jama 2006; 295: 1784-92.

[90] Clarkson TW. The three modern faces of mercury. Environ Health Perspect 2002; 110 Suppl 1: 11-23.

[91] Brown. A. Clear Answers and Smart Advice About Your Baby's Shots. Immunization Action Coalition. 2008;

[92] Pichichero ME, Cernichiari E, Lopreiato J, Treanor J. Mercury concentrations and metabolism in infants receiving vaccines containing thiomersal: a descriptive study. Lancet 2002; 360: 1737-41.

[93] Andrews N, Miller E, Grant A, *et al.* Thimerosal exposure in infants and developmental disorders: a retrospective cohort study in the United kingdom does not support a causal association. Pediatrics 2004; 114: 584-91.

[94] Heron J, Golding J. Thimerosal exposure in infants and developmental disorders: a prospective cohort study in the United kingdom does not support a causal association. Pediatrics 2004; 114: 577-83.

[95] Pichichero ME, Gentile A, Giglio N, *et al.* Mercury levels in newborns and infants after receipt of thimerosal-containing vaccines. Pediatrics 2008; 121: e208-14.

[96] Talbot HK, Keitel W, Cate TR, *et al.* Immunogenicity, safety and consistency of new trivalent inactivated influenza vaccine. Vaccine 2008; 26: 4057-61.

[97] Omer SB, Salmon DA, Orenstein WA, deHart MP, Halsey N. Vaccine refusal, mandatory immunization, and the risks of vaccine-preventable diseases. N Engl J Med 2009; 360: 1981-8.

[98] Thompson WW, Price C, Goodson B, *et al.* Early thimerosal exposure and neuropsychological outcomes at 7 to 10 years. N Engl J Med 2007; 357: 1281-92.

[99] Dorea JG. Exposure to mercury during the first six months via human milk and vaccines: modifying risk factors. Am J Perinatol 2007; 24: 387-400.

[100] Stajich GV, Lopez GP, Harry SW, Sexson WR. Iatrogenic exposure to mercury after hepatitis B vaccination in preterm infants. J Pediatr 2000; 136: 679-81.

[101] Dorea JG, Marques RC, Brandao KG. Neonate exposure to thimerosal mercury from hepatitis B vaccines. Am J Perinatol 2009; 26: 523-7.

[102] Havarinasab S, Haggqvist B, Bjorn E, Pollard KM, Hultman P. Immunosuppressive and autoimmune effects of thimerosal in mice. Toxicol Appl Pharmacol 2005; 204: 109-21.

[103] Hornig M, Chian D, Lipkin WI. Neurotoxic effects of postnatal thimerosal are mouse strain dependent. Mol Psychiatry 2004; 9: 833-45.

[104] Vetvicka V, Vetvickova J. Effects of glucan on immunosuppressive actions of mercury. J Med Food 2009; 12: 1098-104.

[105] Faure AV, Grunwald D, Moutin MJ, *et al.* Developmental expression of the calcium release channels during early neurogenesis of the mouse cerebral cortex. Eur J Neurosci 2001; 14: 1613-22.

[106] Mutkus L, Aschner JL, Syversen T, *et al. In vitro* uptake of glutamate in GLAST- and GLT-1-transfected mutant CHO-K1 cells is inhibited by the ethylmercury-containing preservative thimerosal. Biol Trace Elem Res 2005; 105: 71-86.

[107] Parran DK, Barker A, Ehrich M. Effects of thimerosal on NGF signal transduction and cell death in neuroblastoma cells. Toxicol Sci 2005; 86: 132-40.

[108] Daniels JL, Rowland AS, Longnecker MP, Crawford P, Golding J. Maternal dental history, child's birth outcome and early cognitive development. Paediatr Perinat Epidemiol 2007; 21: 448-57.

[109] Hujoel PP, Lydon-Rochelle M, Bollen AM, *et al.* Mercury exposure from dental filling placement during pregnancy and low birth weight risk. Am J Epidemiol 2005; 161: 734-40.

[110] Drasch G, Schupp I, Hofl H, Reinke R, Roider G. Mercury burden of human fetal and infant tissues. Eur J Pediatr 1994; 153: 607-10.

[111] Castoldi AF, Johansson C, Onishchenko N, *et al.* Human developmental neurotoxicity of methylmercury: impact of variables and risk modifiers. Regul Toxicol Pharmacol 2008; 51: 201-14.

[112] Davidson PW, Myers GJ, Cox C, *et al.* Effects of prenatal and postnatal methylmercury exposure from fish consumption on neurodevelopment: outcomes at 66 months of age in the Seychelles Child Development Study. Jama 1998; 280: 701-7.

[113] Myers GJ, Davidson PW, Cox C, *et al.* Summary of the Seychelles child development study on the relationship of fetal methylmercury exposure to neurodevelopment. Neurotoxicology 1995; 16: 711-16.

[114] Myers GJ, Davidson PW, Shamlaye CF. A review of methylmercury and child development. Neurotoxicology 1998; 19: 313-28.

[115] Myers GJ, Davidson PW. Prenatal methylmercury exposure and children: neurologic, developmental, and behavioral research. Environ Health Perspect 1998; 106 Suppl 3: 841-7.

[116] Daniels JL, Longnecker MP, Rowland AS, Golding J. Fish intake during pregnancy and early cognitive development of offspring. Epidemiology 2004; 15: 394-402.

[117] Nielsen JB, Hultman P. Mercury-induced autoimmunity in mice. Environ Health Perspect 2002; 110 Suppl 5: 877-81.

[118] Via CS, Nguyen P, Niculescu F, *et al.* Low-dose exposure to inorganic mercury accelerates disease and mortality in acquired murine lupus. Environ Health Perspect 2003; 111: 1273-7.

[119] Penkowa M, Camats J, Giralt M, *et al.* Metallothionein-I overexpression alters brain inflammation and stimulates brain repair in transgenic mice with astrocyte-targeted interleukin-6 expression. Glia 2003; 42: 287-306.

[120] Penkowa M, Florit S, Giralt M, *et al.* Metallothionein reduces central nervous system inflammation, neurodegeneration, and cell death following kainic acid-induced epileptic seizures. J Neurosci Res 2005; 79: 522-34.

[121] Charleston JS, Body RL, Mottet NK, Vahter ME, Burbacher TM. Autometallographic determination of inorganic mercury distribution in the cortex of the calcarine sulcus of the monkey Macaca fascicularis following long-term subclinical exposure to methylmercury and mercuric chloride. Toxicol Appl Pharmacol 1995; 132: 325-33.

[122] Shapiro AM, Chan HM. Characterization of demethylation of methylmercury in cultured astrocytes. Chemosphere 2008; 74: 112-8.

[123] Vahter M, Mottet NK, Friberg L, *et al.* Speciation of mercury in the primate blood and brain following long-term exposure to methyl mercury. Toxicol Appl Pharmacol 1994; 124: 221-9.

[124] Shanker G, Syversen T, Aschner M. Astrocyte-mediated methylmercury neurotoxicity. Biol Trace Elem Res 2003; 95: 1-10.

[125] Morken TS, Sonnewald U, Aschner M, Syversen T. Effects of methylmercury on primary brain cells in mono- and co-culture. Toxicol Sci 2005; 87: 169-75.

[126] Charleston JS, Body RL, Bolender RP, *et al.* Changes in the number of astrocytes and microglia in the thalamus of the monkey Macaca fascicularis following long-term subclinical methylmercury exposure. Neurotoxicology 1996; 17: 127-38.

[127] Gajkowska B, Szumanska G, Gadamski R. Ultrastructural alterations of brain cortex in rat following intraperitoneal administration of mercuric chloride. J Hirnforsch 1992; 33: 471-6.

[128] Charleston JS, Bolender RP, Mottet NK, *et al.* Increases in the number of reactive glia in the visual cortex of Macaca fascicularis following subclinical long-term methyl mercury exposure. Toxicol Appl Pharmacol 1994; 129: 196-206.

[129] Warfvinge K, Hua J, Logdberg B. Mercury distribution in cortical areas and fiber systems of the neonatal and maternal adult cerebrum after exposure of pregnant squirrel monkeys to mercury vapor. Environ Res 1994; 67: 196-208.

[130] Bertossi M, Girolamo F, Errede M, *et al.* Effects of methylmercury on the microvasculature of the developing brain. Neurotoxicology 2004; 25: 849-57.

[131] Shanker G, Allen JW, Mutkus LA, Aschner M. Methylmercury inhibits cysteine uptake in cultured primary astrocytes, but not in neurons. Brain Res 2001; 914: 159-65.

[132] James SJ, Slikker W, 3rd, Melnyk S, *et al.* Thimerosal neurotoxicity is associated with glutathione depletion: protection with glutathione precursors. Neurotoxicology 2005; 26: 1-8.

[133] Allen JW, Mutkus LA, Aschner M. Mercuric chloride, but not methylmercury, inhibits glutamine synthetase activity in primary cultures of cortical astrocytes. Brain Res 2001; 891: 148-57.

[134] Choi BH, Lapham LW, Amin-Zaki L, Saleem T. Abnormal neuronal migration, deranged cerebral cortical organization, and diffuse white matter astrocytosis of human fetal brain: a major effect of methylmercury poisoning in utero. J Neuropathol Exp Neurol 1978; 37: 719-33.

[135] Sager PR, Aschner M, Rodier PM. Persistent, differential alterations in developing cerebellar cortex of male and female mice after methylmercury exposure. Brain Res 1984; 314: 1-11.

[136] Choi BH. Methylmercury poisoning of the developing nervous system: I. Pattern of neuronal migration in the cerebral cortex. Neurotoxicology 1986; 7: 591-600.

[137] Blaylock RL, Strunecka A. Immune-glutamatergic dysfunction as a central mechanism of the autism spectrum disorders. Curr Med Chem 2009; 16: 157-70.

[138] Brookes N. Specificity and reversibility of the inhibition by HgCl2 of glutamate transport in astrocyte cultures. J Neurochem 1988; 50: 1117-22.

[139] Juarez BI, Martinez ML, Montante M, *et al.* Methylmercury increases glutamate extracellular levels in frontal cortex of awake rats. Neurotoxicol Teratol 2002; 24: 767-71.

[140] Juarez BI, Portillo-Salazar H, Gonzalez-Amaro R, *et al.* Participation of N-methyl-D-aspartate receptors on methylmercury-induced DNA damage in rat frontal cortex. Toxicology 2005; 207: 223-9.

[141] Miyamoto K, Nakanishi H, Moriguchi S, *et al.* Involvement of enhanced sensitivity of N-methyl-D-aspartate receptors in vulnerability of developing cortical neurons to methylmercury neurotoxicity. Brain Res 2001; 901: 252-8.

[142] Shanker G, Aschner M. Methylmercury-induced reactive oxygen species formation in neonatal cerebral astrocytic cultures is attenuated by antioxidants. Brain Res Mol Brain Res 2003; 110: 85-91.

[143] Sorg O, Horn TF, Yu N, Gruol DL, Bloom FE. Inhibition of astrocyte glutamate uptake by reactive oxygen species: role of antioxidant enzymes. Mol Med 1997; 3: 431-40.

[144] Sorg O, Schilter B, Honegger P, Monnet-Tschudi F. Increased vulnerability of neurones and glial cells to low concentrations of methylmercury in a prooxidant situation. Acta Neuropathol 1998; 96: 621-7.

[145] Behan WM, Stone TW. Enhanced neuronal damage by co-administration of quinolinic acid and free radicals, and protection by adenosine A2A receptor antagonists. Br J Pharmacol 2002; 135: 1435-42.

[146] Brown GC, Borutaite V. Nitric oxide inhibition of mitochondrial respiration and its role in cell death. Free Radic Biol Med 2002; 33: 1440-50.

[147] Pacher P, Beckman JS, Liaudet L. Nitric oxide and peroxynitrite in health and disease. Physiol Rev 2007; 87: 315-424.

[148] Beal MF, Hyman BT, Koroshetz W. Do defects in mitochondrial energy metabolism underlie the pathology of neurodegenerative diseases? Trends Neurosci 1993; 16: 125-31.

[149] Hare MF, McGinnis KM, Atchison WD. Methylmercury increases intracellular concentrations of Ca++ and heavy metals in NG108-15 cells. J Pharmacol Exp Ther 1993; 266: 1626-35.

[150] Konigsberg M, Lopez-Diazguerrero NE, Bucio L, Gutierrez-Ruiz MC. Uncoupling effect of mercuric chloride on mitochondria isolated from an hepatic cell line. J Appl Toxicol 2001; 21: 323-9.

[151] Fujimura M, Usuki F, Sawada M, Takashima A. Methylmercury induces neuropathological changes with tau hyperphosphorylation mainly through the activation of the c-jun-N-terminal kinase pathway in the cerebral cortex, but not in the hippocampus of the mouse brain. Neurotoxicology 2009; 30: 1000-7.

[152] Monnet-Tschudi F. Induction of apoptosis by mercury compounds depends on maturation and is not associated with microglial activation. J Neurosci Res 1998; 53: 361-7.

[153] Eskes C, Honegger P, Juillerat-Jeanneret L, Monnet-Tschudi F. Microglial reaction induced by noncytotoxic methylmercury treatment leads to neuroprotection via interactions with astrocytes and IL-6 release. Glia 2002; 37: 43-52.

[154] Allen JW, Mutkus LA, Aschner M. Methylmercury-mediated inhibition of 3H-D-aspartate transport in cultured astrocytes is reversed by the antioxidant catalase. Brain Res 2001; 902: 92-100.

[155] Saijoh K, Fukunaga T, Katsuyama H, Lee MJ, Sumino K. Effects of methylmercury on protein kinase A and protein kinase C in the mouse brain. Environ Res 1993; 63: 264-73.

[156] Pawlak J, Brito V, Kuppers E, Beyer C. Regulation of glutamate transporter GLAST and GLT-1 expression in astrocytes by estrogen. Brain Res Mol Brain Res 2005; 138: 1-7.

[157] Kim P, Choi BH. Selective inhibition of glutamate uptake by mercury in cultured mouse astrocytes. Yonsei Med J 1995; 36: 299-305.

[158] Mutkus L, Aschner JL, Syversen T, *et al.* Mercuric chloride inhibits the *in vitro* uptake of glutamate in GLAST- and GLT-1-transfected mutant CHO-K1 cells. Biol Trace Elem Res 2006; 109: 267-80.

[159] Kugler P, Schleyer V. Developmental expression of glutamate transporters and glutamate dehydrogenase in astrocytes of the postnatal rat hippocampus. Hippocampus 2004; 14: 975-85.

[160] Chmielnicka J, Komsta-Szumska E, Sulkowska B. Activity of glutamate and malate dehydrogenases in liver and kidneys of rats subjected to multiple exposures of mercuric chloride and sodium selenite. Bioinorg Chem 1978; 8: 291-302.

[161] Inage YW, Itoh M, Wada K, Takashima S. Expression of two glutamate transporters, GLAST and EAAT4, in the human cerebellum: their correlation in development and neonatal hypoxic-ischemic damage. J Neuropathol Exp Neurol 1998; 57: 554-62.

CHAPTER 9

Fluoride and Aluminum: Possible Risk Factors in Etiopathogenesis of Autism Spectrum Disorders

Anna Strunecka[*1] and Russell L. Blaylock[2]

[1]*Department of Physiology, Faculty of Science, Charles University in Prague, Prague, Czech Republic and* [2]*Institute for Theoretical Neuroscience, LLC, and Visiting Professor of Biology, Belhaven University, Ridgeland, MS 39157, USA*

Abstract: Fluoride and aluminum ions (Al^{3+}) are considered as new ecotoxicological factors. While aluminum has been involved among the possible culprits of autism spectrum disorders (ASD), fluoride is rarely considered. Al^{3+} is non-essential for all forms of life and serves no known biological role. Fluoride and Al^{3+} can elicit impairment of homeostasis, growth, development, cognition, and behavior. Several symptoms induced by Al^{3+} and/or fluoride overload can be seen in ASD. Several laboratory studies demonstrate that many effects primarily attributed to fluoride are caused by synergistic action of fluoride plus Al^{3+}. In water solutions, Al^{3+} forms in the presence of fluoride, water soluble aluminofluoride complexes (AlFx). AlFx has been widely used as an analogue of phosphate groups to study heterotrimeric G proteins involvement. AlFx affects numerous receptors and signaling systems. It is evident that the long-term intake of low amounts of fluoride and Al^{3+} can evoke receptor malfunctions. The synergistic interactions of fluoride plus Al^{3+} may thus evoke several histological, neurological, biochemical, and behavioral symptoms of ASD. This chapter brings evidence that AlFx represents a hidden potential danger for pathogenesis of ASD.

INTRODUCTION

The heterogeneity of pathophysiological, histological, neurological, biochemical, clinical, and behavioral symptoms provide us little reason to assume that there is one cause of ASD pathogenesis. Nevertheless, we suggested in our previous reviews [1,2] that multiple risk factors may cause the dysregulation of immune-glutamate pathways and evoke various multiple symptoms. Blaylock coined the term imunoexcitotoxicity as the possible central mechanism in ASD etiopathogenesis [3-6]. The interactions between excitotoxins, free radicals, lipid peroxidation products (LPP), inflammatory cytokines, and disruption of neuronal calcium homeostasis can result in brain changes suggestive of the pathological findings in cases of ASD. There is compelling evidence from a multitude of studies of various designs indicating that environmental and food borne excitotoxins can elevate blood and brain glutamate to levels known to cause neurodegeneration, brain inflammation, and alterations in the developing brain. We will focus in this chapter on the potential contributions of fluoride and Al^{3+}. While aluminum has been involved among the possible culprits of ASD, fluoride is rarely considered.

Despite the abundance of aluminum in nature, it has no biological function in humans. On the contrary, wide ranges of toxic effects of Al^{3+} to hundreds of cellular processes have been demonstrated [7]. There is compelling evidence that an accumulation of aluminum in the body appeared recently as the inevitable consequence of the activities of modern human civilization. There are a number of aluminum sources, such as the drinking water, nutrition, cosmetics, and the widespread use of aluminum in medicine. It means that humans are not able to avoid exposure to aluminum at present. Fluoride exposure is common in fetuses, newborns, and small children as a result of the artificial fluoridation of drinking water and a dramatic increase in the volume of man-made industrial fluoride compounds released into the environment. Many investigations of the long-term administration of fluoride to laboratory animals have demonstrated that fluoride and aluminum can elicit impairment of homeostasis, growth, development, cognition, and behavior.

Numerous epidemiological, ecological, and clinical studies have shown the effects of fluoride and aluminum burden on humans. Fluoride could complex with any pre-existing Al^{3+} within body fluids to produce the aluminofluoride

---

*Address correspondence to: Anna Strunecka, Faculty of Science, Charles University in Prague, Vinicna 7, 128 00 Prague 2, Czech Republic; E-mail: strun@natur.cuni.cz

complexes (AlFx), the most potent activators of G protein regulated systems known [8]. AlFx increases the potential neurotoxic effect of Al^{3+} and fluoride alone particularly in children, whose brains are uniquely sensitive to environmental toxins. The synergistic interactions of fluoride plus Al^{3+} may evoke several histological, neurological, biochemical, and behavioral symptoms of ASD [2]. We suggest that the chronic burden with these new ecotoxicological factors could contribute to an alarming increase in the incidence of ASD.

ALUMINUM AND FLUORIDE BIOAVAILABILITY FROM ENVIRONMENT

Aluminum

Aluminum, the most abundant metal of the earth's lithosphere, is everywhere: in water sources, in nourishment, in different food additives and also in air in the form of dust particles. It has, until relatively recently, existed in forms not generally available to living organisms, and was therefore regarded as non-toxic. On the other hand, it remains one of the great paradoxes of life on the Earth that the most abundant metal in the lithosphere has no biological function.

With the use of aluminum in industry and presence in acid rain, there has been a dramatic increase in the amount of bioavailable Al^{3+} in the environment. Aluminum occurs in only one oxidative state (Al^{3+}). Al^{3+} can be formed in solution and this ion is known to exist in groundwater in concentrations ranging from 0.1 mg to 8.0 mg /L. The water supply industry uses aluminum sulphate to produce a less turbid drinking water. Aluminum sulfate used for water treatment was ranked 43rd among the top 50 additives during the years 1993 and 1994. Al hydroxide or sodium aluminate can be present in tap water as residuals from these processes. At present, there is no Environmental Protection Agency (EPA) health standard for aluminum in drinking water, only a recommended Secondary Maximum Contaminant Level (SMCL), which is 0.05 to 0.2 mg/L. This would represent the safe level of a contaminant in drinking water below which there is no known or expected risk to health. Although water is the most extensively studied aluminum source, it provides only about 1% of normal daily human intake [9].

The primary normal source of aluminum is food, which provides approximately 16-100 fold more Al^{3+} to systemic circulation than drinking water. For example, aluminum is found in a number of baked goods, such as pancakes and biscuits, processed cheeses, frozen foods, and beverages [10, 11]. Several papers report that all tea samples release aluminum (0.70-5.93 mg/L) during a standard infusion period. The results indicated that tea consumption must be considered in any assessment of the total dietary intake of aluminum. Aluminum is added to frozen foods to improve their appearance. Aluminum salts are added to processed cheeses and to beer. The aluminum accumulation continues from cookware, cans, tetrapacks, and antacids. Krewski *et al.* [12] estimated that for the general population, intake of aluminum from food (7.2 mg/day for females and 8.6 mg/day for males) dominated that from drinking water (0.16 mg/day) and inhalation exposure (0.06 mg/day). Antacids and buffered aspirin can contribute on the order of thousands of mg/day to aluminum intake.

This evaluation places the gastrointestinal tract (GIT) as the focal point of understanding human exposure to aluminum. Al^{3+} may form complexes with other dietary acids, for example, malic, oxalic, tartaric, succinic, aspartic, and glutamic acids, which may dramatically increase its GIT absorption.

There are a number of ways in which humans are exposed to aluminum, such as the skin, the lung, the nose, and of course, the intramuscular vaccination. Human can be exposed to aluminum through aerosol and topical application of anti-perspirants, topical application of sunscreens, suntan lotion, skin moisturizers, and smoking [13]. Such items as deodorants, vaginal douches and baby wipe not only have high aluminum content, but are applied to areas where there is far greater tendency to absorption through the skin. However, because inhaled aluminum is approximately seven times more bioavailable than aluminum in drinking water, the contribution of inhaled aluminum exceeds the corresponding contribution from drinking water [12]. Aluminum absorbed across olfactory epithelia would bypass the defense of blood- brain barrier (BBB) and pass directly to the hippocampal region of the brain by way of the olfactory tract.

Other major contributors include aluminum used in medicines: antacids, dialysis solutions, parental, and intravenous nutrition solutions used in pediatrics [9,14,15]. Vaccines, allergy skin tests, human serum albumin, baby skin creams, baby diaper wipes, and antacids, which are frequently given to infants, are extremely high in aluminum.

Recently, the content of aluminum in vaccines has been discussed [16]. According to Yokel and McNamara [10] calculations based on 20 injections in the first 6 years of life and an average weight of 20 kg, children receive 0.07 - 0.4 µg/kg of aluminum. This value is comparable with the amount of aluminum absorbed from food.

The natural barrier systems for Al^{3+} such as low absorption in the GIT, and various physiological ligands, such as citrate, phosphate, and silicic acid, were efficient buffers preventing the increased absorption of this metal in natural conditions. The population is now exposed to more bioavailable aluminum. The skin and mucus-lined epithelia of the GIT and respiratory systems are initially barriers to exposure to aluminum and they are also likely transitory sinks for biologically available Al^{3+}. No one has determined the level of Al^{3+} in the hair and nails.

There are few data addressing the excretion of aluminum from the body. It is assumed that a majority of ingested aluminum is excreted in feces, but the amount of aluminum in feces has not been measured. Several studies have demonstrated that the kidney is a route of excretion of systemic aluminum and that a proportion of aluminum in the blood is continually removed by the kidney [17]. An estimated $t_{1/2}$ in one human subject who received intravenous ^{26}Al was 7 years.

Exley [18] suggests that the lack of an Al-specific mechanism to enable its excretion from the body allows for the accumulation of aluminum and the persistence of any symptoms which might be the consequence of a body burden. We can assume that every compartment of human body including cells, organelles, cytoplasm, and extracellular fluids, contain a few atoms of potentially biologically available Al^{3+}. It seems that humans are not able to avoid exposure to aluminum at present. The question thus arises: What is the consequence of a burgoing body burden of aluminum? Since Al^{3+} has no known function in life, then the first manifestations of the effect on human physiology in its presence will almost certainly be negative and most likely in the form of chronic disease.

Fluoride

It is noteworthy that fluorine, which is the most abundant halogen and the thirteenth most abundant element in the earth's crust, was not involved as a regular component of organic compounds of living organisms [19]. The concentration of available fluoride in the living environment, water, and food chain was very low before the Second World War. In 1942, H. Trendley Dean published his famous 21 City study in which he showed that at 1 ppm fluoride (1milligram per litre) there was a marked decrease in tooth decay [20]. The artificial fluoridation of drinking water as a way of preventing dental caries has been a practice for many years in several countries. While a majority of European countries stop water fluoridation, the U.K. introduced it in 1999. Currently about 6 million people (10% of the U.K. population, mainly in the West Midlands and Newcastle) receive water artificially fluoridated to 1 ppm. Approximately 60-70% of the American people and more than 50% of the population of Australia, Columbia, Ireland, New Zealand, and Singapore are supplied with fluoridated drinking water. The U.S. EPA and the National Research Council (NRC) considered safe levels of fluoride in drinking water ≤ 4 ppm (4 mg/L). The WHO has a guideline for a recommended fluoride concentration of 1.5 ppm of drinking water. The problem of fluoride concentration higher than 2 ppm in ground water is one of the most important health-related environmental issues in many parts of India and China [1]. In central and northern Mexico millions of people are affected by high fluoride content in household-use groundwater.

The past 50 years have seen a dramatic increase in the volume of man-made industrial fluoride compounds released into the environment, so total fluoride intake has become an issue of particular concern [21]. People now get fluoride from many other sources, such as food and beverages, pesticide and fertilizers, industry, dental treatments, fluorinated drugs, and fluorine air pollution. Exposure to airborne fluorides from many diverse manufacturing processes such as phosphate fertilizer production, aluminum smelting, uranium enrichment facilities, coal-burning and nuclear power plants, petroleum refining and vehicle emissions, can be considerable. Several studies reported a high content of fluoride in most of tested teas. Tea is very high in fluoride because tea leaves accumulate more fluoride from pollution of soil and air than other plant species. Fluoride content in tea has risen dramatically over the last 20 years due to industrial contamination. Recent analyses have revealed that consuming tea infusions may lead to exposure to a high amount of fluoride [22-25].

Overexposure to fluoride among infants is a widespread problem in most major American cities. The investigation of the Environmental Working Group of the Fluoride Action Network found that up to 60% of formula-fed babies in

US cities were exceeding the upper tolerable limit for fluoride. Using fluoridated water, a bottle-fed baby will receive up to 250 times more fluoride than from the mother's milk (http://www.fluoridealert.org).

Under most conditions, fluoride is rapidly and extensively absorbed from the GIT. The rate of gastric absorption is inversely related to the pH of the gastric contents. High concentrations of Ca^{2+} and Al^{3+} can reduce the uptake of fluoride at this stage and the complexes or insoluble fluoride usually exit the body in the feces. Fluoride removal from plasma occurs by calcified tissue uptake and urinary excretion. A great part of the body burden of fluoride is associated with calcified tissues, and most of it is not exchangeable. In a healthy adult, about 50% of the fluoride, which enters plasma, is excreted by the kidney. Numerous studies document the increase of fluoride in blood, bones, and urine in children and adults occurring during the last two decades (for a review see [1]).

SYMPTOMS OF ALUMINUM AND FLUORIDE INTOXICATION

Aluminum

The studies showing the increased bioavailability of Al^{3+} due to acid rains reveal that aluminum may be toxic for many organisms. Toxic effects of aluminum concentration, which under legislation of both EPA and EU is allowed in drinking water (0.200 mg/L), have already been observed in some animal and plant species living in lakes with acidic water. Experiments with rats showed that the toxicity of 0.5 ppm aluminum in the drinking water was significantly greater than at 5 or even 50 ppm [26]. The reason for this paradoxical concentration effect is obscure.

The human body does not normally use aluminum, yet human physiology does not preclude its participation in wide and varied functions. Berend *et al.* [17] described the large content of aluminum in the lungs, liver, bone, myocardium, and parathyroid cells. The amount of aluminum that is not eliminated in the urine is retained in the body and removed very slowly.

Despite intensive research, we do not know exactly the conditions under which biologically reactive Al^{3+} has produced an aberrant physiological response in the affected cell or tissue. Exley [18,27] suggested that aluminum may be stored, thus being a silent "visitor" in tissues such as hair, skin, and bone. Its toxicity may be via a mechanism of free radical generation; in addition, aluminum may be recycled and used over and over again in promoting oxidative damage.

Much has already been written about human diseases attributed to exposure to Al^{3+}. These clinical observations demonstrate that severity and development of the symptoms increased long-lasting burden depend on a person's age, genetic background, nutrition status, kidney function, and many other factors. Aluminum exposure is linked to neurodegenerative diseases such as Alzheimer's disease, Parkinson disease, and multiple sclerosis [7,27,28]. A new distinct neurological disease, dialysis encephalopathy, which is associated with speech disturbances, personality changes, seizures and myoclonus, has been described in dialysis patients [29,30].

A new emerging syndrome, *macrophagic myofasciitis,* was first described in France [31]. Victims of this syndrome suffer severe muscle and joint pains and severe weakness. Subsequent studies indicate widespread, severe brain injury as well—confirmed by MRI scanning. This brain syndrome has been described in American children as well. Recently, Exley *et al.* [16] reported the coincidence of macrophagic myofasciitis, chronic fatigue syndrome, and aluminum overload in an individual. This condition developed progressively following five vaccinations over a period of four weeks. Each of these vaccinations included an Al-based adjuvant and, three years later, the persistence of aluminum salt at an injection site was confirmed by muscle biopsy in the diagnosis of macrophagic myofasciitis. Aluminum overload was diagnosed four years post vaccination though the prevelance of this condition is unknown. This case has highlighted potential dangers associated with Al-containing adjuvants and authors have elucidated a possible mechanism whereby Al could trigger a cascade of immunological events, which are associated with autoimmune conditions [32]. Because vaccine adjuvants are designed to produce prolonged immune stimulation, they pose a particular hazard to the developing nervous system. Studies have shown that immune activation following vaccination can last up to two years. This means that the brain's microglial cells are also primed for the same length of time or possibly longer.

Another interesting question is connected with the potential of Al^{3+} to act as a potential antigen. It is because there is an immune response against Al^{3+} that it is used as a vaccine adjuvant. Aluminum is used as adjuvant in a wide range of common vaccination. This is one of the reasons why vaccination has been assumed as the possible trigger for ASD. The Al - based adjuvants are known as long-lived depots of antigen but it is known that these Al-adjuvants activate innate immune signals, even in the absence of an adsorbed antigen. Research in rabbits show that labelled ^{26}Al, when injected into muscle as an adjuvant, was present after one hour in the blood sample. The distribution profile of ^{26}Al demonstrated its presence in various tissues; kidney > spleen > liver > heart > lymph node > brain [33]. Adjuvant aluminum is now implicated in a wide spectrum of human diseases, including adverse skin reactions, macrophagic myofasciitis, Alzheimer's disease, Parkinson' s disease, ALS, cutaneous lymphoid hyperplasia, vaccine-related hypersensitivity to Al, immunotherapy-related hypersensitivity to Al, Gulf War illness, and Guillain-Barre syndrom [18].

The burgeoning use of Al-based adjuvants in vaccinations, which cover the full spectrum of human diseases and are administered throughout life, from new born babies through to the elderly, may already have created cohorts of individuals which are hypersensitive to aluminum exposure. Bergfors *et al.* [34] followed the group of 76, 000 vaccinees with Al- adsorbed diphtheria-tetanus/acellular pertussis vaccines and observed that 645 children (0, 8%) had persistent itching nodules at the vaccination site. The itching was intense and long-lasting. So far, 75% still have symptoms after a median duration of 4 years. Contact hypersensitivity to Al was demonstrated in 77% of the children with itching nodules and in 8% of the symptomless siblings who had received the same vaccines (P<0.001). Identifiable diseases may then be the manifestation of immune-mediated responses to the burgeoning body burden of aluminum.

It has been known for a long time that local administration or application of aluminum to the brain can cause experimental animals to develop seizures [17].

Fluoride

The apparent side-effect of fluoride overload appears to be dental fluorosis. WHO recently estimated that dental fluorosis is endemic in at least 25 countries across the globe [1]. Millions of people live in endemic fluorosis area. WHO estimated that 2.7 million people have skeletal fluorosis in China, and over 6 millions suffer this crippling bone disease in India.

The toxic action of fluoride has been attributed to the fact that fluoride acts as enzymatic poison, inhibiting activities of many important enzymes such as enolase, lipase, phosphofructokinase, pyruvate kinase, glycogen synthase, succinate dehydrogenase, cytochrome oxidase, various phosphatases, ATP-ases, urease, and cholinesterases, to name a few. Strunecka *et al.* [1] review found that fluoride has been found to inhibit 22 various enzymes. On the other hand, 20 enzymes including adenylyl cyclase, lactate dehydrogenase, and glycogen phosphorylase, are stimulated by fluoride in milimolar concentrations. The use of fluoride in laboratory investigations helped in the discovery of glycolytic and Krebs-cycle pathways and provided key evidence of fluoride effects on the biochemical and physiological processes.

During the last decade numerous animal studies have been published, which have raised the level of concern about the impacts of increasing fluoride exposure on the brain [35-41]. These studies further highlight that it is not just the teeth, but the brain, that may be impacted by too much fluoride during infancy and childhood. Reduction of children's intelligence, various psychiatric symptoms in adults, such as memory impairment, and difficulties with concentration and thinking were reported [42,43].

Several studies appeared from China, which indicated a lowering of IQ in children associated with fluoride exposure [42,44,45]. Their conclusions have been criticized because of the possibility of unaccounted confounding variables. However, the later study by Xiang *et al.* [46] controlled for parental economic status and education, as well as exposure to iodide and lead. These authors found that IQ scores below 80 were significantly associated with higher serum fluoride level and estimated that children's IQ would be lowered at 1.8 ppm fluoride in their drinking water. Such a finding represents little margin of safety considering the potentially serious outcome for infants drinking fluoridated water. Rocha-Amador *et al.* [47] found that the increased content of fluoride in urine was associated with reduced performance, verbal, and full IQ scores. The effects on individuals indicated that for each mg increase of fluoride in urine, a decrease of 1.7 point in full IQ might be expected. Fluoride may belong to the class of developmental neurotoxicants such as arsenic, lead, and methylmercury.

Elevated fluoride content was found in embryonic brain tissues obtained from required abortions in areas where fluorosis was prevalent. These studies showed poor differentiation of brain nerve cells and delayed brain development [48]. The fetal BBB is immature and readily permeable to fluoride [35]. Mullenix *et al.* [49] demonstrated that fluoride accumulated in the brain of rats exposed to sodium fluoride in drinking water. The accumulations of fluoride were found in all the regions of the brain, with the highest levels in the hippocampus, one of the most sensitive areas of the brain to neurotoxicity. Mullenix and co-workers compared behavior, body weight, plasma, and brain fluoride levels after NaF exposures during late gestation, at weaning or in adults. Rats exposed prenatally had dispersed behaviors typical of hyperactivity, whereas rats exposed as adults displayed behavior-specific changes typical of cognitive deficits.

Histopathological studies of fluoride-exposed animals have demonstrated damage to CA1 and CA4 areas of the hippocampus and to the dentate gyrus, which is also consistent with excitotoxicity [50,51]. In their study using aluminum fluoride and sodium fluoride, Varner *et al.* [26] found damage in the superficial layers of the cortex, amygdala, and cerebellum—all areas endowed with abundant GluRs. Others have described a loss of Purkinje cells with chronic fluoride exposure, a cell type containing abundant AMPA GluRs. Also consistent with excitotoxicity is the finding of elevated levels of reactive oxygen species (ROS), reactive nitrogen species (RNS), and LPP in the brain, following fluoride exposure, both *in vitro* and *in vivo* [40,52]. Also of interest is the finding of elevation in nitric oxide (NO) via induced nitric oxide synthase (iNOS), again a critical component of excitotoxicity. At least two studies have shown that fluoride compounds can activate immune pathways that can lead to, or enhance, autoimmunity. A growing number of studies have shown that inflammatory cytokines and chemokines can markedly enhance excitotoxicity. This could represent fluoride alterations in cerebral glutamate levels, which are known to play a vital role in neuron migration and pruning of synaptic connections and dendrites [50,51].

The endocrine glands such as the thyroid and the pineal gland, are extremely sensitive to fluoride. It was shown that normal healthy individuals had thyroid function lowered when consuming water at 2.3 ppm [53]. The thyroid gland appears to be the most sensitive tissue in the body to fluoride burden, which is able to increase the concentration of *thyroid-stimulating hormone* (TSH) and decrease the concentration of T_3 and T_4 hormones, thereby producing hypothyroidism [54]. Up until the late 1950's, the doses of fluoride 2.3 - 4.5 mg/day were recommended in Europe to reduce the activity of the thyroid gland of those suffering from hyperthyroidism [55]. The search for a mechanism to explain how fluoride might lower thyroid activity has a very long and elusive history. A possible explanation has come from Tezelman *et al.* [56] who have suggested that overproduction of cAMP leads to a feedback mechanism resulting in a desensitization of the TSH receptor, thus ultimately leading to reduced activity of the gland.

Disturbance of thyroid hormone production has been found in correlation with lowered IQ in children in China [46]. A decreased level of T_3 was found in residents of Villa Ahumada, Mexico, where fluoride concentration in drinking water averages 5.3 ppm [57]. Susheela *et al.* [58] compared the production of thyroid hormones and TSH of 90 children living in fluoride endemic, non-iodine deficient areas of Delhi, India, along with 21 children from non-endemic areas. The data indicated an association of excess of fluoride intake and thyroid hormone disturbances leading to manifestation of iodine deficiency disorders (IDD). This study clearly documents that the primary cause of IDD may not always be iodine deficiency, but rather an excess of fluoride might induce it. Susheela *et al.* [58] suggests that iodine metabolism is being disturbed through fluoride's effect on deiodinases, the three enzymes, which regulate the conversion of T_4 to T_3 in target tissues. The role of excess of fluoride in development of IDD has been largely unnoticed at present, despite the fact that millions of children suffer with IDD. More targeted research is needed, considering the globally increasing problem of IDD.

The Synergistic Effects of Aluminum and Fluoride

Several laboratory studies have demonstrated that many effects primarily attributed to fluoride are caused by synergistic action of fluoride plus Al^{3+}. In water solution, Al^{3+} forms, in the presence of fluoride, water soluble aluminofluoride complexes (AlFx) whose average stoichiometry depends on the excess concentration of fluoride ions and the pH of the solution (for a review see [1]). AlFx has been widely used an analogue of phosphate groups to study phosphoryl transfer reactions and heterotrimeric G proteins involvement [59]. Numerous laboratory studies demonstrated that AlFx interacts with all known G protein-activated effector enzymes. Fluorides in the presence of Al^{3+} affect the levels of second messenger molecules, including cAMP, inositol phosphates, and Ca^{2+}. These studies brought a great deal of understanding concerning the involvement of G proteins in cell signaling. Moreover, they bring evidence that AlFx influences various functions and biochemical reactions of many cells and tissues of animals or human organisms. Fluoride in the presence of trace amounts of Al^{3+} affects blood elements, endothelial cells and blood circulation, the function of lymphocytes and cells of the immune system, bone cells, fibroblasts and keratinocytes, ion transport, calcium influx and mobilization, processes of neurotransmission, metabolism of the liver, growth and differentiation, protein phosphorylation, and cytoskeletal proteins. These effects are not surprising in respect to the extensive role of G proteins in the cell. Physiological agonists of G protein-coupled receptors (GPCR) include neurotransmitters and hormones, such as dopamine, epinephrine, norepinephrine, serotonin, acetylcholine, glucagon, vasopressin, melatonin, TSH, neuropeptides, opioids, excitatory amino acids, prostanoids, purines, photons, and odorants.

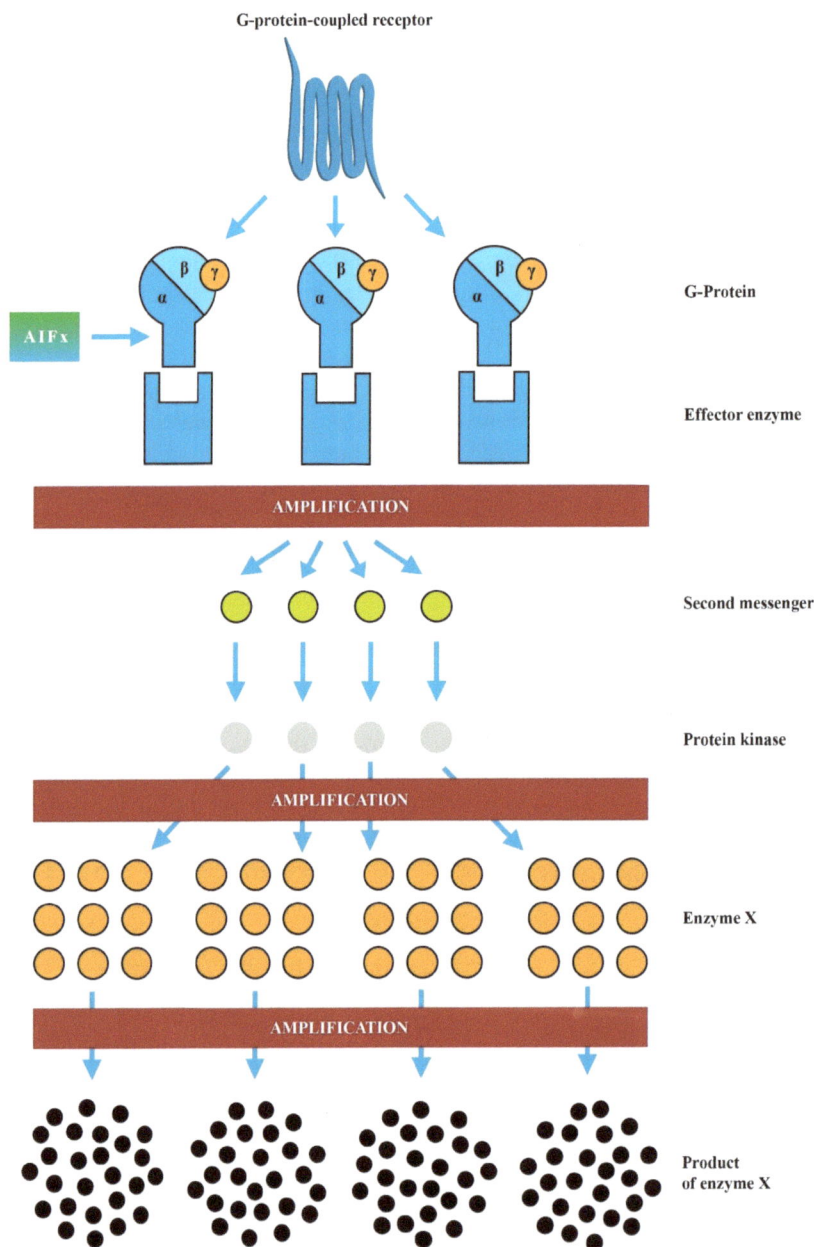

Figure 1: AlFx acts as the messenger of false information. Its message is greatly amplified during the conversion into the functional response of a cell. The second messenger molecule could be cAMP, Ins(1,4,5)P$_3$, and DAG.

The synergistic action of fluoride and Al^{3+} has the important implication for pathology. The effects of fluoride or Al^{3+} alone substantially differ from the effects of AlFx. While during the burgeoning accumulation of these elements, cell or tissue manage to cope with them until a threshold concentration is reached. Al^{3+} in micromolar concentrations avidly binds with fluoride to form AlFx. AlFx has a potency that allows it to activate hundreds of GPCRs. This means that the effects of AlFx result in pathophysiological consequences at several times lower concentrations than either Al^{3+} and fluoride acting alone. Moreover, the effects of AlFx are amplified by processes of signal transduction (Fig. **1**). The principle of amplification of the initial signal during its conversion into a functional response has been a widely accepted tenet in cell physiology. It is evident that AlFx is a molecule giving a false message.

Toxicological potential of fluoride is thus markedly increased in the presence of trace amounts of Al^{3+}. It has been also observed that Al-induced neural degeneration in rats is greatly enhanced when the animals were fed low doses of fluoride. The presence of fluoride caused more Al^{3+} to cross the BBB and be deposited in the brain of rat [26].

THE POTENTIAL ROLE OF ALUMINUM AND FLUORIDE IN ETIOPATHOGENESIS OF ASD

Several symptoms induced by aluminum and/or fluoride overload can be seen in ASD. ASD is strongly correlated with neurodevelopmental alterations in prenatal, as well as the postnatal period. A considerable amount of evidence supports the conclusion that fluorides belongs to a group of other recognized causes of neurodevelopmental disorders and subclinical brain dysfunctions. Currently, fluoride exposure is common in fetuses, newborns, and small children mostly as a result of the artificial fluoridation of drinking water and a dramatic increase in the volume of man-made industrial fluoride compounds released into the environment. With over half of mothers using infant formula reconstituted using tap water, infant exposure to significant levels of AlFx has become a major problem. Chronic exposure of humans to Al and fluoride begins during the first trimester in the womb. While we have known for a long time that fluoride might cross the placenta and that the immature fetal BBB is readily permeable to both fluoride and Al^{3+} [12, 35], their impacts on the development of human fetal brain are not yet fully recognized. The association between embryonic errors and the development of autism has been reported in the literature. The implication is that the origin of ASD can be much earlier [60] in embryologic development than has been frequently reported.

A study by Du [48] revealed adverse effects of fluoride overload on the brains of 15 aborted fetuses between the 5-8th months of gestation from an endemic fluorosis area in China compared with those from a non-endemic area. Stereological study of the brains showed that the numerical density and the volume of the neurons were abnormal and the presence of undifferentiated neuroblasts as well as the increase in the nucleus-cytoplasm ratio of neurons indicated disordered neurodevelopment. The overall mean volume of the neurons was reduced. These results indicated that chronic fluoride overload in the course of intrauterine fetal life may produce certain harmful effects on the developing brain of the fetus. The recent study found more premature births in fluoridated than non-fluoridated upstate New York communities (http://apha.confex.com/apha/137am/webprogram/Paper197468.html). Gillberg and Gillberg [61] reported significantly more incidences of prematurity and postmaturity at birth of ASD patients. May-Benson *et al.* [62] investigated 467 children with ASD and found that the incidence of premature birth was 16%. Brimacombe *et al.* [63] studied a cohort of 164 families of autistic children referred to The Autism Center at New Jersey Medical School. Prevalence rates in this cohort for prematurity were higher than comparable rates reported nationally and in New Jersey.

Fluoridated drinking water contains up to 200 times more fluoride than breast milk (1000 ppb in fluoridated tap water vs 5-10 ppb in breast milk). As a result, babies consuming formula made with fluoridated tap water are exposed to much higher levels of fluoride than are breast-fed infants. Although there are some other reasons for encouraging breastfeeding, several authors are warning against ingestion of excessive fluoride from high quantities of intake of fluoridated water used to reconstitute concentrated infant formula early in infancy [64-66]. Newborn and premature infants also are particularly at risk for aluminum neurotoxicity. High permeability of the immature BBB to aluminum, the increased uptake via poorly developed GIT, and immature function of the kidney increases the risk of Al^{3+} intoxication. While aluminum content of breast milk is very small (about 20 µg/L), formulas based milk contain about ten times as much [17].

Tanoue and Oda [67] statistically compared weaning times of 145 children diagnosed as autistic with those of 224 normal children in the same catchment area: 24.8% of the patients and 7.5% of the controls were weaned by the end of first week, a significant difference. Early weaning because of the mother's rather than the child's condition occurred with 17.9% of the patients and 5.8% of the controls, also a significant difference. Historical studies on infantile autism revealed that the disorder developed more prevalently in the socioeconomic status where the incidence of breast-feeding was less frequent. These results suggest that early weaning may contribute to the etiology of infantile autism.

Some symptoms of ASD such as the sleep problems and the early onset of puberty suggest abnormalities in melatonin physiology and dysfunctions of the pineal gland. Luke [68,69] reported that fluoride accumulates in the pineal gland and that mongolian gerbils fed higher doses of fluoride excreted less melatonin metabolite in their urine and took a shorter time to reach puberty. When Luke had the pineal glands from 11 human corpses analyzed, the fluoride in the apatite crystals averaged about 9,000 ppm and in one case went as high as 21,000 ppm. Many studies indicate clearly that nocturnal production of melatonin is reduced in ASD [70]. Melatonin is responsible for regulating numerous life processes, including development and aging [71]. It is also known that production of melatonin by the pineal is controlled by mGluR and that excess aspartate or glutamate activity can inhibit melatonin release. Being a G protein type receptor, AlFx could also activate mGluR. Melatonin has been shown to have powerful neutralizing effects on ROS and LPP and to increase the levels of several of the antioxidant enzymes in the brain. A study of Tauman *et al.* [72] revealed that babies with the lowest melatonin production had the most neurobehavioral problems. In the light of these findings it is interesting to note that the Newburgh-Kingston fluoridation trial (1945-55) found that the girls in fluoridated Newburgh were menstruating on average 5 months earlier than girls in unfluoridated Kingston [73]. Considering the importance of melatonin in ASD, this issue warrants further study.

Mullenix *et al.* [49] demonstrated significant behavioral changes in rats exposed neonatally to fluoride in drinking water, with the effects dependent on the timing of the dose and sex of the animals. They found males were more affected with prenatal exposure than females if the exposure occurred after weaning or adulthood.

The contribution of aluminum to symptoms of ASD has been discussed for a long time; mainly the contribution of Al from vaccination in development of ASD pathology. Al-adjuvants potentiate the immune response, thereby ensuring the potency and efficacy of typically sparingly available antigen. While we do not understand fully how Al-based adjuvants work; it could be assumed that their efficacy is based upon similar principles in that their injection into tissue results in a threshold concentration of aluminum being reached instantaneously [32]. Thus, aluminum from vaccines is redistributed to numerous organs including the brain where it can accumulate. Each vaccine adds to this tissue level of aluminum. A total dose of 30.6 mg (and not the 0.85 mg considered safe by the FDA) is available when we calculate the total aluminum dose available from 36 vaccinations [4-6]. Of course, not all of this aluminum ends up in the tissues. However, aluminum can accumulate in substantial amounts from ingesting foods containing aluminum and from drinking water.

Recent research has also demonstrated sensitization to food allergens following their co-administration with Al salts such as antacid preparations, via childhood vaccinations, and an ongoing aluminum overload. While the body may cope robustly with a mild but persistent immune response to aluminum overload, the coping mechanism could possibly be suddenly and dramatically overwhelmed by a new exposure to Al-adjuvant. The latter, will not only enhance the antigenicity of itself, but it will raise the level of the immune response against all significant body stores of aluminum. Under these conditions, an individual's everyday exposure to aluminum will continue to fuel the response and myriad symptoms of associated autoimmunity will take over the life of the affected individual [32]. The individual will now respond adversely to aluminum exposures which previously were not sufficient to elicit a biological response and the only solution will be to treat the aluminum overload and to reduce everyday exposure to aluminum.

Moreover, we show that fluoride could complex with any pre-existing Al^{3+} within body fluids to produce the AlFx and this could lead to a combination of chronic activation of G protein regulated systems, dysregulation of calcium homeostasis, and sustained activation of receptor functions. A number of studies have shown that AlFx can evoke all symptoms of ASD. We have reviewed studies that indicate that fluoride and Al^{3+} can exacerbate the pathological and clinical problems by worsening excitotoxicity, immune activation, and microglial priming. Moreover, AlFx accumulation in various compartments of the body may enhance the subclinical pathological alterations and/or the genetic susceptibility [1]. A number of ASD-related symptoms including intellectual disability, seizures, persistence of primitive reflexes, stereotypies, self-injurious behavior, among others, are known as potentially being emergent. We can assume that AlFx might play a role of trigger of these emergent phenomena [2].

AMELIORATION OF FLUORIDE AND ALUMINUM EFFECTS

Laboratory studies have revealed that withdrawal of fluoride resulted in some recovery of symptoms of fluoride toxicity. In 2006, a 12-person U.S. NRC committee reviewed the health risks associated with fluoride in the water and concluded:

> *"After reviewing the collective evidence, including studies conducted since the early 1990s, the committee concluded unanimously that the concentration of 4 mg/L for fluoride should be lowered. Such exposure clearly puts children at risk of developing severe enamel fluorosis, a condition that is associated with enamel loss and pitting."* (http://www.epa.gov/safewater/contaminants/index.html).

Many scientists are calling for a stop to water fluoridation namely in the USA. On the other hand, authorities from the CDC claim the safety and effectiveness of fluoride at levels used in community water fluoridation. The Director of CDC J. L.Geberding wrote:

> *"Experts have weighed the findings and the quality of the available evidence and found that the weight of peer-reviewed scientific evidence does not support an association between water fluoridation and any adverse health effect or systemic disorder. Current evidence supports water fluoridation as being safe and effective for all population groups; nevertheless, CDC and other health organizations constantly review the scientific literature and safety evidence for information that might indicate a need for closer examination or additional research."* (Personal letter to AS from April 29, 2008).

Most recently, proponents of water fluoridation have put their efforts into introducing mandatory fluoridation on a statewide level. They did this recently in the USA and are trying to do it presently in New Jersey. New Jersey has the highest overall prevalence of ASD among the US states - 9.9 autistic children per 1,000 children aged 8 years. New Jersey was the first state, where fluoride burden occurred during the Second Word War [21] due to the production of enriched uranium for atomic bomb. In this area ASD has recently been diagnosed for one from every 67 boys (14.8 ASD cases per 1000 healthy boys). (http://www.cdc.gov/mmwr/preview/mmwrhtml/ss5601a1.htm).

Since tea is considered as the source of aluminum, it is interesting to know that adding milk to the tea can form insoluble Al-phosphate, thereby reducing absorption of aluminum in GIT. Yet, addition of lemon, which contains substantial amounts of the organic acid citrate, forms an Al-citrate complex, which has been shown to increase aluminum absorption from the gut 6-fold. [17]. Citrate is known to increase both GI absorption as well as tissue accumulation of aluminum. The bone aluminum content in rats treated with Al-hydroxide plus citrate was 41 times higher than in the bones of control rats treated with Al-hydroxide alone. However, in one study with human volunteers administered aluminum and citrate in the drinking water, no difference was observed between the plasma concentrations of Al^{3+} in subjects receiving citrate with the drinking water and those that did not. While increasing aluminum absorption, citrate may also enhance urinary excretion in humans, resulting in no significant increase in Al^{3+} plasma concentration in individuals with normal kidney function [17].

Ameliorative effects of vitamins C, E, and D alone, and in combination, were reported in laboratory animals [74]. Vitamins C and E act as antioxidants scavengers of free radicals and peroxides, which accumulate after fluoride exposure. Vitamin E channels the conversion of oxidized glutathione to reduced glutathione, which in turn helps compression of mono- and dehydroascorbic acid to maintain ascorbic acid levels. Oral administration of vitamin C (50 mg/kg body weight/day) and vitamin E (2 mg/0.2 ml olive oil/animal/day) from day 6 to 19 of gestation along with NaF (40 mg/kg body weight) significantly ameliorates NaF-induced total percentage of skeletal and visceral abnormalities in rats. Vitamin E was comparatively less effective than vitamin C [75]. Vitamin D is known to promote GI absorption of Ca^{2+} and phosphate. Cotreatment with vitamins C, D, and E ameliorates NaF-induced reduction in serum Ca^{2+} and phosphorus [76]. Ekambaram and Paul [77] reported that calcium carbonate prevents not only fluoride-induced hypocalcemia but also the locomotor behavioral and dental toxicities of fluoride by decreasing bioavailability of fluoride in rats. Toxic effects of fluoride were reversible if its exposure was withdrawn for two months. Recovery was also possible by feeding antioxidants (superoxide dismutase, glutathione, β-carotene, and some herbal extracts) [78]. Liu *et al.* [79] reported that synthetic catalytic scavengers of ROS proved beneficial in mouse brain for reversal of age related learning deficits and oxidative stress in mice.

Reversal of fluoride induced cell injury and fluorosis through the elimination of fluoride and consumption of a diet containing essential nutrients and antioxidants, have been shown in humans [80]. Increasing dietary proteins, calcium, and vitamins may help in its prevention especially in pregnant and nursing women and children [81]. Treatments of vitamins C, D, and Ca^{2+} showed significant improvement in skeletal, clinical, and biochemical parameters in children consuming water containing 4.5 ppm of fluoride.

There is a need for a non-invasive method to both reduce the absorption of aluminum in the GIT and facilitate the excretion of systemic aluminum in the urine. Exley and co-workers are currently working on solutions to reduce the human body burden of Al [82,83]. Based on the knowledge that silicon is the natural antagonist to aluminum, researchers have shown that silicon-rich mineral waters can be used to reduce the body burden of aluminum in individuals with Alzheimer's disease, macrophagic myofasciitis, and chronic fatigue syndrome. In one case report they demonstrated that regular drinking of a silicon-rich mineral water over a three month period dramatically reduced the body burden of aluminum from aluminum overload, to a normal range [16].

The iron chelating drug desferrioxamine (DFO) has been used extensively to treat individuals with suspected aluminum overload and, in spite of the significant side effects associated with its use, DFO remains, as yet, the only accepted course of treatment for the removal of systemic aluminum. Chelation with DFO may increase urinary aluminum excretion. A new oral agent, Feralex-G, in experimental tests, have shown less toxicity and greater removal of aluminum from tissues than more toxic intramuscular agents [84]. Several *in vitro* tests have also show a number of flavonoids to be excellent aluminum chelators, such as curcumin, quercetin and fisetin [85-87].

CONCLUSIONS

Fluoride and Al^{3+} exposure is common in fetuses, newborns, and small children as a result of the dramatic increase in the volume of man-made industrial fluoride and aluminum compounds released into the environment and food chain. The long-term fluoride burden has several health effects with a striking resemblance to ASD. These include hypocalcemia, hypomagnesemia, hypothyroidism, sleep-pattern disturbance, and IQ deficits. Conceivably, fluoride inhibits the release of pineal melatonin by elevating glutamate levels. Fluoride interferes with a number of glycolytic enzymes, resulting in a significant suppression of cellular energy production, which has been shown to dramatically increase excitotoxicity. Several studies demonstrate that autistic children have altered energy metabolism. Fluoride and Al^{3+} also reduce the antioxidant potential in the cells. This could be another significant source of priming of microglia as well as excitotoxicity. In addition, both fluoride and Al^{3+} have effects on cell signaling that can affect neurodevelopment and neuronal function.

The discovery of synergistic action of fluoride plus Al^{3+} expanded our understanding of mechanisms of their effects on living organism. The widespread use of AlFx as a general activator of heterotrimeric G proteins provided evidence that AlFx is a molecule giving false messages, which are amplified by processes of signal transduction. It is evident that AlFx might evoke receptor malfunction. Importantly, mGluRs operate by G-proteins. AlFx accumulation in various compartments of the body may enhance the subclinical pathological alterations and/or genetic susceptibility to various pathologies. This mechanism could explain the emergence phenomena in etiopathogenesis of ASD on a molecular and cellular level. Signaling disorders represent a major cause for the etiopathology of ASD. A number of studies have shown that AlFx can affect learning and behavior, and induce a loss of cerebrovascular integrity both in experimental animals and humans.

Scientific studies have already provided accumulating evidence demonstrating that environmental and dietary excitotoxins, as well as fluoride and aluminum, can exacerbate the pathological and clinical problems by worsening excitotoxicity. The awareness of increasing load of fluoride and Al^{3+} as a new ecotoxicological phenomenon could contribute to the qualified assessment of their widespread use in water sources, food chains, and medicine.

REFERENCES:

[1] Strunecka A, Patocka J, Blaylock R, Chinoy N. Fluoride interactions: From molecules to disease. Current Signal Transduction Therapy 2007; 2: 190-213.

[2] Blaylock RL, Strunecka A. Immune-glutamatergic dysfunction as a central mechanism of the autism spectrum disorders. Curr Med Chem 2009; 16: 157-70.

[3] Blaylock R. Chronic microglial activation and excitotoxicity secondary to excessive immune stimulation: possible factors in Gulf War Syndrome and autism. J Am Phys Surg 2004; 9: 46-51.

[4] Blaylock RL. A possible central mechanism in autism spectrum disorders, part 1. Altern Ther Health Med 2008; 14: 46-53.

[5] Blaylock RL. A possible central mechanism in autism spectrum disorders, part 2: immunoexcitotoxicity. Altern Ther Health Med 2009; 15: 60-7.

[6] Blaylock RL. A possible central mechanism in autism spectrum disorders, part 3: the role of excitotoxin food additives and the synergistic effects of other environmental toxins. Altern Ther Health Med 2009; 15: 56-60.

[7] Strunecká A, Patočka J. Aluminofluoride Complexes in the Etiology of Alzheimer's disease. In: Roesky C, editor. Structure and Bonding. New Developments in Biological Aluminum Chemistry - Book 2.: Springer-Verlag; 2003. p. 139-81.

[8] Strunecka A, Strunecky O, Patocka J. Fluoride plus aluminum: useful tools in laboratory investigations, but messengers of false information. Physiol Res 2002; 51: 557-64.

[9] Yokel RA, Florence RL. Aluminum bioavailability from the approved food additive leavening agent acidic sodium aluminum phosphate, incorporated into a baked good, is lower than from water. Toxicology 2006; 227: 86-93.

[10] Yokel RA, McNamara PJ. Aluminium toxicokinetics: an updated minireview. Pharmacol Toxicol 2001; 88: 159-67.

[11] Yokel RA, Rhineheimer SS, Brauer RD, *et al.* Aluminum bioavailability from drinking water is very low and is not appreciably influenced by stomach contents or water hardness. Toxicology 2001; 161: 93-101.

[12] Krewski D, Yokel RA, Nieboer E, *et al.* Human health risk assessment for aluminium, aluminium oxide, and aluminium hydroxide. J Toxicol Environ Health B Crit Rev 2007; 10 Suppl 1: 1-269.

[13] Nicholson S, Exley C. Aluminum: a potential pro-oxidant in sunscreens/sunblocks? Free Radic Biol Med 2007; 43: 1216-7.

[14] Flarend R, Bin T, Elmore D, Hem SL. A preliminary study of the dermal absorption of aluminium from antiperspirants using aluminium-26. Food Chem Toxicol 2001; 39: 163-8.

[15] Speerhas RA, Seidner DL. Measured versus estimated aluminum content of parenteral nutrient solutions. Am J Health Syst Pharm 2007; 64: 740-6.

[16] Exley C, Swarbrick L, Gherardi RK, Authier FJ. A role for the body burden of aluminium in vaccine-associated macrophagic myofasciitis and chronic fatigue syndrome. Med Hypotheses 2009; 72: 135-9.

[17] Berend K, Voet BVd, Wolff Fd. Acute aluminum intoxication. Structure and Bonding 2003; 104: 2-58.

[18] Exley C. Aluminum and medicine. In: Merce A, editor. Molecular and Supramolecular Bioinorganic Chemistry: Nova Science Publishers, Inc.; 2008. p. 1-24.

[19] O'Hagan D, Schaffrath C, Cobb SL, Hamilton JT, Murphy CD. Biochemistry: biosynthesis of an organofluorine molecule. Nature 2002; 416: 279.

[20] Dean HT. The investigation of physiological effects by the " epidemiological method". In: Moulton FR, editor. Fluorine and dental health. Washington: AAAS; 1942. p. 23.

[21] Bryson C. The fluoride deception. In:. editor. The fluoride deception.: Seven Stories Press US; 2004. p. 1-272.

[22] Cao J, Zhao Y, Li Y, *et al.* Fluoride levels in various black tea commodities: measurement and safety evaluation. Food Chem Toxicol 2006; 44: 1131-7.

[23] Lung SC, Cheng HW, Fu CB. Potential exposure and risk of fluoride intakes from tea drinks produced in Taiwan. J Expo Sci Environ Epidemiol 2008; 18: 158-66.

[24] Malinowska E, Inkielewicz I, Czarnowski W, Szefer P. Assessment of fluoride concentration and daily intake by human from tea and herbal infusions. Food Chem Toxicol 2008; 46: 1055-61.

[25] Emekli-Alturfan E, Yarat A, Akyuz S. Fluoride levels in various black tea, herbal and fruit infusions consumed in Turkey. Food Chem Toxicol 2009; 47: 1495-8.

[26] Varner JA, Jensen KF, Horvath W, Isaacson RL. Chronic administration of aluminum-fluoride or sodium-fluoride to rats in drinking water alterations in neuronal and cerebrovascular integrity. Brain Res 1998; 784: 284-98.

[27] Exley C. The aluminium-amyloid cascade hypothesis and Alzheimer's disease. Subcell Biochem 2005; 38: 225-34.

[28] Exley C, Mamutse G, Korchazhkina O, *et al.* Elevated urinary excretion of aluminium and iron in multiple sclerosis. Mult Scler 2006; 12: 533-40.

[29] Arnow PM, Bland LA, Garcia-Houchins S, Fridkin S, Fellner SK. An outbreak of fatal fluoride intoxication in a long-term hemodialysis unit. Ann Intern Med 1994; 121: 339-44.

[30] Berend K, van der Voet G, Boer WH. Acute aluminum encephalopathy in a dialysis center caused by a cement mortar water distribution pipe. Kidney Int 2001; 59: 746-53.

[31] Gherardi RK, Coquet M, Cherin P, *et al.* Macrophagic myofasciitis: an emerging entity. Groupe d'Etudes et Recherche sur les Maladies Musculaires Acquises et Dysimmunitaires (GERMMAD) de l'Association Francaise contre les Myopathies (AFM). Lancet 1998; 352: 347-52.

[32] Exley C, Siesjo P, Eriksson H. The immunobiology of aluminium adjuvants: how do they really work? Trends Immunol 2010; 31: 103-9.

[33] Flarend RE, Hem SL, White JL, *et al. In vivo* absorption of aluminium-containing vaccine adjuvants using 26Al. Vaccine 1997; 15: 1314-8.

[34] Bergfors E, Trollfors B, Inerot A. Unexpectedly high incidence of persistent itching nodules and delayed hypersensitivity to aluminium in children after the use of adsorbed vaccines from a single manufacturer. Vaccine 2003; 22: 64-9.

[35] Woodbury DM. Maturation of the blood-brain and blood-CSF barriers. In: Vernadakis A, Weiner N, editors. Drugs and the developing brain. New York: Plenum Press; 1974. p. 259-80.

[36] Paul V, Ekambaram P, Jayakumar AR. Effects of sodium fluoride on locomotor behavior and a few biochemical parameters in rats. Environ Toxicol Pharmacol 1998; 6: 187-91.

[37] Shao Q, Wang Y, Guan Z. Influence of free radical inducer on the level of oxidative stress in brain of rats with fluorosis. Chung-Hua Yu Fang I Hsueh Tsa Chih [Chin J Preventive Med]   2000; 34: 330-2.

[38] Sun ZR, Liu FZ, L.N. W. Effects of high fluoride drinking water on the cerebral functions of mice. Chin J Endemiol 2000; 19: 262-3.

[39] Bhatanagar M, Rao P, Jain S, Bhatnagar R. Neurotoxicity of fluoride: Neurodegeneration in hippocampus of female mice. Indian J Exp Biol 2002; 40: 546-54.

[40] Shivarajashankara YM, Shivarajashankara AR, Bhat PG, Rao SH. Brain lipid peroxidation and antioxidant systems of yound rats in chronic fluoride intoxication. Fluoride 2002; 35: 197-203.

[41] Shivarajashankara YM, Shivarajashankara AR, Gopalakrishna Bhat P, Muddanna Rao S, Hanumanth Rao S. Histological changes in the brain of young fluoride-intoxicated rats. Fluoride 2002; 35: 12-21.

[42] Lu Y, Sun ZR, Wu LN, *et al.* Effect of high fluoride water on intelligence in children. Fluoride 2000; 33: 74-8.

[43] Spittle B. Fluoride and intelligence. Fluoride 2000; 33: 49-52.

[44] Li XS, Zhi JL, Gao RO. Effect of fluorine exposure on intelligence in children. Fluoride 1995; 28: 189-92.

[45] Zhao LB, Liang GH, Zhang DN, Wu XR. Effect of a high fluoride water supply on children's intelligence. Fluoride 1996; 29: 190-2.

[46] Xiang QY, Liang YX, Chen LS, *et al.* Effect of fluoride in drinking water on intelligence in children. Fluoride 2003; 36: 84-94.

[47] Rocha-Amador D, Navarro ME, Carrizales L, Morales R, Calderon J. Decreased intelligence in children and exposure to fluoride and arsenic in drinking water. Cad Saude Publica 2007; 23 Suppl 4: S579-87.

[48] Du L. The effect of fluorine on the developing human brain. Zhonghua Bing Li Xue Za Zhi 21:218-220, 1992. Zhonghua Bing Li Xue Za Zhi 1992; 21: 218-20.

[49] Mullenix PJ, Denbesten PK, Schunior A, Kernan WJ. Neurotoxicity of sodium fluoride in rats. Neurotoxicol Teratol 1995; 17: 169-77.

[50] Blaylock R. Excitotoxicity: a possible central mechanism in fluoride neurotoxicity. Fluoride 2004; 37: 264-77.

[51] Blaylock R. Fluoride neurotoxicity and excitotoxicity/microglial activation: critical need for more research. Fluoride 2007; 40: 89-92.

[52] Shivarajashankara YM, Shivarajashankara AR, Bhat PG, Rao SH. Effect of fluoridation intoxication on lipid peroxidation and antioxidant systems. Fluoride 2001; 34:

[53] Bachinskii PP, Gutsalenko OA, Naryzhniuk ND, Sidora VD, Shliakhta AI. [Action of the body fluorine of healthy persons and thyroidopathy patients on the function of hypophyseal-thyroid the system]. Probl Endokrinol (Mosk) 1985; 31: 25-9.

[54] Ge Y, Ning H, Wang S, Wang J. DNA thyroid damage in rats exposed to high F and low iodine. Fluoride 2005; 38: 318-23.

[55] Gallerti P. On the use of fluoride to treat overactive thyroid. Fluoride 1976; 9: 105-15.

[56] Tezelman S, Shaver JK, Grossman RF, *et al.* Desensitization of adenylate cyclase in Chinese hamster ovary cells transfected with human thyroid-stimulating hormone receptor. Endocrinology 1994; 134: 1561-9.

[57] Ruiz-Payan A, Duarte-Gardea M, Ortiz M, R. H. Chronic effects of fluoride on growth, blood chemistry, and thyroid hormones in adolescents residing in three communities in Northern Mexico. Fluoride 2005; 38: 246.

[58] Susheela AK, Bhatnagar M, Vig K, Mondal NK. Excess fluoride ingestion and thyroid hormone derangements in children living in Delhi, India. Fluoride 2005; 38: 98-108.

[59] Wittinghofer A. Signaling mechanistics: aluminum fluoride for molecule of the year. Curr Biol 1997; 7: R682-5.

[60] Ploeger A, Raijmakers ME, van der Maas HL, Galis F. The association between autism and errors in early embryogenesis: what is the causal mechanism? Biol Psychiatry 2010; 67: 602-7.

[61] Gillberg C, Gillberg IC. Infantile autism: a total population study of reduced optimality in the pre-, peri-, and neonatal period. J Autism Dev Disord 1983; 13: 153-66.

[62] May-Benson TA, Koomar JA, Teasdale A. Incidence of pre-, peri-, and post-natal birth and developmental problems of children with sensory processing disorder and children with autism spectrum disorder. Front Integr Neurosci 2009; 3: 31.

[63] Brimacombe M, Ming X, Lamendola M. Prenatal and birth complications in autism. Matern Child Health J 2007; 11: 73-9.

[64] Levy SM, Kohout FJ, Guha-Chowdhury N, *et al.* Infants' fluoride intake from drinking water alone, and from water added to formula, beverages, and food. J Dent Res 1995; 74: 1399-407.

[65] Fomon SJ, Ekstrand J. Fluoride intake by infants. J Public Health Dent 1999; 59: 229-34.

[66] Levy SM, Guha-Chowdhury N. Total fluoride intake and implications for dietary fluoride supplementation. J Public Health Dent 1999; 59: 211-23.

[67] Tanoue Y, Oda S. Weaning time of children with infantile autism. J Autism Dev Disord 1989; 19: 425-34.

[68] Luke J. The effect of fluoride on the physiology of the pineal gland. Guildford: University of Surrey; 1997.

[69] Luke J. Fluoride deposition in the aged human pineal gland. Caries Res. 2001; 35: 125-8.

[70] Tordjman S, Anderson GM, Pichard N, Charbuy H, Touitou Y. Nocturnal excretion of 6-sulphatoxymelatonin in children and adolescents with autistic disorder. Biol Psychiatry 2005; 57: 134-8.

[71] Pandi-Perumal SR, Srinivasan V, Maestroni GJ, *et al.* Melatonin: Nature's most versatile biological signal? FEBS J 2006; 273: 2813-38.

[72] Tauman R, Zisapel N, Laudon M, Nehama H, Sivan Y. Melatonin production in infants. Pediatr Neurol 2002; 26: 379-82.

[73] Schlesinger ER, Overton DE, Chase HC, Cantwell KT. Newburgh-Kingston caries-fluorine study. XIII. Pediatric findings after ten years. J Am Dent Assoc 1956; 52: 296.

[74] Chinoy NJ, Sharma AK. Amelioration of fluoride toxicity by vitamin E and D in reproductive function of male mice. Fluoride 1998; 31: 203-16.

[75] Verma RJ, Sherlin DM. Vitamin C ameliorates fluoride-induced embryotoxicity in pregnant rats. Hum Exp Toxicol 2001; 20: 619-23.

[76] Guna Sherlin DM, Verma RJ. Vitamin D ameliorates fluoride-induced embryotoxicity in pregnant rats. Neurotoxicol Teratol 2001; 23: 197-201.

[77] Ekambaram P, Paul V. Modulation of fluoride toxicity in rats by calcium carbonate and by withdrawal of fluoride exposure. Pharmacol Toxicol 2002; 90: 53-8.

[78] Chinoy NJ. Studies on fluoride, aluminium and arsenic toxicity in mammals and amelioration by some antidotes. In: Tripathi G, editor. Modern Trends in Environmental Biology. New Delhi: CBS Publishers; 2002. p. 164-96.

[79] Liu R, Liu IY, Bi X, *et al.* Reversal of age related learning deficits and brain oxidative stress in mice with superoxide dismutase/ catalase mimetics. Proc Natl Acad Sci USA 2003; 128: 15103.

[80] Susheela AK. Fluorosis management programme in India. Curr Sci 1999; 70: 298-304.

[81] Zang ZY, Fan JY, Yen W, *et al.* The effect of nutrition on the development of endemic osteomalacia in patients with skeletal fluorosis. Fluoride 1996; 29: 20-4.

[82] Exley C, Korchazhkina O, Job D, *et al.* Non-invasive therapy to reduce the body burden of aluminium in Alzheimer's disease. J Alzheimers Dis 2006; 10: 17-24; discussion 9-31.

[83] Exley C. Silicon in life: whither biological silicification? Prog Mol Subcell Biol 2009; 47: 173-84.

[84] Shin RW, Kruck TP, Murayama H, Kitamoto T. A novel trivalent cation chelator Feralex dissociates binding of aluminum and iron associated with hyperphosphorylated tau of Alzheimer's disease. Brain Res 2003; 961: 139-46.

[85] Cornard JP, Merlin JC. Spectroscopic and structural study of complexes of quercetin with Al(III). J Inorg Biochem 2002; 92: 19-27.

[86] Dimitric Markovic JM, Markovic ZS, Veselinovic DS, Krstic JB, Predojevic Simovic JD. Study on fisetin-aluminium(III) interaction in aqueous buffered solutions by spectroscopy and molecular modeling. J Inorg Biochem 2009; 103: 723-30.

[87] Kumar A, Dogra S, Prakash A. Protective effect of curcumin (Curcuma longa), against aluminium toxicity: Possible behavioral and biochemical alterations in rats. Behav Brain Res 2009; 205: 384-90.

CHAPTER 10

The Role of Melatonin in Etiopathogenesis and Therapy of Autism Spectrum Disorders

Anna Strunecka

Institute of Medical Biochemistry, Laboratory of Neuropharmacology, 1st Faculty of Medicine, Charles University in Prague, Prague, Czech Republic

Abstract: Pineal melatonin, an endogenous signal of darkness, is believed to be an important regulator of circadian and seasonal rhythms. The changing melatonin levels serve as hands of a bio-clock and dates of the bio-calendar in vertebrates including humans. Circulating melatonin regulates and influences the sleep wake cycle, sexual development, as well as various immune, endocrine, and metabolic functions. Initially it was thought to be produced exclusively by pineal gland. Subsequently, it was shown that melatonin is also produced in several other tissues. Substantial amounts of melatonin are found in the gut as well as the brain. Moreover, melatonin is one of the most powerful scavengers of free radicals. Sleep disorders and low melatonin levels are frequently observed in people with autism spectrum disorders (ASD). The role of abnormal melatonin biosynthesis in the gastrointestinal system relating to the development of ASD symptoms warrants further studies. The important multiple role of melatonin in prevention and amelioration of ASD symptoms is not fully recognized at present. Melatonin appears to be promising as one of the efficient and seemingly safe adjunctive treatments in children and adults with ASD.

INTRODUCTION

Melatonin was first purified and characterized from the bovine pineal gland extract by Aron Lerner and co-workers in 1958 [1]. Since this discovery the knowledge of the function of the pineal gland and the role of melatonin in humans has tremendously increased. It is generally known that melatonin is produced in the pineal gland, particularly at night [2-4]. The environmental light detected by the retina adjusts the biological clock in the suprachiasmatic nuclei (SCN), which innervates the pineal gland through sympathetic fibers. The nocturnal hormone melatonin circulates throughout the entire body and adjusts several bodily functions according to the presence and duration of darkness. The circulating melatonin level has been shown to be a good biomarker of circadian dysregulation. It can be reliably measured directly and indirectly through its metabolites in urine, blood, and saliva. Urinary melatonin has been shown to be stable over time, making it useful in many clinical and epidemiologic studies. Melatonin administration has been used as effective treatment of sleep disturbances during the last decades. Melatonin as a food supplement is widely used in the USA and several other countries.

In 1993, melatonin was discovered to function as a direct free radical scavenger *in vitro* [5]. Since then, an excess of one thousand publications have confirmed the ability of melatonin and its metabolites to reduce oxidative stress *in vivo*. Over the last thirty years, a great number of reports have documented a relationship between melatonin/pineal gland and the immune system in various species, including humans. This evidence implicates melatonin in a broad range of effects with a significant regulatory influence over many of the body's physiological functions (see Fig. **1**).

At present, several considerations seem to indicate that chronic sleep disorders in children with pervasive developmental disorder are associated with a disturbance in melatonin secretion. The pineal secretion of melatonin can be inhibited by accumulation of fluoride and aluminum in this gland, by accumulation of glutamate and aspartate, and by light exposure at night [6,7]. In this chapter we focus on melatonin biosynthesis and the physiological role of melatonin in pineal gland and gastrointestinal tract (GIT). This knowledge may be useful in understanding the etiopathogenesis of ASD as well as in the search for efficient and safe therapy.

*Address correspondence to: **Anna Strunecka**, Laboratory of Neuropharmacology, 1st Faculty of Medicine, Charles University in Prague, Albertov 4, 128 00 Prague 2, Czech Republic; E-mail: strun@natur.cuni.cz

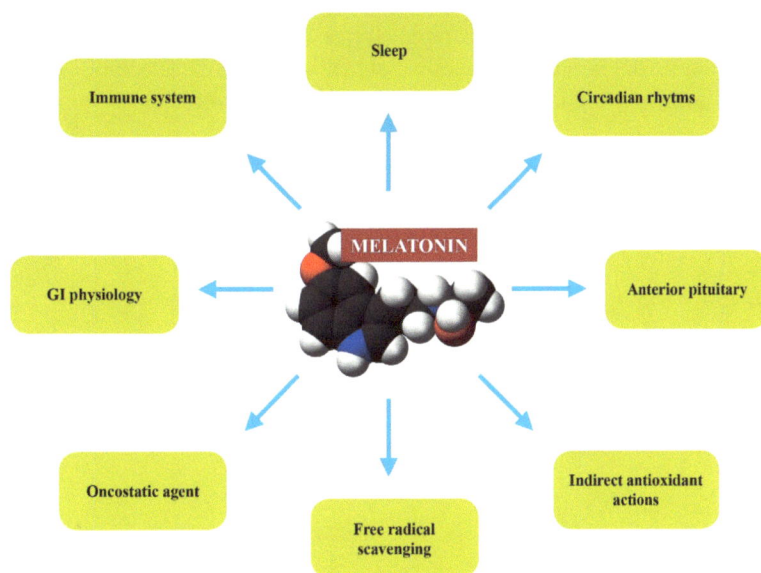

Figure 1: Most important physiological actions of melatonin. A file melatonin has been used from the <u>Wikimedia Commons</u> according GNU Free Documentation License.

BIOSYNTHESIS AND THE ROLE OF MELATONIN IN HEALTH

Melatonin Biosynthesis in Human Body

Melatonin was extracted by Lerner *et al.* (1958) from the bovine pineal gland. It had not been know to exist in biological tissue although it had been isolated as a urinary excretion product in rats after administration of 5-hydroxytryptamine (serotonin). Initially it was thought to be produced exclusively by pineal gland [8]. Subsequently it was shown that melatonin is also produced in other tissues including the retina, ciliary body, lens, Harderian gland, brain, thymus, airway epithelium, bone marrow, ovary, testicle, placenta, skin, cells of the immune system, the pancreas, the hepatobiliary system, and the human appendix [9]. Melatonin was confirmed immunohistologically in all segments of the GIT where it is produced in the enteroendocrine cells of the GIT mucosa. The concentrations of melatonin in the GIT are 10-100 times higher than in the plasma [10-12].

The biosynthesis of melatonin is initiated by the uptake of the essential amino acid tryptophan. Tryptophan is the least abundant of essential amino acids in normal diets. It is converted to another amino acid, 5-hydroxytryptophan, through the action of the enzyme tryptophan hydroxylase and then to 5-hydroxytryptamine (serotonin) by the enzyme aromatic amino acid decarboxylase. Serotonin is converted to melatonin through the action of two enzymes: serotonin N-acetyltransferase (NAT) and hydroxyindole O-methyltransferase (HIOMT) (Fig. **2**). NAT is often stated to be the rate-limiting enzyme in melatonin biosynthesis. HIOMT transfers a methyl group from S-adenosylmethionine to the 5-hydroxyl of the N-acetylserotonin. These reactions yield melatonin (N-acetyl-5-methoxytryptamine).

Here we do not deal with the detailed description of all other steps of metabolism of hydroxyindoles. It is important to know that the conversion of serotonin to melatonin is regulated by light/darkness cycle in the pineal gland. The activities of NAT and HIOMT rise soon after the onset of darkness because of the enhanced release of norepinephrine from sympathetic neurons terminating on the pineal parenchymal cells.

The biosynthetic steps of melatonin transforming tryptophan to melatonin have been also detected in cells of the upper GIT wall. There are substantial differences in the mode of melatonin secretion in the GIT. Whereas night time levels of melatonin in blood are mostly of pineal origin, day time melatonin concentrations in blood are produced mostly in the GIT [13]. Because of high levels of melatonin in the GIT tissues and the large size of the GIT, melatonin in the GIT was shown to be generated in about 500 times larger amount than is produced in the pineal gland [14,15]. After oral application of tryptophan, the plasma melatonin increases in a dose-dependent manner both in intact and pinealectomized animals. However, extrapineal melatonin, despite its large quantity, is not subjected to light/dark regulation and it does not serve as a chemical signal of light/dark adaptation. Surgical pinealectomy or constant light exposure, all markedly diminish the circulating melatonin level in vertebrates [9].

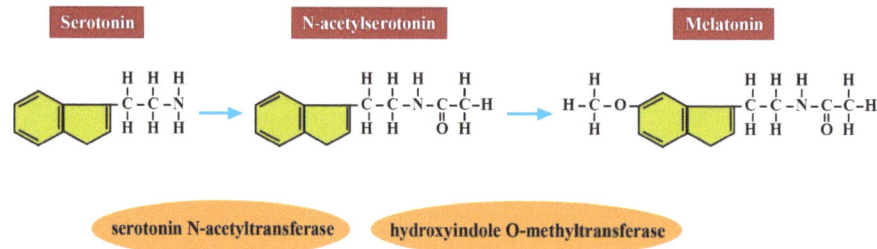

Figure 2: Serotonin is converted to melatonin through the action of two enzymes: serotonin N-acetyltransferase and hydroxyindole O-methyltransferase.

It is generally accepted that a major urinary metabolite of melatonin in vertebrates is 6-hydroxymelatonin sulfate, which is catabolized either in the liver or in other organs and tissues. Tan *et al.* [9] suggest that the original melatonin metabolite may be N1-acetyl-N2-formyl-5-methoxykynuramine (AFMK) rather than its commonly measured urinary excretory product 6-hydroxymelatonin sulfate. AFMK is present in mammals including humans. Numerous pathways for AFMK formation have been identified both *in vitro* and *in vivo*. These include enzymatic and pseudo-enzymatic pathways, interactions with reactive oxygen species /reactive nitrogen species (ROS/RNS) and with ultraviolet irradiation.

The Physiological Role of Pineal Melatonin

In its role as a pineal hormone, melatonin is a pleiotropic, nocturnally peaking and systemically acting chronobiotic. Melatonin provides vertebrates with a circulating signal of time and is essential for optimal integration of physiological functions with environmental lighting on a daily and seasonal basis. Thus, the daily and seasonally changing melatonin rhythms are involved in signaling time of day and time of year and, thus, they serve as signals of bio-clock and a bio-calendar in vertebrates including humans [9,16]. The melatonin rhythm in mammals is driven by a circadian clock located in the SCN, which is connected to the pineal gland by a dense network of catecholamine-containing sympathetic fibers [4,17]. The pineal gland is subordinate to the eye/SCN system: Activation of the SCN-pineal pathway occurs at night and results in the release of norepinephrine from the sympathetic fibers into the pineal perivascular space. The level of melatonin can be used as a biomarker of circadian regulation that could be used in studies of the effects of circadian disruption in humans. Circulating melatonin influences various immune, endocrine and metabolic functions. Dysfunction of the endogenous melatonin secretion is associated with mood and behavioral disorders and includes body weight regulation. The effects of the hormone melatonin are largely explained by actions via G protein-coupled receptors (GPCR) (see below).

The recent study of pineal demonstrated the functional categorization of the highly expressed and/or night/day differentially expressed genes identified clusters in pineal gland that are markers of specialized functions, including photodetection, melatonin synthesis, the immune/inflammation response, thyroid hormone signaling, and diverse aspects of cellular signaling and cell biology [18]. This study produce a paradigm shift in our understanding of the 24-h dynamics of the pineal gland from one focused on melatonin synthesis to one revealing the multiple integrative regulatory networks inside the small tissue which is the pineal gland.

The Potential Role of Melatonin in the GIT

There are many indications that melatonin has a variety of beneficial effects in the GIT. Melatonin functions in the gut generally seem to be protective of the mucosa from erosion and ulcer formation and to possibly influence movement of the GI contents through the digestive system [19]. In the GIT, melatonin exhibits endocrine, paracrine, autocrine, and luminal actions. Generally, the episodic secretion of melatonin from the GIT is related to the intake and digestion of food and to the prevention of tissue damage caused by hydrochloric acid and digestive enzymes. Some actions, such as the scavenging of hydroxyl free radicals, immunoenhancement and antioxidant effects are of general nature, whereas others, such as an increase of mucosal blood flow, the reduction of peristalsis and the regulation of fecal water content, are specific to the tubular GIT.

Generally, melatonin actions oppose those of serotonin. Laboratory and clinical studies indicate that the utilization of melatonin can prevent or treat pathological conditions such as esophageal and gastric ulcers, pancreatitis, colitis, irritable bowel disease, and colon cancer. Tan and co-workers speculate that the locally generated melatonin is consumed by the tissues in which it is produced as a protective mechanism of oxidative stress [9,20]. This might occur especially in the GIT, which is continuously exposed to the hostile outside environments such as food pollutants, bacteria, parasites, and toxins. The function of locally produced high levels of melatonin may be to help them cope with these stressors as an antioxidant and anti-inflammatory agent.

MECHANISMS OF MELATONIN ACTION

Melatonin Receptors and Signaling Pathways

As a highly lipophilic compound, melatonin diffuses easily though the plasma membranes to reach, within a short time, almost every cell in the body. Melatonin exerts many of its physiological actions by interacting with two receptors, MT1 and MT2, which have been cloned in mammals [21,22]. Both belong to the family of GPCRs. The heptahelical MT1 and MT2 can form homo- and heterodimers, which are expressed in various parts of the CNS (SCN, hippocampus, cerebellar cortex, prefrontal cortex, basal ganglia, substantia nigra, ventral tegmental area, nucleus accumbens, and retinal horizontal, amacrine and ganglion cells) and in peripheral organs (blood vessels, mammary gland, GI tract, liver, kidney and bladder, ovary, testis, prostate, skin, and the immune system) [23]. RT-PCR analysis showed that in rats MT1 is highly expressed in the hypothalamus, lung, kidney, adrenal gland, stomach, and ovary, while MT2 is highly expressed in the hippocampus, kidney, and ovary [24].

Melatonin receptors mediate a plethora of intracellular effects. It has been shown that the MT1 receptor subtype is coupled to different G proteins that mediate adenylyl cyclase (AC) inhibition and frequently decreases cAMP, but also activates phospholipase C (PLC) β and protein kinase C (PKC), acts via the MAP kinase and PI3 kinase/Akt pathways, modulates large conductance Ca^{2+}-activated K^{+}, and voltage-gated Ca^{2+} channels. The MT2 receptor is also coupled to inhibition of AC and additionally it inhibits the soluble guanylyl cyclase pathway [22]. Whereas MT1 receptors are necessary for the acute inhibitory action of melatonin or melatonin receptor agonists on neuronal activity in the SCN, MT2 receptors mediate the phase shift of circadian rhythm of neuronal activity in the SCN.

Melatonin as Free Radical Scavenger and Antioxidant

In excess of one thousand publications have confirmed the ability of melatonin and its metabolites to scavenge both oxygen and nitrogen-based reactants, to stimulate antioxidative enzymes, and to reduce oxidative stress. Melatonin's antioxidant capacity involves the direct, receptor-independent scavenging of free radicals and reactive oxygen species. The indirect antioxidative actions may depend on cellular melatonin receptors. Reiter [25] and Tan *et al.* [9] suggested that to act as a receptor-independent free radical scavenger and as a broad-spectrum antioxidant, is a primitive and primary function of melatonin. Melatonin is phylogenetically very old molecule. Melatonin is estimated to have evolved 2.5-3 billion years ago, coincident with the development of oxygen-based metabolism.

It has been documented that the free radical scavenging capacity of melatonin extends to its secondary, tertiary and quaternary metabolites. It means that melatonin's interaction with ROS/RNS is a prolonged scavenging cascade that involves many of melatonin derivatives. Considering the cascade of reactions that includes melatonin metabolite AFMK, a melatonin molecule can scavenge up to ten ROS/RNS molecules [9]. This cascade reaction is a specific property of melatonin, which makes melatonin action different from other conventional antioxidants. This mechanism also makes melatonin highly effective, even at physiological concentrations, in protecting organisms from oxidative stress.

Melatonin, at least in pharmacological concentrations, has the capability of increasing either mRNA levels or the activities of the major antioxidative enzymes, such as glutathione peroxidase, glutathione reductase, and catalase. Evidence from several laboratories shows that glutathione (GSH) and GSSG levels are significantly changed in autistic individuals (see Chapter 7). Reduced levels of GSH greatly increase the vulnerability of neurons and astrocytes to excitotoxicity and oxidative stress. When combined with reduced mitochondrial energy production, low antioxidant enzymes, low glutathione, and reduced secretion of melatonin, one can reasonably expect an acceleration of damage to brain's elements (see Chapter 4). Beeing amphiphilic, melatonin crosses the BBB, the placenta, and cell membranes. Melatonin thus protects the brain and the fetus from damage induced by free radicals and toxins (Fig. **3**).

Intensive oxidative stress results in a rapid drop of circulating melatonin levels. This melatonin decline is not related to its reduced synthesis but to its rapid consumption, i.e. circulating melatonin is rapidly metabolized by interaction with ROS/RNS induced by stress. Rapid melatonin consumption during elevated stress may serve as a protective mechanism of organisms in which melatonin is used as a first-line defensive molecule against oxidative damage. The oxidative status of organisms thus modifies melatonin metabolism. It has been reported that the higher the oxidative state, the more AFMK is produced. The ratio of AFMK and another melatonin metabolite, cyclic 3-hydroxymelatonin (3-OHM), may serve as an indicator of the level of oxidative stress in organisms [25].

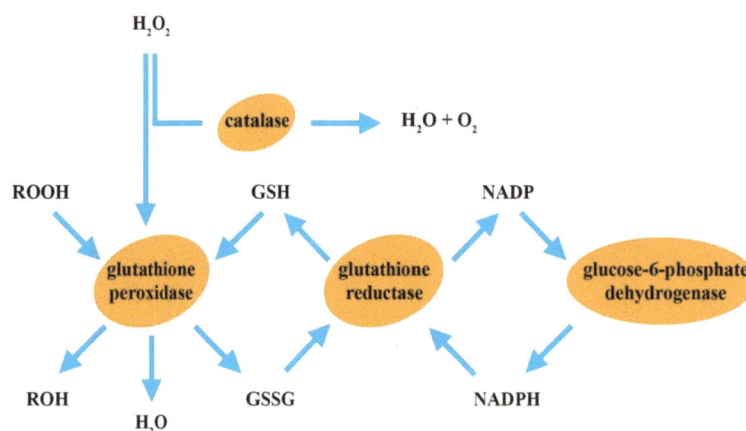

Figure 3: Melatonin is an efficient direct and indirect antioxidant. Hydrogen peroxide (H_2O_2), the immediate precursor of hydroxyl radical, is metabolized via the action of catalase and glutathione peroxidase. Glutathione peroxidase also metabolizes reactive hydroperoxides (ROOH).Glutathion peroxidase also oxidizes reduced glutathione (GSH) to its disulfide form (GSSH), which is recycled back to GSH by the action of glutathione reductase (see Chapter 7). A cofactor for glutathione reductase is NADPH, which is supplied by the action of glucose-6-phosphate dehydrogenase. Melatonin stimulates all these enzymes. Additionally, melatonin increases GSH production by stimulating its synthesis (Modified according to [25].)

Physiological Link between Melatonin and the Immune System

A tight, physiological link between the pineal gland and the immune system is emerging from a series of experimental studies [26-28]. The localization of melatonin biosynthesis in lymphoid organs such as the bone marrow, thymus, and lymphocytes reflects its immunoneuroendocrine role. Melatonin has been shown to be involved in the regulation of both cellular and humoral immunity. In both normal and leukaemic mice, melatonin administration results in quantitative and functional enhancement of natural killer (NK) cells, whose role is to mediate defenses against virus-infected cells and cancer cells. Melatonin appears to regulate cell dynamics, including the proliferative and maturational stages of virtually all hemopoietic and immune cells lineages involved in host defense - not only NK cells but also T and B lymphocytes, granulocytes and monocytes - in both bone marrow and tissues [27]. Melatonin not only stimulates the production of natural killer cells, monocytes and leukocytes, but also alters the balance of T helper (Th)-1 and Th-2 cells mainly towards Th-1 responses and increases the production of relevant cytokines such as interleukin IL-2, IL-6, IL-12, and interferon-γ. The regulatory function of melatonin on immune mechanisms is seasonally dependent [28]. It is possible to recognize reciprocal influences between cytokine action and pineal endocrine activity, suggesting the existence of feedback mechanisms responsible for a central regulation of cytokine network [29-30].

THE ROLE OF MELATONIN IN ETIOPATHOGENESIS OF ASD

Regulation of melatonin secretion appears to be abnormal in children with ASD but the underlying cause of this deficit is unknown. The absence of nocturnal secretion and elevated daytime levels has been observed [31]. Some symptoms of ASD such as the sleep problems and the early onset of puberty suggest abnormalities in melatonin physiology and dysfunctions of the pineal gland [32]. A study of Tauman *et al.* [33] revealed that babies with the lowest melatonin production had the most neurobehavioral problems.

Melke with co-workers [34] studied the *ASMT* gene, encoding the last enzyme of melatonin synthesis. Interestingly, this gene is located on the pseudo-autosomal region 1 of the sex chromosomes, deleted in several individuals with ASD. Melke *et al.* [34] sequenced all *ASMT* exons and promoters in 250 individuals with ASD and compared the allelic frequencies with 255 controls. Non-conservative variations of *ASMT* were identified, including a splicing mutation present in two families with ASD, but not in controls. Two polymorphisms located in the promoter (rs4446909 and rs5989681) were more frequent in ASD compared to controls and were associated with a dramatic decrease in *ASMT* transcripts in blood cell lines. Biochemical analyses performed on blood platelets and/or cultured cells revealed a highly significant decrease in *ASMT* activity and melatonin level in individuals with ASD. These results indicate that a low melatonin level, caused by a primary deficit in *ASMT* activity, is a risk factor for ASD. They also support *ASMT* as a susceptibility gene for ASD.

Luke [35] reported that fluoride accumulates in the pineal gland and that mongolian gerbils fed higher doses of fluoride excreted less melatonin metabolite in their urine and took a shorter time to reach puberty. It is also known that production of melatonin by the pineal is controlled by metabotropic glutamate receptors (mGluRs) and that excess aspartate or glutamate activity can inhibit melatonin release [36]. Pinealocytes also secrete glutamate and this suppresses further melatonin secretion, a possible feedback mechanism. Yamada and co-workers, using rat pinealocytes, demonstrated that suppression of melatonin secretion occurred when mGluR3 were activated [36].

Melatonin and Pregnancy

Melatonin works in its variety of ways in human pregnancy and it appears to be essential for successful pregnancy. Laure-Kamionowska *et al.* [37] demonstrated that during its development the pineal gland is very susceptible to injury. The failure of normal pineal gland development and subsequent impaired production of melatonin decrease resistance of newborns and children to various environmental harmful agents.

The data available raise the intriguing possibility that the fetal SCN and fetal tissues act as peripheral clocks commanded by separate maternal signals. Lipophilic melatonin crosses the placenta freely without being altered. Maternal melatonin enters the fetal circulation with ease providing photoperiodic information to the fetus as a circadian rhythm modulator, endocrine modulator, immunomodulator, direct free radical scavenger and indirect antioxidant. The nighttime serum concentration of melatonin shows an incremental change toward the end of pregnancy [38]. The SCN develops throughout gestation but is still immature for some time after. The maternal environment affects the phase of rhythms and the fetal response of the circadian timing system to light pulses. The fetal SCN could be thus vulnerable to maternal influences, such as poor nutrition, stress and drugs, all of which can affect neuronal development [39]. Oral and *i.p.* administered monosodium glutamate has been shown to damage the SCN.

Sleep Disturbances

Alteration in melatonin rhythm may be responsible for sleep-onset and maintenance difficulties in individuals with ASD. The sleep-promoting and sleep/wake rhythm regulating effects of melatonin are attributed to its action on MT1 and MT2 melatonin receptors present in the SCN of the hypothalamus [40]. Melatonin primarily regulates the length of time needed to fall asleep and on the total length of sleep, and that it has many advantages compared to commonly used medications such as the benzodiazepines since it has no side effects when used for short periods and does not induce dependence or withdrawal syndromes.

Tordjman *et al.* [31] measured the nocturnal urinary excretion of 6-sulphatoxymelatonin by radioimmunoassay in groups of 49 children and adolescents with ASD and 88 normal control individuals matched on age, sex, and Tanner stage of puberty. Nocturnal 6-sulphatoxymelatonin excretion rate was significantly and substantially lower in patients with autism than in normal controls and was significantly negatively correlated with severity of autistic impairments in verbal communication and play.

Few studies have been published, which could document the advantages of melatonin in the treatment of sleep disorders in children [41,42], adolescents [43], and adults [44] with ASD. A randomized, placebo-controlled double-blind crossover trial of melatonin taken by 11 children with ASD provided evidence of effectiveness of melatonin in children with sleep difficulties [41]. An Italian study [42] followed 25 children, aged 2.6-9.6 years with autism after 1-3-6-month melatonin treatment and one month after discontinuation. During treatment sleep patterns of all children improved. After discontinuation 16 children returned to pre-treatment score, re-administration of melatonin was again effective. Treatment gains were maintained at 12 and 24-month follow-ups. No adverse side effects were reported.

Galli-Carminatti with co-workers observed six adults with autism. Melatonin was initiated at a daily dose of 3 mg at nocturnal bedtime. If this proved ineffective, the melatonin dose was titrated over the following 4 weeks at increments of 3 mg/2 weeks up to a maximum of 9 mg, unless it was not tolerated [44].

Andersen *et al.* investigated one hundred seven children (2-18 years of age) with a confirmed diagnosis of ASD who received melatonin by reviewing the electronic medical records of a single pediatrician [45]. Clinical response to melatonin was based on parental report. The melatonin dose varied from 0.75 to 6 mg. After initiation of melatonin, parents of 27 children (25%) no longer reported sleep concerns at follow-up visits. Parents of 64 children (60%) reported improved sleep, although continued to have concerns regarding sleep. Parents of 14 children (13%) continued to report sleep problems as a major concern, with only one child having worse sleep after starting melatonin (1%), and one child having undetermined response (1%). Only three children had mild side-effects after starting melatonin, which included morning sleepiness and increased enuresis. There was no reported increase in seizures after starting melatonin in children with pre-existing epilepsy and no new-onset seizures. The majority of children were taking psychotropic medications. Melatonin has been generally reported as a safe and well-tolerated treatment for insomnia in children with ASD.

The Link between Glutamate and Melatonin Synthesis in the Pineal Gland

Rat pinealocytes use L-glutamate as a modulator for melatonin synthesis. It has been observed that aspartate and glutamate inhibit melatonin synthesis [36,46,47]. In addition to noradrenergic (NA) innervation that stimulates melatonin synthesis in a cAMP-mediated manner, pinealocytes secrete L-glutamate through an exocytic mechanism. The released glutamate inhibits melatonin synthesis [36]. Upon binding of L-glutamate to the class II mGluRs, NA-dependent formation of cAMP was inhibited, resulting in decreased NAT activity and melatonin output. Although L-glutamate at 1 mM caused 90% inhibition of melatonin synthesis, about 30% of the NAT activity remained, suggesting the presence of another target for L-glutamate. Yamada *et al.* found that L-glutamate also inhibits HIOMT [36]. The inhibition is reversible and dose-dependent, with the maximal inhibition being obtained with more than 0.4 mM L-glutamate. These results indicated that HIOMT is another target for L-glutamate due to its inhibition of melatonin synthesis, and the signaling pathway toward the inhibition is distinct from that of NAT.

The link between glutamate and melatonin synthesis could play a role in the pathogenesis of ASD. The dysregulation of glutamate homeostasis in the pineal gland will decrease nocturnal melatonin biosynthesis. Kim *et al.* suggested that glutamate transporters mediate synchronized elevation of L-glutamate and thereby efficiently down-regulate melatonin secretion *via* previously identified inhibitory mGluR in the mammalian pineal gland [48]. It has been well documented that glutamate transporters are quite sensitive to mercury toxicity and that inhibition of glutamate transport can dramatically increase extracellular glutamate levels.

The ability of mercury to induce glutamate accumulation in the brain has also been demonstrated in freely moving mice [49]. At a dose of 10 μM they found a 9.8-fold rise in extracellular glutamate and at 100 μM dose they observed a 2.4-fold rise, indicating that a lower dose of mercury was significantly more excitotoxic.

The effect of glutamate transporters on the membrane potential depends on the subtype of the transporter [48]. However, in the presence of mercury, expressing glutamate transporter-1 (GLT-1) allows pinealocytes to be depolarized upon EAA stimulation because of reverse transport of glutamate from the astrocyte. Depolarization of the membrane caused by GLT-1 dysfunction leads to Ca^{2+} elevation intraneuronally. A significant rise in cytosolic calcium level $[Ca^{2+}]_i$ occurs with 10 μM L-aspartate, and larger rises occur at higher substrate concentrations. The increased glutamate and or aspartate level in the blood could also inhibit melatonin synthesis, since the pineal gland belongs to the so called circumventricular organs (CVO). This means that pinealocytes are not protected by BBB and both glutamate and aspartate can freely penetrate into pinealocytes from the systemic circulation.

Melatonin Synthesis in the GIT

Bubenik with co-workers detected melatonin in the mucous membrane of the entire GIT [50]. The possibility of extrapineal melatonin synthesis was accepted when melatonin-synthesizing enzymes, NAT and HIOMT were also detected in the GIT and confirmed by transcription polymerase chain reaction [12]. Moreover, beside production of melatonin by enteroendocrine cells of the GIT, Bubenik with co-workers demonstrated that the GIT can extract melatonin from the circulation. At any day or night time, GIT tissues contain at least 400 times more melatonin than the pineal gland. Concentrations of melatonin in the GIT exceed those in plasma by 10 to 100 times. Melatonin produced in the GIT is forwarded to the general circulation via hepatic portal vein [12]. In later study, melatonin was found in the liver in levels 15 times higher than its concentrations in the blood and degradation of melatonin in the liver was reported to be the main pathway for its catabolism [14]. The concentration of melatonin in bile is enormous; it exceeds melatonin concentrations in the GIT by some 10-40 times. It has been speculated, that melatonin in the bile may protect the mucosa of GIT from oxidative stress [51].

Melatonin has been tested as a clinical remedy for prevention or treatment of diseases of the GIT, ulceration of esophagus, duodenum, pancreas, and colon [12]. Administration of melatonin increased the number and size of Payer´s patches, the main component of the GIT immune system. Melatonin was found to be effective agent against experimental colitis in more than a dozen studies [52].

It is of interest that the addition of tryptophan to the breakfast of infants and young students (0-8 years) was found to increase the quality of sleep, as well as improve mental health and appetite, probably due to its conversion to serotonin and melatonin [53]. Interestingly, administering melatonin in a dose of 3 mg before bedtime for two weeks has been shown to significantly reduce abdominal pain and rectal pain sensitivity in autistic children, a common affliction [54].

Melatonin Bioavailability in Humans

Compared with rodents, the bioavailability of melatonin in humans is poor [9]. A clear study using deuterium labeled melatonin as a reference standard revealed that the bioavailability of orally consumed melatonin was as low as 1% in some subjects. The low bioavailability of melatonin is attributed to its first pass effect through the liver. In addition to its low bioavailability, substantial individual variations in melatonin bioavailability have been observed. This variation can be as high as 37-fold. The average individual variations calculated from available data in terms of melatonin's bioavailability in humans differ by about 18-fold.

Currently, the most popular melatonin formula commercially available is a 3 mg tablet. For some subjects, this dose when taken to benefit sleep may induce drowsiness the day after; for others, whose bioavailability is low, this dose may not be sufficient to treat insomnia The melatonin dose has been increased in some subjects with autism up to a maximum of 9 mg, unless it was not tolerated [44].

To obtain the optimal effects of melatonin treatment, individualization of dose is suggested based on the serum or salivary melatonin levels after melatonin administration or adjusting the dose depending on the responses of the subjects. Drug interactions also influence the bioavailability of melatonin. For example, co-administration of melatonin with CYP1A2 inhibitor, fluvoxamine (also a serotonin reuptake inhibitor), in healthy subjects, results in a 17-fold increase in melatonin blood levels [9].

CONCLUSIONS

The discovery of the versatile role of melatonin in human body opens the possibility of assessing the impact of altered melatonin on the pathophysiology and behavioral expression of ASD. Melatonin has a gamut of actions on organismal physiology. The pineal gland, represented by melatonin, is truly a "regulator of regulators". The absence of nocturnal secretion of melatonin was found in children and adolescents with autism. Some symptoms of ASD such as the sleep problems and the early onset of puberty suggest abnormalities in melatonin physiology and dysfunctions of the pineal gland [31,32]. Melatonin has an important role in the embryonic and fetal development from very early stages. The fetal pineal is very susceptible to injury and vulnerable to maternal influences [39,37].

Excitotoxic factors such as glutamate, aspartate, mercury, fluoride and aluminum could influence the biosynthesis of pineal melatonin as well as its metabolism in other tissues and organs. Oxidative stress and immunoexcitotoxicity play key roles in the pathogenesis of ASD; therefore, the effect of melatonin as the most powerful scavenger of free radicals is of great importance in amelioration and therapy of ASD symptoms. Melatonin protects the brain and the fetus from damage induced by free radicals. Further research on the role of melatonin in the GIT of autistic subjects is warranted in order to understand the mechanisms of its protective effects and links with immunity as regards GI pathology of ASD.

Melatonin as a food supplement is widely used in the USA and in several other countries. According to estimates by the Mayo Clinic, more than 20 million Americans now regularly take melatonin supplements, spending $200-$350 million every year (http://your-doctor.com/patient_info/nutrition_supplements/melatonin.html). However, the pharmacokinetics of melatonin, especially its bioavailability in humans, has not drawn sufficient attention. The available data show the low utility of melatonin administration in various individuals with autism.

The powerful regulator of pineal melatonin biosynthesis is light. It has been documented by many animal and human studies that light exposure at night severely compromises the circadian production of melatonin [2-4,55]. It seems that the danger of sleeping with light in the bedroom (like a night light, or the computer or a TV that's left on) for disturbances of melatonin biosynthesis and therefore the importance of a sleeping in a dark room for the restoration of melatonin rhythms has not been fully recognized. The nutritional supplement with amino acid tryptophan – the precursor of melatonin biosynthesis is also recommended for patients with ASD.

This work was supported by VZ MSM 0021620806.

REFERENCES

[1]   Lerner AB, Case LD, Lee TH, Mori V. Isolation of melatonin, the pineal factor that lightens melanocytes. J Am Chem Soc 1958; 80: 2587-89.
[2]   Maitra SK, Huesgen A, Vollrath L. The effects of short pulses of light at night on numbers of pineal "synaptic" ribbons and serotonin N-acetyltransferase activity in male Sprague-Dawley rats. Cell Tissue Res 1986; 246: 133-6.
[3]   Vollrath L, Seidel A, Huesgen A, *et al.* One millisecond of light suffices to suppress nighttime pineal melatonin synthesis in rats. Neurosci Lett 1989; 98: 297-8.
[4]   Foulkes NS, Whitmore D, Sassone-Corsi P. Rhythmic transcription: the molecular basis of circadian melatonin synthesis. Biol Cell 1997; 89: 487-94.
[5]   Hardeland R, Pandi-Perumal SR, Cardinali DP. Melatonin. Int J Biochem Cell Biol 2006; 38: 313-6.
[6]   Strunecka A, Patocka J, Blaylock R, Chinoy N. Fluoride interactions: from molecules to disease. Curr Signal Transduct Ther 2007; 2: 190-213.
[7]   Blaylock RL, Strunecka A. Immune-glutamatergic dysfunction as a central mechanism of the autism spectrum disorders. Curr Med Chem 2009; 16: 157-70.
[8]   Lerner AB, Nordlund JJ. Melatonin: clinical pharmacology. J Neural Transm Suppl 1978; 339-47.
[9]   Tan DX, Manchester LC, Terron MP, Flores LJ, Reiter RJ. One molecule, many derivatives: a never-ending interaction of melatonin with reactive oxygen and nitrogen species? J Pineal Res 2007; 42: 28-42.
[10] Bubenik GA. Localization, physiological significance and possible clinical implication of gastrointestinal melatonin. Biol Signals Recept 2001; 10: 350-66.
[11] Bubenik GA. Gastrointestinal melatonin: localization, function, and clinical relevance. Dig Dis Sci 2002; 47: 2336-48.

[12]   Bubenik GA. Thirty four years since the discovery of gastrointestinal melatonin. J Physiol Pharmacol 2008; 59 Suppl 2: 33-51.

[13]   Bubenik GA, Pang SF, Cockshut JR, *et al.* Circadian variation of portal, arterial and venous blood levels of melatonin in pigs and its relationship to food intake and sleep. J Pineal Res 2000; 28: 9-15.

[14]   Messner M, Huether G, Lorf T, Ramadori G, Schworer H. Presence of melatonin in the human hepatobiliary-gastrointestinal tract. Life Sci 2001; 69: 543-51.

[15]   Konturek SJ, Konturek PC, Brzozowska I, *et al.* Localization and biological activities of melatonin in intact and diseased gastrointestinal tract (GIT). J Physiol Pharmacol 2007; 58: 381-405.

[16]   Reiter RJ, Korkmaz A. Clinical aspects of melatonin. Saudi Med J 2008; 29: 1537-47.

[17]   Karolczak M, Korf HW, Stehle JH. The rhythm and blues of gene expression in the rodent pineal gland. Endocrine 2005; 27: 89-100.

[18]   Bailey MJ, Coon SL, Carter DA, *et al.* Night/day changes in pineal expression of >600 genes: central role of adrenergic/cAMP signaling. J Biol Chem 2009; 284: 7606-22.

[19]   Reiter RJ, Tan DX, Mayo JC, *et al.* Neurally-mediated and neurally-independent beneficial actions of melatonin in the gastrointestinal tract. J Physiol Pharmacol 2003; 54 Suppl 4: 113-25.

[20]   Tan DX, Manchester LC, Hardeland R, *et al.* Melatonin: a hormone, a tissue factor, an autocoid, a paracoid, and an antioxidant vitamin. J Pineal Res 2003; 34: 75-8.

[21]   Wu YH, Zhou JN, Balesar R, *et al.* Distribution of MT1 melatonin receptor immunoreactivity in the human hypothalamus and pituitary gland: colocalization of MT1 with vasopressin, oxytocin, and corticotropin-releasing hormone. J Comp Neurol 2006; 499: 897-910.

[22]   Jockers R, Maurice P, Boutin JA, Delagrange P. Melatonin receptors, heterodimerization, signal transduction and binding sites: what's new? Br J Pharmacol 2008; 154: 1182-95.

[23]   Pandi-Perumal SR, Trakht I, Srinivasan V, *et al.* Physiological effects of melatonin: role of melatonin receptors and signal transduction pathways. Prog Neurobiol 2008; 85: 335-53.

[24]   Ishii H, Tanaka N, Kobayashi M, Kato M, Sakuma Y. Gene structures, biochemical characterization and distribution of rat melatonin receptors. J Physiol Sci 2009; 59: 37-47.

[25]   Reiter RJ. Melatonin: Lowering the high price of free radicals. News Physiol Sci 2000; 15: 246-250

[26]   Maestroni GJ. The immunotherapeutic potential of melatonin. Expert Opin Investig Drugs 2001; 10: 467-76.

[27]   Miller SC, Pandi-Perumal SR, Esquifino AI, Cardinali DP, Maestroni GJ. The role of melatonin in immuno-enhancement: potential application in cancer. Int J Exp Pathol 2006; 87: 81-7.

[28]   Srinivasan V, Spence DW, Trakht I, *et al.* Immunomodulation by melatonin: its significance for seasonally occurring diseases. Neuroimmunomodulation 2008; 15: 93-101.

[29]   Lissoni P. The pineal gland as a central regulator of cytokine network. Neuro Endocrinol Lett 1999; 20: 343-9.

[30]   Lissoni P, Rovelli F, Brivio F, Fumagalli L, Brera G. A study of immunoendocrine strategies with pineal indoles and interleukin-2 to prevent radiotherapy-induced lymphocytopenia in cancer patients. *In Vivo* 2008; 22: 397-400.

[31]   Tordjman S, Anderson GM, Pichard N, Charbuy H, Touitou Y. Nocturnal excretion of 6-sulphatoxymelatonin in children and adolescents with autistic disorder. Biol Psychiatry 2005; 57: 134-8.

[32]   Pandi-Perumal SR, Smits M, Spence W, *et al.* Dim light melatonin onset (DLMO): a tool for the analysis of circadian phase in human sleep and chronobiological disorders. Prog Neuropsychopharmacol Biol Psychiatry 2007; 31: 1-11.

[33]   Tauman R, Zisapel N, Laudon M, Nehama H, Sivan Y. Melatonin production in infants. Pediatr Neurol 2002; 26: 379-82.

[34]   Melke J, Goubran Botros H, Chaste P, *et al.* Abnormal melatonin synthesis in autism spectrum disorders. Mol Psychiatry 2008; 13: 90-8.

[35]   Luke J. Fluoride deposition in the aged human pineal gland. Caries Res 2001; 35: 125-8.

[36]   Yamada H, Yatsushiro S, Ishio S, *et al.* Metabotropic glutamate receptors negatively regulate melatonin synthesis in rat pinealocytes. J Neurosci 1998; 18: 2056-62.

[37]   Laure-Kamionowska M, Maslinska D, Deregowski K, Czichos E, Raczkowska B. Morphology of pineal glands in human foetuses and infants with brain lesions. Folia Neuropathol 2003; 41: 209-15.

[38]   Tamura H, Nakamura Y, Terron MP, *et al.* Melatonin and pregnancy in the human. Reprod Toxicol 2008; 25: 291-303.

[39]   Kennaway DJ. Programming of the fetal suprachiasmatic nucleus and subsequent adult rhythmicity. Trends Endocrinol Metab 2002; 13: 398-402.

[40]   Srinivasan V, Pandi-Perumal SR, Trahkt I, *et al.* Melatonin and melatonergic drugs on sleep: possible mechanisms of action. Int J Neurosci 2009; 119: 821-46.

[41]   Garstang J, Wallis M. Randomized controlled trial of melatonin for children with autistic spectrum disorders and sleep problems. Child Care Health Dev 2006; 32: 585-9.

[42] Giannotti F, Cortesi F, Cerquiglini A, Bernabei P. An open-label study of controlled-release melatonin in treatment of sleep disorders in children with autism. J Autism Dev Disord 2006; 36: 741-52.

[43] Hayashi E. Effect of melatonin on sleep-wake rhythm: the sleep diary of an autistic male. Psychiatry Clin Neurosci 2000; 54: 383-4.

[44] Galli-Carminati G, Deriaz N, Bertschy G. Melatonin in treatment of chronic sleep disorders in adults with autism: a retrospective study. Swiss Med Wkly 2009; 139: 293-6.

[45] Andersen IM, Kaczmarska J, McGrew SG, Malow BA. Melatonin for insomnia in children with autism spectrum disorders. J Child Neurol 2008; 23: 482-5.

[46] Yamada H, Yamaguchi A, Moriyama Y. L-aspartate-evoked inhibition of melatonin production in rat pineal glands. Neurosci Lett 1997; 228: 103-6.

[47] Ishio S, Yamada H, Hayashi M, *et al.* D-aspartate modulates melatonin synthesis in rat pinealocytes. Neurosci Lett 1998; 249: 143-6.

[48] Kim MH, Uehara S, Muroyama A, *et al.* Glutamate transporter-mediated glutamate secretion in the mammalian pineal gland. J Neurosci 2008; 28: 10852-63.

[49] Juarez BI, Martinez ML, Montante M, *et al.* Methylmercury increases glutamate extracellular levels in frontal cortex of awake rats. Neurotoxicol Teratol 2002; 24: 767-71.

[50] Bubenik GA, Brown GM, Grota LJ. Immunohistological localization of melatonin in the rat digestive system. Experientia 1977; 33: 662-3.

[51] Tan DX, Manchester LC, Reiter RJ, Plummer BF. Cyclic 3-hydroxymelatonin: a melatonin metabolite generated as a result of hydroxyl radical scavenging. Biol Signals Recept 1999; 8: 70-4.

[52] Terry PD, Villinger F, Bubenik GA, Sitaraman SV. Melatonin and ulcerative colitis: evidence, biological mechanisms, and future research. Inflamm Bowel Dis 2009; 15: 134-40.

[53] Harada T, Hirotani M, Maeda M, Nomura H, Takeuchi H. Correlation between breakfast tryptophan content and morning-evening in Japanese infants and students aged 0-15 yrs. J Physiol Anthropol 2007; 26: 201-7.

[54] Song GH, Leng PH, Gwee KA, Moochhala SM, Ho KY. Melatonin improves abdominal pain in irritable bowel syndrome patients who have sleep disturbances: a randomised, double blind, placebo controlled study. Gut 2005; 54: 1402-7.

[55] Brainard GC, Sliney D, Hanifin JP, *et al.* Sensitivity of the human circadian system to short-wavelength (420-nm) light. J Biol Rhythms 2008; 23: 379-86.

CHAPTER 11

The Search for Plausible Role of Oxytocin in Etiology and Therapy of Autism Spectrum Disorders

Anna Strunecka

Institute of Medical Biochemistry, Laboratory of Neuropharmacology, 1st Faculty of Medicine, Charles University in Prague, Prague, Czech Republic

Abstract: The discoveries of novel sites of oxytocin receptor (OTR) expression in central nervous system have greatly expanded the classical biological roles of oxytocin (OT), which are stimulation of uterine smooth muscle contraction at parturition and milk ejection during lactation. Central actions of OT range from the modulation of the response on individual synapses to the establishment of complex social and bonding behaviors. While there are currently no animal models reflecting the broad range of the autism spectrum disorders (ASD) behavioral and neurological phenotypes, studies of vole pair bonding and sexual behavior provided important clues useful for understanding the neurobiological mechanisms of social behavior. It has been suggested that OT dysfunction might contribute to the development of social deficits in autism, a core symptom domain and potential target for intervention. The intranasal administration of OT has been reported to reduce repetitive behaviors in adult patients with ASD. The studies of the possible role of OT in ASD etiology have raised interest but also brought some contradictory findings. Though there are preliminary promising findings with OT intranasal delivery in autistic adults, further studies are needed to replicate these findings on a large scale. Studies concerning the safety of these potential treatments are needed as well.

INTRODUCTION

Oxytocin (OT) was the first peptide hormone to have its structure determined (Fig. **1**) and the first to be chemically synthesized in biologically active form [1]. The nonapeptide OT is synthesized in the brain and released from neurohypophyseal terminals into the blood and within defined brain regions. Immunohistochemical studies revealed that magnocellular neurons of the hypothalamic paraventricular and supraoptic nuclei are the neurons of origin for the OT released from the posterior pituitary.

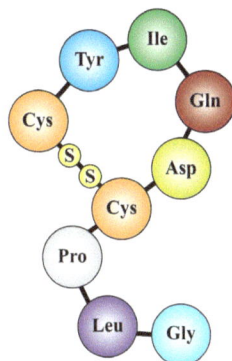

Figure 1: Oxytocin is a peptide of nine amino acids acids (a nonapeptide). The sequence is cysteine - tyrosine - isoleucine - glutamine - asparagine - cysteine - proline - leucine - glycine. The cysteine residues form a sulfur bridge.

OT is a very abundant neuropeptide. This became obvious in a study where the most prevalent hypothalamic-specific mRNAs were analyzed. OT was found to be the most abundant of 43 transcripts identified [2]. However, OT is also synthesized in peripheral tissues, e.g. uterus, placenta, amnion, corpus luteum, testis, and heart. Today, we recognize that OT exerts a wide spectrum of central and peripheral effects [3]. Its main physiological role is the

*Address correspondence to: Anna Strunecka, Laboratory of Neuropharmacology, 1st Faculty of Medicine, Charles University in Prague, Albertov 4, 128 00 Prague 2, Czech Republic; E-mail: strun@natur.cuni.cz

ejection of milk from the lactating breast and the regulation of the contraction of uterine smooth muscle at parturition.

Expression and presence of OT receptors (OTRs) have been revealed in various peripheral tissues as well as in the brain. Recently, neuropeptide OT has attracted intense attention due to the discovery of its amazing variety of behavioral functions. Brain OT is implicated in learning, anxiety, pain perception, stress responses, and aggression. Brain OT modulates emotional and social behaviors, including parental care, feeding, pair bonding and sexual behavior, grooming, social memory and support. Interest of many researchers attracted the observations concerning OTR distribution in the brains of two closely related species of vole (genus *Microtus)* with remarkably different reproductive behavior strategies [4-6]. These animal models contributed enormously to the hypothesis that OT might have a substantial role in the etiology of ASD.

THE EFFECTS OF OT IN SOCIAL BEHAVIOR

OT Involvement in Studies of Pair Bonding in Voles

Voles are hamster-sized rodents that vary widely in their social organization across species. The prairie voles (Microtus ochrogaster) are socially monogamous; males and females form long-term pair bonds, establish a nest site and rear their offsprings together (Fig. **2**). In contrast, montane and meadow voles (Microtus montanus and Microtus pennsylvanicus) do not form bonds with a mate and only the females take part in rearing the young [4,7,8]. Carter and coworkers revealed a role for OT and vasopressin in behavioral and endocrine changes during social interactions, and changes that are associated with the absence of social interactions (i.e. social isolation). These researchers also reported that OT infusions can hasten the formation of a partner preference. These results implicated OT in the formation of adult heterosexual social bonds.

Figure 2: The prairie voles (*Microtus ochrogaster*): Rearing conditions. (Courtesy Todd Ahearn, Larry Young Lab, Neuroscience Program, Emory University, Atlanta, GA, USA.)

Insel and Shapiro [5,6] demonstrated that species of voles selected for differences in social affiliation show contrasting patterns of OTR expression in the brain. They used *in vitro* receptor autoradiography with an iodinated OT analogue and found that in the prairie vole, OTRs density was significantly higher in the prelimbic cortex, bed nucleus of the stria terminalis, nucleus accumbens, midline nuclei of the thalamus, and the lateral aspects of the amygdala. These brain areas showed little binding in the montane vole, in which OTRs were localized to the lateral septum, ventromedial nucleus of the hypothalamus, and cortical nucleus of the amygdala. Furthermore, in the montane voles, which show little affiliative behavior, except during the postpartum period, brain OTR distribution changed within 24 hr of parturition concurrent with the onset of maternal behavior. Moreover, OT had little effect on aggressive behavior when administered before mating but had profound effects on the aggression of male prairie voles when administered after mating. OT had relatively modest effects on the behavior of montane voles, and neither the behavior nor the peptide effects were affected by mating experience. The data indicate that differences in peptide binding in these two species of vole may be functionally related to difference in social behavior [9].

Recent study demonstrates that OT regulates partner preference formation and alloparental behavior in the socially monogamous prairie vole by activating OTR in the nucleus accumbens of females. Ross *et al.* [10] show for the first time that extracellular concentrations of OT are increased in the nucleus accumbens of female prairie voles during unrestricted interactions with a male. They further demonstrate that the nucleus accumbens OT-immunoreactive fibers likely originate from paraventricular and supraoptic hypothalamic neurons. If correct, this may serve to coordinate peripheral and central release of OT with appropriate behavioral responses associated with reproduction, including pair bonding after mating, and maternal responsiveness following parturition and during lactation.

These animal studies have generated testable hypotheses regarding the motivational systems and underlying molecular neurobiology involved in social engagement and social bond formation that may have important implications for the core social deficits characterizing ASD [8,11]. While there are currently no animal models reflecting the broad range of the ASD behavioral and neurological phenotypes, studies into the neurobiology of normal social cognition, engagement and bonding in animals may provide important clues useful for understanding the neurobiological mechanisms underlying social deficits in ASD.

A Crucial Role of OT in Human Social Bonding

Recent human studies implicate a crucial role of OT in human social bonding, namely social attachment behavior and social interactions [12-14]. Intranasal infusion of OT appears to enhance the buffering of social support in humans, as indicated by decreased stress response during a socially stressful situation. Kosfeld with co-workers conducted a famous study with male university students playing "the trust game" [15]. Healthy male participants were given intranasal OT or placebo. Their results revealed that intranasal infusion of OT significantly increased trust among participants thereby greatly increasing the benefits from social interactions. The effect of OT on trust is not due to a general increase in the readiness to bear risks. On the contrary, OT specifically affects an individual's willingness to accept social risks arising through interpersonal interactions. These results concur with animal research suggesting an essential role for OT as a biological basis of prosocial approach behavior.

Kirsch *et al.* [16] used functional magnetic resonance imaging (fMRI) to investigate amygdala activity in response to fear-inducing stimuli of a social (angry and fearful faces) and non-social (threatening scenes) nature. OT reduced amygdala activation to both kinds of stimuli but, interestingly, had a more pronounced effect on the social stimuli. Moreover, results revealed that OT reduced the functional connection between the amygdala and structures in the upper brain stem (i.e., the periaqueductal gray and reticular formation), which have been implicated in autonomic and behavioral fear responses. This study is consistent with previous studies documenting the anxiolytic effects of OT and, as the authors' suggest, point to the possibility that OT might increase trust and prosocial behavior more generally by dampening amygdala responsively to the potential dangers inherent in social situations.

Domes *et al.* [17] investigated the effects of OT in the Reading the Mind in the Eyes Test (RMET). The RMET test was originally developed and revised to measure severe impairment in mind-reading capability in adults with ASD [18,19]. Domes with co-workers tested 30 healthy male volunteers aged 21 to 30 years in a double-blind, placebo controlled study after intranasal administration of 24 IU OT. Compared with placebo, OT improved performance on the RMET in 20 of 30 participants. It means that OT improved the ability to infer the mental state of others from social cues of the eye region. The authors state that reading the mind of an interactive partner is a cornerstone of all human interactions, which would also pertain to parenting. The definition of sensitive parenting explicitly includes the reading of the child's attachment needs from subtle facial or other non-verbal signals as a first and important step to responding in a prompt and adequate manner [20].

The oxytocinergic system has recently been placed amongst the most promising targets for various psychiatric treatments due to its role in prosocial behavior and anxiety reduction. Recently, Fischer-Shofty *et al.* [21] investigated the effects of intranasal administration of OT on emotion recognition in a double-blind placebo-controlled crossover study. A single dose of OT or a placebo was administered intranasally to 27 healthy male subjects 45 min prior to task performance. The results showed that a single intranasal administration of OT, as opposed to the placebo, improved the subjects' ability to recognize fear, but not other emotions.

THE SEARCH FOR THE ROLE OF OT IN THE ETIOLOGY, PATHOGENESIS, AND THERAPY OF ASD

A number of researchers have suggested that OT might be implicated in the etiology of ASD, which is characterized by severe social impairment [17, 22-34]. Baron-Cohen *et al.* [18] showed amygdala activation in response to the RMET using fMRI. Some areas of the prefrontal cortex also showed activation when using social intelligence. In contrast, patients with autism activated the fronto-temporal regions but not the amygdala when making mentalistic inferences from the eyes. Hollander noticed that 60 % of the autistic patients had OT treatment during labor and postulated hypothesis that excess OT, possibly through OT administration at birth, could contribute to the development of ASD [32,33]. Interestingly, Hollander and his colleagues are the first to have used both intravenous and intranasal delivery to study the behavioral effects of OT in ASD. Infusions of synthetic OT and Pitocin significantly reduced repetitive behaviors in adult patients with autistic and Asperger syndrome. The questions thus have risen: Does OT evoke or ameliorate ASD? Do autistic children have too much or too little OT in their brains?

Increased OT during Birth

The first speculations regarding the contribution of OT in ASD appeared in connection with the use of OT at birth. The knowledge of the classical physiological role of OT to stimulate uterine smooth muscle contraction at parturition led to the widely spread administration of OT to induce labor or to spur labor that's going slowly. OT is administered under the trade names such as Pitocin or Syntocinin for labor induction and/or labor augmentation during childbirth. Hattori *et al.* [22] reported that children born in one Japanese hospital routinely using a combination of anesthetics and synthetic OT - nonapeptid Pitocin - had significantly higher rate of autism than children born in three other hospitals, which rarely used general anesthesia. Laila Y. Al-Ayadhi [35] also found a higher incidence of Pitocin-induced labor among autistics children from Riyadh – Saudi Arabia as compared to normal. No indication for a relationship between labor induction and autism can be found in Fein's study [36]. These researchers examined 478 preschool children with autism (51 high functioning non-verbal IQ>79, 123 low functioning), language disorders (197) and low-IQ (107). Amongst the high functioning autistic children, 21.6 % were found OT-induced, amongst the low functioning autistic 24 %, amongst the language disordered 19.3 % and amongst the low IQ 27.1 % were found induced [37]. When compared to the U. S. national average of labor induction, which has been used in 20 % of mothers, no indication for a relationship between labor induction and autism can be found.

No correlation between Pitocin at labor and prevalence of ASD was found by Gale *et al.* [38]. These researchers examined the rates of labor induction using Pitocin in 41 children with autism and age-matched controls (15 typically developing and ten with mental retardation). There were no differences in Pitocin induction rates as a function of either diagnostic group (autism versus control) or IQ level (average versus sub average range), failing to support an association between exogenous exposure to OT and neurodevelopmental abnormalities.

Neurologists Woodward, in a personal letter to ARI notes that it seems improbable that a single dose of OT, an endogenous hormone naturally present during labor, would lead to cerebral maldevelopment [39]. Two barriers are commonly thought of as preventing OT from potentially accessing the infant's brain: The maternal placenta barrier and the blood-brain barrier (BBB) of the infant. However, it should be considered that during labor, a unique stress situation exists for both mother and child, which could result in the release of cytokines and create an oxidative stress situation which has been shown to render the BBB more permeable than usual. Also, the BBB in the infant might not be as fully developed as compared to adults and might therefore be more permeable towards small lipid insoluble molecules. Wahl [37] offered the hypothesis in support of Hollander's hypothesis that excess OT, possibly through OT administration at birth, could contribute to the development of ASD by proposed down regulation of the OTR. He proposed that to test this hypothesis, the OT concentrations in the fetal blood and brain should be routinely monitored right after birth. Furthermore, at a molecular level in a large clinical test study, OT could be administered in a radiolabeled form, preferably [^{3}H]-, but also [^{125}I]- or [^{131}I]-OT, and its occurrence/non-occurrence in the infant's brain should be determined by radiotracing. Therefore, the issue appears to remain open for further research and debate.

The evidence that OT not only has a critical role in birth and lactation but also in the emergence of an intimate bond with offspring is related to OT's reported effects on animal behavior, favoring social bondage, notably in sheep, voles, rats, and especially mice.

Although research on the neurobiological foundation of social affiliation has implicated the neuropeptide OT in processes of maternal bonding in mammals, there is little evidence to support such links in humans. Feldman *et al.* [40] sampled plasma OT and cortisol of 62 pregnant women during the first trimester, last trimester, and first postpartum month. OT was assayed using enzyme immunoassay, and free cortisol was calculated. After the infants were born, their interactions with their mothers were observed, and the mothers were interviewed regarding their infant-related thoughts and behaviors. OT was stable across time, and OT levels at early pregnancy and the postpartum period were related to a clearly defined set of maternal bonding behaviors, including gaze, vocalizations, positive affect, and affectionate touch; to attachment-related thoughts; and to frequent checking of the infant. This study demonstrated that OT level may play a role in the emergence of behaviors and mental representations typical of bonding in the human mother across pregnancy and the postpartum period. Experimental research showing that OT improves 'mind reading' suggests that OT may facilitate parental sensitivity at any stage in parents' lives and not only during the period around birth [20].

Circumstantial evidence for the potentially important role of OT in human parenting may be derived from experimental studies administering OT to patients with autism, which enhanced their social cognitions and empathic feelings [41]. Further studies relating autism to variations in the OTR gene will also document that the hypothesis of Pitocin administration during labor as the cause of ASD cannot give a satisfying explanation for direct involvement of OT in the etiology of ASD.

OT has been coined as the hormone of love and cuddle. Can we suggest that the mothers with low endogenous production of OT need primarily exogenous OT to support labor? Such simplification could be seemingly in concordance with the oldest Kanner's hypothesis. Leo Kanner was calling attention to what he saw as a lack of parental warmth and attachment to their autistic children. In his 1949 paper, he attributed autism to a "genuine lack of maternal warmth" and the "Refrigerator Mother" theory of autism was born [42].

Jane Strathearn and colleagues [43] examined 30 first-time new mothers to test whether differences in attachment, using a modified version of the Adult Attachment Interview, a semistructured interview that assesses a person's childhood interactions and bonding experience with their primary caregivers (generally parents). On viewing their own infant's smiling and crying faces during fMRI scanning, mothers with secure attachment showed greater activation of brain reward regions, including the ventral striatum, and the OT-associated hypothalamus/pituitary region. Peripheral OT response to infant contact at 7 months was also significantly higher in secure mothers, and was positively correlated with brain activation in both regions. Insecure/dismissing mothers showed greater insular activation in response to their own infant's sad faces. These results show that individual differences in maternal attachment may be linked with development of the oxytocinergic neuroendocrine systems.

Decreased Plasma Levels of OT in Autistic Children

Although making inferences to central OT functioning from peripheral measurement is difficult, the data suggest that OT abnormalities may exist in autism. Modahl *et al.* [29] found significantly lower levels of plasma OT in 29 children diagnosed with autism compared with 30 age-matched healthy control children. OT increased with age in the normal but not the autistic children. Elevated OT was associated with higher scores on social and developmental measures for the normal children, but was associated with lower scores for the autistic children. A follow-up study of these same subjects revealed that the differences in plasma OT levels were associated with an increase in incompletely processed OT fragments, suggesting that peptide processing may be dysregulated in the autistic patients [31]. The hypothalamic synthetic pathway of nonapeptide OT involves the synthesis of carboxy-extended forms that serve as intermediate prohormones. A prohormone is sequentially processed to peptides. These peptides are the bioactive amidated form and the carboxy- extended peptides, OT-Gly, OT-Gly-Lys and OT-Gly-Lys-Arg. Using OT antisera with different specificity for the peptide forms, Green *et al.* measured plasma OT and carboxy-extended peptides in each group of subjects. T tests showed that there was a decrease in plasma OT and an increase in extended peptides compared with control subjects. Green with co-workers [31] concluded that observed deficits in OT peptide processing in children with autism may be important in the development of this syndrome. However, it is not known whether the carboxy-extended OT prohormones have some biological activity and could compete with OT for receptor binding in human brain. In peripheral tissues they only serve as a substrate for OT synthesis and do not compete with OT for OTR binding [44].

The second study found lower level of OT in autistic children in Central Saudi Arabia [35]. Seventy-seven autistic children with an age ranging from 3.5-14 years from Riyadh area participated in the study, with the confirmed diagnosis according to DSM-IV diagnostic criteria of autism. Results showed a statistically significant lower plasma level of OT in autistic group (0.074 ± 0.01 ng/ml), as compared to the control group (0.107 ± 0.01 ng/ml). Further more, vasopressin plasma level was significantly lower in autistic children. There was no significant correlation between the degree of autism, or the age of the affected child and plasma levels of OT or vasopressin.

Marchini and Stock [45] investigated OT profile in 26 healthy newborn (1-day-old) infants using the pulse detection program PULSAR. The plasma OT concentrations were determined by specific radioimmunoassay. They found that 42 % of the infants presented one peak in the OT level during a 4-min period. The peak constituted a 111 ± 66 % increment of the baseline value. These researchers suggested that the release of OT during basal conditions occurs in a pulsatile way in newborn infants and that these hormone pulses reflect fluctuations in the activity of the hypothalamic neurosecretory cells.

However, measurements of OT concentration in blood using a sensitive enzyme immunoassay (EIA) are infrequently performed. Carter *et al.* [46] described the method for measurements of biologically relevant changes in salivary OT. Their results confirm the biological relevance of changes in salivary OT with stressors and support saliva as a noninvasive source to monitor central neuroendocrine function. In the near future it may become easier to measure salivary OT as a biomarker for affiliative behavior in humans, which would enable direct tests of the association between OT and ASD.

The Effects of OT Administration in Adult Autistic Patients

Several animal studies, mostly with rodents, investigated the effect of OT administered centrally or peripherally, and found that OT facilitated social recognition [47]. Fergusson *et al.* [48-50] observed that male mice with a null mutation in the gene coding for OT ("OT knockout mice") failed to recognize a conspecific even over repeated exposures and that a single intracerebroventricular injection of OT before the initial encounter with the conspecific enabled social memory acquisition. No studies have yet examined the pharmacological influences of OT on the social deficits in autism; however, infusions of synthetic OT and Pitocin significantly reduced repetitive behaviors in adult patients with autistic and Asperger syndrome. Hollander *et al.* [32] investigated the influence of OT infusion on repetitive behaviors in autism. In a double-blind cross-over design, 15 ASD adult patients showed a significant reduction in repetitive behaviors — need to know, repeating, ordering, need to tell/ask, self-injury, and touching — after OT infusion versus placebo infusion.

Inspired by these findings, Hollander with coworkers [33] investigated further whether increased levels of OT would facilitate social information processing for adult individuals diagnosed with autism or ASD. They focused on auditory processing of social stimuli—and specifically participants' ability to assign affective significance to speech—because this deficit is present in most autistic people, and it has been hypothesized that its disruption could be central to the social and speech deficits in autism. In this randomized, placebo-controlled, double-blind crossover investigation, 15 adults (14 men; mean age 32.9 years, range 19–56 years) diagnosed with autism (n = 6) or Asperger syndrome (n = 9) received OT and placebo challenges during visits separated by a minimum of one week. Participants were medically healthy and medication-free for at least two weeks before and throughout the study. Each subject served as his or her own control and completed both OT and placebo challenges on separate days; synthetic OT (Pitocin) or placebo was continuously infused over a 4-hour period. Eight subjects received OT first, and seven received placebo first. Comprehension of affective speech was tested at baseline, just before the intravenous OT/placebo infusion, and at 30, 60, 120, 180, and 240 min over the course of the infusion. In this study, participants were presented with four sentences of neutral content: "The boy went to the store," "The game ended at 4 o'clock," "Fish can jump out of the water," and "He tossed the bread to the pigeons." The sentences were pre-recorded and played for participants on a tape player; the voice reading the sentences was unknown to participants and was the same for all participants. Each sentence was presented with one of four emotional intonations (happy, indifferent, angry, and sad) with the pairing of emotional expression and sentences in one of six counterbalanced orders. Participants were then asked to indicate the emotional mood of the speaker by pointing to the word that corresponded to the emotion that they believed matched the one they heard on the tape; the examiner recorded participants' responses for all trials. As it turned out, this task was relatively easy for the adult participants in this study and, consequently, the findings were negatively skewed. In an effort to reduce the negative skew of the variable and to better balance the difficulty of the task, the outcome measure was scored dichotomously as 1 (all items correct) and 0 (not all items correct) [33]. This study found that OT administration facilitated the processing and retention of social information in adults diagnosed with autism or Asperger syndrome: compared with subjects who received placebo first, subjects who received OT first showed increased retention of affective speech comprehension after a delay. These authors pointed that it would be interesting to explore the effects of OT on other aspects of social cognition, to investigate whether different methods of administration (i.e., intranasal versus intravenous) yield different results and to investigate whether cerebrospinal fluid OT levels differ depending on the method of administration used.

Treatment studies with standardized measures to assess "real-life" improvements in social functioning are also needed to demonstrate the practical utility of OT in the treatment of autism. Finally, studies are needed to investigate the effects of OT administration in younger children who could potentially benefit from early intervention.

The more recent studies have brought demonstration that intranasal OT improves emotion recognition for youth with ASD [51]. In a double-blind, randomized, placebo-controlled, crossover design, OT nasal spray (18 or 24 IU) or a placebo was administered to 16 male youth aged 12 to 19 who were diagnosed with autistic or Asperger syndrome. Participants then completed the RMET, a widely used and reliable test of emotion recognition. These researchers found that OT intranasal administration improved performance on the RMET. This effect was also shown when analysis was restricted to the younger participants aged 12 to 15 who received the lower dose. These findings suggest the potential of earlier intervention and further evaluation of OT nasal spray as a treatment to improve social communication and interaction in young people with ASD.

Recently, Andari *et al.* [52] investigated the behavioral effects of OT in 13 individuals with high-functioning autism or with Asperger syndrome in a simulated ball game where participants interacted with fictitious partners. These researchers found that after OT inhalation, patients exhibited stronger interactions with the most socially cooperative partner and reported enhanced feelings of trust and preference. Andari with colleagues thus support a therapeutic potential of OT through its action on a core dimension of autism.

INVESTIGATIONS OF OTR IN ASD

It is interesting to note that OT genes belong to the oldest ones; they should be even older than 500-700 million years [3,53]. Despite the great diversity of the proposed functions of OT and the oxytocinergic system, only one type of OTR has been identified both in laboratory animals as well as in humans [54-56].

Biochemistry of OTR

The encoded OTR is a 388-amino-acid polypeptide, which belongs to the family of G protein-coupled receptors (GPCR) [57]. All GPCR display seven heptihelical domains that are hydrophobic and span lipid bilayer (see Fig. **3**). Extracellular domains and core of bundle of seven transmembrane segments act in signal discrimination and ligand binding [58,59]. The affinity of the receptor for ligands is strongly dependent on the presence of divalent cations and cholesterol that both act like positive allosteric modulators. Notably, some evidence is provided that OTR are also present in the form of dimeric or oligomeric complexes at the cell surface.

Intracellular domains of OTR function in signal propagation to heterotrimeric G proteins. Heterotrimeric G proteins are constructed of three types of subunits, an α-subunit uniquely capable of binding and degrading GTP and a tightly knit complex of β- and γ-subunits. The nomenclature now popularly known as Gq, Gs, and Gi classes determines the interaction of the α-subunit with various effectors molecules [60]. Gs means that the α-subunit of heterotrimeric G protein interacts with adenylyl cyclase (AC) and stimulates the production of cyclic adenosine monophosphate (cAMP), while Gi inhibits the production of cAMP. Gq has been used for a class of G proteins, which activate phospholipase C (PLC). OTR is able to couple to different G proteins with a subsequent stimulation of various signaling cascades [61-64] (see Fig. **3**). In spite of the fact that there is only one type of OTR, its stimulation can produce an array of cell functions, and moreover, sometimes in an opposite manner. Because of dependence on G protein coupling, OTR can give rise to opposite effects on the same cellular function. For example, OTR coupling to Gq induces the contraction in myometrial cells, while OTR activation of Gi decreases preterm labor [65].

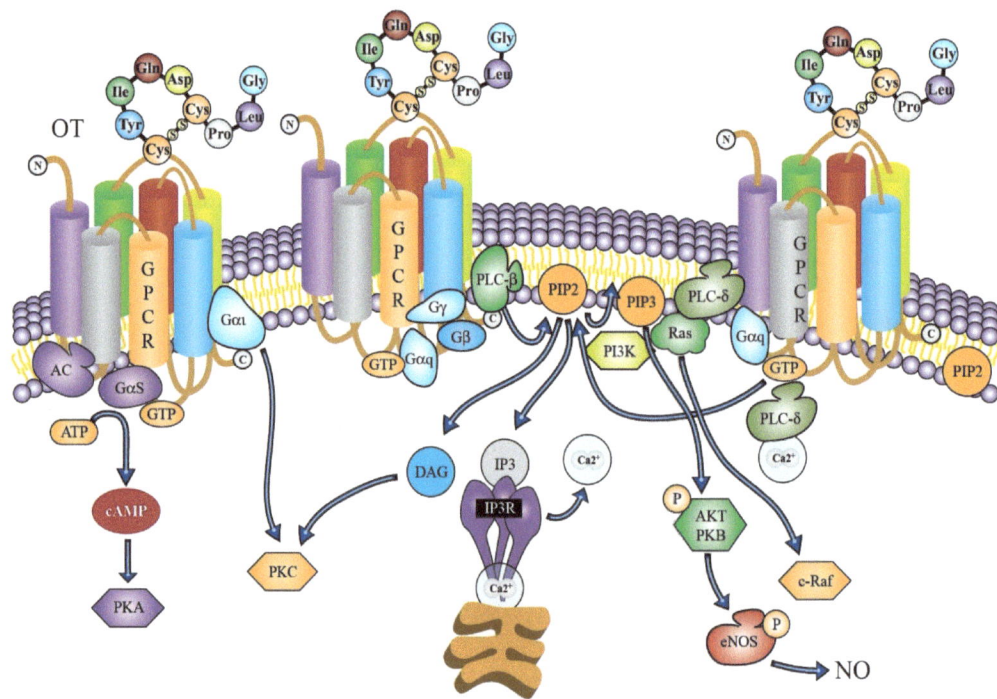

Figure 3: Various signaling pathways used by OTR.

Interesting findings concerning the coupling of OTR with various G proteins were reported in the brain. To investigate the potential role of Gq signaling in behavior, Wettschureck *et al.* [66] generated mice, which lack the α-subunits of the two main members of the Gq/11 family, selectively in the forebrain. These authors found that forebrain Gq/11-deficient females did not display any maternal behavior such as nest building, pup retrieving, crouching, or nursing. However, olfaction, motor behavior, and mammary gland function were normal in forebrain Gq/11-deficient females. It seems therefore that Gq/11 signaling is indispensable to the neuronal circuit that connects the perception of pup-related stimuli to the initiation of maternal behavior.

Several findings indicate that the OTR is regulated in a very complex manner. For example, it has been shown that OTR in plasma membrane mediates the inhibition of cell proliferation while OTR localized in caveolin-enriched microdomains (lipid rafts) mediates a mitogenic effect [63,67]. The search of new OT analogs could bring a class of highly selective compounds with therapeutic relevance in obstetrics, oncology, psychiatry, and some other areas of medicine. On the other hand, it is difficult to predict the effects of exogenous administration of OT or its synthetic analogues.

Genetic Variants in the OTR Gene Associated with ASD

Animal models and linkage data from genome screens indicate that the *OTR* gene is an excellent candidate for research concerning psychiatric disorders, particularly those involving social impairments, such as autism. Some modest evidence suggesting a possible association of the *OTR* gene with autism has been already presented.

The *OTR* gene is a single-copy gene consisting of four exons and three introns, localized at chromosome 3 of the human genome (3p25-3p26.2). Specifically, a combined analysis of Autism Genetic Resource Exchange (AGRE) and a sample of Finnish families of probands with autism implicated chromosome locus 3p24–26 as a candidate autism locus [68]. This region contains 40 genes, including the *OTR* gene. However, this study failed to identify an association between specific polymorphisms and autism, although a limited number of polymorphisms were analyzed.

Several authors have shown the association of polymorphisms in OTR with autism. Wu *et al.* [69] genotyped four single nucleotide polymorphisms (SNPs) located within the *OTR* gene of 195 Chinese Han autism trios (parents and child), using polymerase chain reaction (PCR)-restriction fragment length polymorphism analysis. Technical details may not be of interest here but the family-based association test (FBAT) revealed a significant genetic association between autism and two of the SNPs tested (rs2254298 and rs53576).

Jacob and co-workers [70] tested whether these associations replicated in a Caucasian sample with strictly defined ASD and genotyped the two previously associated SNPs (rs2254298, rs53576) in 57 Caucasian autism trios. Significant association was detected at rs2254298. However, in their Caucasian autism trios and the CEU Caucasian HapMap samples the frequency of A was less than that reported in the Chinese Han samples. The haplotype test of association did not reveal excess transmission from parents to affected offspring. However, these findings also provide support for association of OTR with ASD in a Caucasian population. Authors concluded that overtransmission of different alleles in different populations may be due to a different pattern of linkage disequilibrium between the marker rs2254298 and an as yet undetermined susceptibility variant in OTR.

The third association study, in a third ethnic group from Israel, confirmed that SNPs and haplotypes in the *OTR* gene confer risk for ASD [71]. Lerer and co-workers at Hebrew University, Jerusalem, undertook a comprehensive study of all 18 tagged SNPs across the entire *OTR* gene region identified using HapMap data and the Haploview algorithm. Altogether 152 subjects diagnosed with ASD from 133 families were genotyped (parents and affected siblings). In particular, a five-locus haplotype block (rs237897-rs13316193-rs237889-rs2254298-rs2268494) was significantly associated with ASD.

Association of the *OTR* gene polymorphisms with ASD in the Japanese population has been recently reported by Liu *et al.* [72]. These researchers investigated the associations between OTR and ASD by analyzing 11 SNPs using both FBAT and population-based case-control test. No significant signal was detected in the FBAT test. However, significant differences were observed in allelic frequencies of four SNPs, including rs2254298 between patients and controls. The risk allele of rs2254298 was 'A', which was consistent with the previous study in China [69], and not with the observations in Caucasian. The difference in the risk allele of this SNP in previous studies might be attributable to an ethnic difference in the linkage disequilibrium structure between the Asians and Caucasians. In addition, haplotype analysis exhibits a significant association between a five-SNP haplotype and ASD, including rs2254298. Liu with co-workers suggest that OTR has a significant role in conferring the risk of ASD in the Japanese population."

Tansey *et al.* [73] examined 18 SNPs at the *OTR* gene for association in three independent autism samples from Ireland, Portugal, and the United Kingdom. These authors investigated cis-acting genetic effects on OTR expression in lymphocytes and amygdala region of the brain using an allelic expression imbalance (AEI) assay and by investigating the correlation between RNA levels and genotype in the amygdala region. Their results performed in the lymphoblast cell lines highlighted two SNPs associated with relative allelic abundance in *OTR* (rs237897 and rs237895). Two SNPs were found to be effecting cis-acting variation through AEI in the amygdala. One was weakly correlated with total gene expression (rs13316193) and the other was highlighted in the lymphoblast cell lines (rs237895). However, Tansey with co-workers concluded that their data does not support the role of common genetic variation in OTR in the etiology of ASD in Caucasian samples.

Kelemenova and co-workers investigated polymorphism of candidate genes including *OTR* gene in 90 autistic boys from Slovakia [74]. These researchers have examined only one polymorphism in *OTR* gene (rs2228485) and no association with autism has been found. Taken together, the authors cannot conclude that these genes are not connected with autism in Slovakia. In contrast, these results could imply genetic heterogeneity of autism in various geographic regions (see Table **1**).

Table 1: Polymorphism of *OTR* Gene in Autistic Patients in Various Geographic Regions.

SUBJECTS	OBSERVED ASSOCIATION BETWEEN ASD AND GENETIC CHANGE	REF
195 Chinese Han autism trios	SNPs rs2254298 and rs53576	[69]
Japanese population	5 SNPs including rs2254298	[71]
57 Caucasian autism trios	rs2254298	[70]
152 subjects with ASD; Hebrew University, Israel	rs237897; rs13316193; rs237889; rs2254298; rs2268494	[72]
Ireland, Portugal, and the UK	no association	[73]
90 Autism boys, Slovakia	no association with rs2228485	[74]

Gregory *et al.* [75] studied genomic and epigenetic alterations of OTR gene within 119 probands from multiplex autism families. The involvement of this gene was suggested since their analysis revealed a genomic deletion containing the OTR gene. Gregory with coworkers used high-resolution genome-wide tilepath microarrays and comparative genomic hybridization to identify copy number variants. They also carried out DNA methylation analysis by bisulfite sequencing in a proband and his family, expanding this analysis to methylation analysis of peripheral blood and temporal cortex DNA of autism cases and matched controls from independent datasets. The analysis of a group of unrelated autistic subjects did not show an OTR gene deletion, but rather hypermethylation of the gene promoter, with a reduced mRNA expression. Further DNA methylation analysis of the CpG island known to regulate OTR gene expression identified several CpG dinucleotides that show independent statistically significant increases in the DNA methylation status in the peripheral blood cells and temporal cortex in independent datasets of individuals with autism as compared to control samples. The nature of this epigenetic dysregulation is unknown but, if proved to be true, might explain the failure to identify sequence alterations in a host of candidate genes.

Gregory *et al.* [75] also assessed *OTR* gene expression within the temporal cortex tissue by quantitative real-time PCR. Associated with the increase in methylation of the CpG dinucleotides is the finding that *OTR* gene mRNA showed decreased expression in the temporal cortex tissue of autism cases matched for age and sex compared to controls.

CONCLUSIONS

The possible implication of oxytocinergic system in ASD pathology and therapy has attracted intense attention of several groups of researchers. While the plasma OT data are consistent with the hypothesis that disruptions in the OT system contribute to the social behavioral phenotype in ASD, it is also equally plausible that these differences in OT levels may be the consequence of altered cognitive processing in autistic patients and/or their mothers. A crucial role of brain OT in parenting, by enhancing the reward value of social interactions and intimate bonds with the offspring is related to OT's reported effects on animal behavior, notably in voles, rats, and especially mice.

Two related studies in adults found that OT treatments resulted in an increased retention of affective speech in adults with autism, decreased repetitive behaviors, and improved interpretation of emotion [32,33], but these preliminary results do not yet necessarily apply to children. Though the preliminary promising findings with OT intranasal delivery in autistic adults further studies replicated in large scale and the study about the safety of these potential treatments are needed. Animal models and linkage data from genome screens indicate that the *OTR* gene is an excellent candidate for research concerning psychiatric disorders, particularly ASD involving social impairments. Nevertheless, it is evident that the majority of OT actions are due to the activation of the single OTR subtype. Several researchers have reported that autism is correlated with genomic deletion of the *OTR* gene. Studies involving Caucasian, Finnish, Jewish, and Chinese Han samples families with autistic children provide support for the relationship of OTR with autism. Autism may also be associated by an aberrant methylation of OTR. It has been suggested that methylation-modifying drugs also may be a new avenue for ASD treatments. Nevertheless, it is becoming clear, that the great diversity of OT actions in the brain and peripheral organs is paralleled by activation of a diversity of signaling pathways. The studies of OTR-linked signaling cascades could also bring interesting knowledge, opening new avenues for research in molecular pharmacology of ASD. However, the research has not yet brought plausible answers regarding the role of OT in autism pathology and its efficiency in therapy.

This work was supported by VZ MSM 0021620806.

REFERENCES

[1]   Du Vigneaud V, Ressler C, Trippett S. The sequence of amino acids in oxytocin, with a proposal for the structure of oxytocin. J Biol Chem 1953; 205: 949-57.

[2]   Gautvik KM, de Lecea L, Gautvik VT, *et al.* Overview of the most prevalent hypothalamus-specific mRNAs, as identified by directional tag PCR subtraction. Proc Natl Acad Sci U S A 1996; 93: 8733-8.

[3]   Gimpl G, Fahrenholz F. The oxytocin receptor system: structure, function, and regulation. Physiol Rev 2001; 81: 629-83.

[4]   Carter CS, Williams JR, Witt DM, Insel TR. Oxytocin and social bonding. Ann N Y Acad Sci 1992; 652: 204-11.

[5]   Insel TR, Shapiro LE. Oxytocin receptors and maternal behavior. Ann N Y Acad Sci 1992; 652: 122-41.

[6]   Insel TR, Shapiro LE. Oxytocin receptor distribution reflects social organization in monogamous and polygamous voles. Proc Natl Acad Sci U S A 1992; 89: 5981-5.

[7]   Ahern TH, Young LJ. The impact of early life family structure on adult social attachment, alloparental behavior, and the neuropeptide systems regulating affiliative behaviors in the monogamous prairie vole (microtus ochrogaster). Front Behav Neurosci 2009; 3: 17.

[8]   Carter CS, Grippo AJ, Pournajafi-Nazarloo H, Ruscio MG, Porges SW. Oxytocin, vasopressin and sociality. Prog Brain Res 2008; 170: 331-6.

[9]   Winslow JT, Shapiro L, Carter CS, Insel TR. Oxytocin and complex social behavior: species comparisons. Psychopharmacol Bull 1993; 29: 409-14.

[10]   Ross HE, Cole CD, Smith Y, *et al.* Characterization of the oxytocin system regulating affiliative behavior in female prairie voles. Neuroscience 2009; 162: 892-903.

[11]   Hammock EA, Young LJ. Oxytocin, vasopressin and pair bonding: implications for autism. Philos Trans R Soc Lond B Biol Sci 2006; 361: 2187-98.

[12]   Neumann ID. Oxytocin: the neuropeptide of love reveals some of its secrets. Cell Metab 2007; 5: 231-3.

[13]   Heinrichs M, von Dawans B, Domes G. Oxytocin, vasopressin, and human social behavior. Front Neuroendocrinol 2009; 30: 548-57.

[14]   Domes G, Lischke A, Berger C, *et al.* Effects of intranasal oxytocin on emotional face processing in women. Psychoneuroendocrinology 2010; 35: 83-93.

[15]   Kosfeld M, Heinrichs M, Zak PJ, Fischbacher U, Fehr E. Oxytocin increases trust in humans. Nature 2005; 435: 673-6.

[16]   Kirsch P, Esslinger C, Chen Q, *et al.* Oxytocin modulates neural circuitry for social cognition and fear in humans. J Neurosci 2005; 25: 11489-93.

[17]   Domes G, Heinrichs M, Michel A, Berger C, Herpertz SC. Oxytocin improves "mind-reading" in humans. Biol Psychiatry 2007; 61: 731-3.

[18]   Baron-Cohen S, Ring HA, Wheelwright S, *et al.* Social intelligence in the normal and autistic brain: an fMRI study. Eur J Neurosci 1999; 11: 1891-8.

[19]   Baron-Cohen S, Wheelwright S, Hill J, Raste Y, Plumb I. The "Reading the Mind in the Eyes" Test revised version: a study with normal adults, and adults with Asperger syndrome or high-functioning autism. J Child Psychol Psychiatry 2001; 42: 241-51.

[20]   Bakermans-Kranenburg MJ, van Ijzendoorn MH. Oxytocin receptor (OXTR) and serotonin transporter (5-HTT) genes associated with observed parenting. Soc Cogn Affect Neurosci 2008; 3: 128-34.

[21]   Fischer-Shofty M, Shamay-Tsoory SG, Harari H, Levkovitz Y. The effect of intranasal administration of oxytocin on fear recognition. Neuropsychologia 2010; 48: 179-84.

[22]   Hattori R, Desimaru M, Nagayama I, Inoue K. Autistic and developmental disorders after general anaesthetic delivery. Lancet 1991; 337: 1357-8.

[23]   Modahl C, Fein D, Waterhouse L, Newton N. Does oxytocin deficiency mediate social deficits in autism? J Autism Dev Disord 1992; 22: 449-51.

[24]   Panksepp J. Oxytocin effects on emotional processes: separation distress, social bonding, and relationships to psychiatric disorders. Ann N Y Acad Sci 1992; 652: 243-52.

[25]   Panksepp J. Commentary on the possible role of oxytocin in autism. J Autism Dev Disord 1993; 23: 567-9.

[26]   Waterhouse L, Fein D, Modahl C. Neurofunctional mechanisms in autism. Psychol Rev 1996; 103: 457-89.

[27]   Insel TR, Young L, Wang Z. Central oxytocin and reproductive behaviours. Rev Reprod 1997; 2: 28-37.

[28]   McCarthy MM, Altemus M. Central nervous system actions of oxytocin and modulation of behavior in humans. Mol Med Today 1997; 3: 269-75.

[29] Modahl C, Green L, Fein D, *et al*. Plasma oxytocin levels in autistic children. Biol Psychiatry 1998; 43: 270-7.

[30] Insel TR, O'Brien DJ, Leckman JF. Oxytocin, vasopressin, and autism: is there a connection? Biol Psychiatry 1999; 45: 145-57.

[31] Green L, Fein D, Modahl C, *et al*. Oxytocin and autistic disorder: alterations in peptide forms. Biol Psychiatry 2001; 50: 609-13.

[32] Hollander E, Novotny S, Hanratty M, *et al*. Oxytocin infusion reduces repetitive behaviors in adults with autistic and Asperger syndromes. Neuropsychopharmacology 2003; 28: 193-8.

[33] Hollander E, Bartz J, Chaplin W, *et al*. Oxytocin increases retention of social cognition in autism. Biol Psychiatry 2007; 61: 498-503.

[34] Panksepp J. Primary process affects and brain oxytocin. Biol Psychiatry 2009; 65: 725-7.

[35] Al-Ayadhi LY. Altered OT and vasopressin levelsin autistic children in Central Saudi Arabia. Neurosciences. 2005; 10: 47-50.

[36] Fein D, Allen D, Dunn M, *et al*. Pitocin induction and autism. Am J Psychiatry 1997; 154: 438-9.

[37] Wahl RU. Could oxytocin administration during labor contribute to autism and related behavioral disorders?--A look at the literature. Med Hypotheses 2004; 63: 456-60.

[38] Gale S SO, J. Lainhart. Pitocin induction in autistic and nonautistic individuals. J Autism Dev Disord 2003; 33: 205-8.

[39] Woodward K. Oxytocin and autism. Autism Research Review International 1996; 10: 2.

[40] Feldman R, Weller A, Zagoory-Sharon O, Levine A. Evidence for a neuroendocrinological foundation of human affiliation: plasma oxytocin levels across pregnancy and the postpartum period predict mother-infant bonding. Psychol Sci 2007; 18: 965-70.

[41] Bartz JA, Hollander E. The neuroscience of affiliation: forging links between basic and clinical research on neuropeptides and social behavior. Horm Behav 2006; 50: 518-28.

[42] Kanner L. Problems of nosology and psychodynamics of early infantile autism. Am J Orthopsychiatry 1949; 19: 416-26.

[43] Strathearn L, Fonagy P, Amico J, Montague PR. Adult attachment predicts maternal brain and oxytocin response to infant cues. Neuropsychopharmacology 2009; 34: 2655-66.

[44] Mitchell BF, Fang X, Wong S. Role of carboxy-extended forms of oxytocin in the rat uterus in the process of parturition. Biol Reprod 1998; 59: 1321-7.

[45] Marchini G, Stock S. Pulsatile release of oxytocin in newborn infants. Reprod Fertil Dev 1996; 8: 163-5.

[46] Carter CS, Pournajafi-Nazarloo H, Kramer KM, *et al*. Oxytocin: behavioral associations and potential as a salivary biomarker. Ann N Y Acad Sci 2007; 1098: 312-22.

[47] Popik P, Vetulani J, van Ree JM. Low doses of oxytocin facilitate social recognition in rats. Psychopharmacology (Berl) 1992; 106: 71-4.

[48] Ferguson JN, Young LJ, Hearn EF, *et al*. Social amnesia in mice lacking the oxytocin gene. Nat Genet 2000; 25: 284-8.

[49] Ferguson JN, Aldag JM, Insel TR, Young LJ. Oxytocin in the medial amygdala is essential for social recognition in the mouse. J Neurosci 2001; 21: 8278-85.

[50] Ferguson JN, Young LJ, Insel TR. The neuroendocrine basis of social recognition. Front Neuroendocrinol 2002; 23: 200-24.

[51] Guastella AJ, Howard AL, Dadds MR, Mitchell P, Carson DS. A randomized controlled trial of intranasal oxytocin as an adjunct to exposure therapy for social anxiety disorder. Psychoneuroendocrinology 2009; 34: 917-23.

[52] Andari E, Duhamel JR, Zalla T, *et al*. Promoting social behavior with oxytocin in high-functioning autism spectrum disorders. Proc Natl Acad Sci U S A 2010; 107: 4389-94.

[53] Acher R, Chauvet J, Chauvet MT. Man and the chimaera. Selective versus neutral oxytocin evolution. Adv Exp Med Biol 1995; 395: 615-27.

[54] Kimura T, Tanizawa O, Mori K, Brownstein MJ, Okayama H. Structure and expression of a human oxytocin receptor. Nature 1992; 356: 526-9.

[55] Inoue T, Kimura T, Azuma C, *et al*. Structural organization of the human oxytocin receptor gene. J Biol Chem 1994; 269: 32451-6.

[56] Kimura T, Saji F, Nishimori K, *et al*. Molecular regulation of the oxytocin receptor in peripheral organs. J Mol Endocrinol 2003; 30: 109-15.

[57] Bockaert J, Pin JP. Molecular tinkering of G protein-coupled receptors: an evolutionary success. EMBO J 1999; 18: 1723-9.

[58] Gimpl G, Postina R, Fahrenholz F, Reinheimer T. Binding domains of the oxytocin receptor for the selective oxytocin receptor antagonist barusiban in comparison to the agonists oxytocin and carbetocin. Eur J Pharmacol 2005; 510: 9-16.

[59] Gimpl G, Reitz J, Brauer S, Trossen C. Oxytocin receptors: ligand binding, signalling and cholesterol dependence. Prog Brain Res 2008; 170: 193-204.

[60] Morris AJ, Malbon CC. Physiological regulation of G protein-linked signaling. Physiol Rev 1999; 79: 1373-430.

[61]   Fanelli F, Barbier P, Zanchetta D, de Benedetti PG, Chini B. Activation mechanism of human oxytocin receptor: a combined study of experimental and computer-simulated mutagenesis. Mol Pharmacol 1999; 56: 214-25.

[62]   Chini B, Fanelli F. Molecular basis of ligand binding and receptor activation in the oxytocin and vasopressin receptor family. Exp Physiol 2000; 85 Spec No: 59S-66S.

[63]   Rimoldi V, Reversi A, Taverna E, *et al.* Oxytocin receptor elicits different EGFR/MAPK activation patterns depending on its localization in caveolin-1 enriched domains. Oncogene 2003; 22: 6054-60.

[64]   Chini B, Manning M. Agonist selectivity in the oxytocin/vasopressin receptor family: new insights and challenges. Biochem Soc Trans 2007; 35: 737-41.

[65]   Zhou XB, Lutz S, Steffens F, Korth M, Wieland T. Oxytocin receptors differentially signal via Gq and Gi proteins in pregnant and nonpregnant rat uterine myocytes: implications for myometrial contractility. Mol Endocrinol 2007; 21: 740-52.

[66]   Wettschureck N, Moers A, Hamalainen T, *et al.* Heterotrimeric G proteins of the Gq/11 family are crucial for the induction of maternal behavior in mice. Mol Cell Biol 2004; 24: 8048-54.

[67]   Herbert Z, Botticher G, Aschoff A, *et al.* Changing caveolin-1 and oxytocin receptor distribution in the ageing human prostate. Anat Histol Embryol 2007; 36: 361-5.

[68]   Ylisaukko-oja T, Rehnstrom K, Auranen M, *et al.* Analysis of four neuroligin genes as candidates for autism. Eur J Hum Genet 2005; 13: 1285-92.

[69]   Wu S, Jia M, Ruan Y, *et al.* Positive association of the oxytocin receptor gene (OXTR) with autism in the Chinese Han population. Biol Psychiatry 2005; 58: 74-7.

[70]   Jacob S, Brune CW, Carter CS, *et al.* Association of the oxytocin receptor gene (OXTR) in Caucasian children and adolescents with autism. Neurosci Lett 2007; 417: 6-9.

[71]   Lerer E, Levi S, Salomon S, *et al.* Association between the oxytocin receptor (OXTR) gene and autism: relationship to Vineland Adaptive Behavior Scales and cognition. Mol Psychiatry 2008; 13: 980-8.

[72]   Liu X, Kawamura Y, Shimada T, *et al.* Association of the oxytocin receptor (OXTR) gene polymorphisms with autism spectrum disorder (ASD) in the Japanese population. J Hum Genet 2010; 55: 137-41.

[73]   Tansey KE, Brookes KJ, Hill MJ, *et al.* Oxytocin Receptor (OXTR) does not play a major role in the Aetiology of Autism: Genetic and Molecular Studies. Neurosci Lett 2010; 474: 163-7.

[74]   Kelemenova S, Schmidtova E, Ficek A, *et al.* Polymorphisms of candidate genes in Slovak autistic patients. Psychiatr Genet 2010; 20: 137-9.

[75]   Gregory SG, Connelly JJ, Towers AJ, *et al.* Genomic and epigenetic evidence for oxytocin receptor deficiency in autism. BMC Med 2009; 7: 62.

CHAPTER 12

Regulation of Cortisol Levels in Autistic Individuals and their Mothers

Anna Strunecka

Institute of Medical Biochemistry, Laboratory of Neuropharmacology, 1st Faculty of Medicine, Charles University in Prague, Prague, Czech Republic

Abstract: The significant evidence has been collected that individuals with autism spectrum disorders (ASD) have disturbed regulation of the hypothalamic-pituitary-adrenal (HPA) axis. Disturbance in daily rhythm, the increased level of pituitary adrenocorticotropin hormone (ACTH), and the decreased concentration of cortisol in plasma and saliva has been found. The hypoactivity of HPA axis and decreased level of saliva cortisol has also been found in parents of autistic individuals. Adverse maternal stress during gestation is involved in abnormal behavior, mental and cognition disorders in offspring. Hippocampus is the principal target site for corticosteroids in the brain as it has the highest concentration of receptor sites for glucocorticoids. Several studies found that repeated prenatal glucocorticoid exposure has profound influences on HPA function both in animals and humans. Lower cortisol is seen in conditions of chronic stress and in social situations characterized by unstable social relationships. The presence of social support has been associated with decreased stress responsiveness and with attenuated free cortisol concentrations in saliva. Several studies document that oxytocin significantly reduced salivary cortisol levels. Findings from several laboratories, which investigated both autistic children and parents, further support the knowledge that mothers may share some metabolic characteristics with their autistic children and that there is the need to intervention of both – autistic children and their mothers. It has been suggested that paired-like homeodomain transcription factor 1(*PITX1*), a key regulator of hormones within the HPA axis, may be implicated in the etiology of autism. The role of 11β-hydroxysteroid dehydrogenase type 1 (11β-HSD1) in the pathogenesis of ASD is discussed.

INTRODUCTION

The glucocorticoid cortisol is a hormone, which is usually referred to as the stress hormone. It is produced in the adrenal gland by zona fasciculata of the adrenal cortex. Cortisol is essential to life and has a wide range of physiological functions throughout the body. The primary function of cortisol is to increase blood glucose and store sugar in the liver as glycogen. Cortisol acts to counteract many of the effects of stress as it is involved in response to stress and anxiety. Moreover, glucocorticoids are immunosupressive, anti-inflammatory, anti-allergic, and also possess mineralocorticoid activity. High levels of cortisol suppress the normal immune response to infection. Cortisol causes a fall in antibody production and in the number of circulating lymphocytes [1].

Interestingly, glucocorticoids also influence the central nervous system (CNS). Although glucocorticoid receptors are highly expressed in the prefrontal cortex, the hippocampus remains the predominant focus in studies examining relationships between cortisol and the brain [2, 3]. Hippocampus thus serves as an important glucocorticoid sensor in relation to the stress response. The hippocampus is also a key structure in the formation of spatial memory. For example, cognitive deficits, especially those closely related to hippocampus function, appear to be related to cortisol secretion in depressed patients [4]. Activation of glucocorticoid receptors enhances mood disorders ranging from mild euphoria to hypomania [5] (Fig. **1**).

The synthesis and secretion of glucocorticoids is under the control of ACTH secreted by the anterior pituitary in response to hypothalamic corticotropin-releasing hormone (CRH). ACTH may dramatically stimulate cortisol from the low baseline value. Stimulation resulting in a greater than 14-fold increase in serum concentration over 30 minutes has been reported, although more typically serum cortisol levels will double or triple from baseline. Glucocorticoid secretion is regulated by a typical negative feedback system [1]. Rising cortisol level acts on the anterior pituitary and probably also on hypothalamus to inhibit ACTH and CRH release, thereby reducing the rate of

---

*Address correspondence to: **Anna Strunecka,** Laboratory of Neuropharmacology, 1st Faculty of Medicine, Charles University in Prague, Albertov 4, 128 00 Prague 2, Czech Republic; E-mail: strun@natur.cuni.cz

secretion of cortisol. The HPA axis modulates the ability to react to change, a feature of which is a dramatic increase in cortisol upon waking, called the Cortisol Awakening Response (CAR).

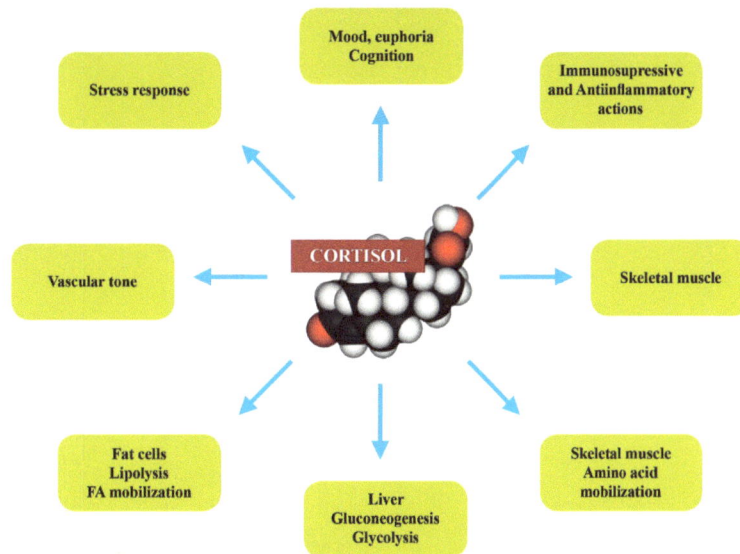

Figure 1: The principal physiological actions of the glucocorticoid hormone, cortisol. (**Cortisol was used from** Benjah-bmm27).

ACTH secretion shows a distinct circadian rhythm related to sleep-wakefulness cycle and this is reflected in the pattern of cortisol secretion. The concentration of cortisol in the plasma is minimal at around midnight and rises to a maximum between 6 and 8 a. m. before falling slowly during the rest of day [1]. Superimposed on this cycle pattern is a pulsatile secretion of cortisol followed by a transient refractory period when the HPA axis appears to be inhibited.

Interestingly, there are differences in the adrenocortical response to stress between male and female subjects. In females pulses of cortisol secretion occur approximately once per hour with variation in pulse amplitude underlying a diurnal rhythm. Males show smaller pulses of secretion which become widely spaced. Pulsatility is altered by genetic programming, early life experience and reproductive status [6, 7].

Several studies have shown that there may be dysfunction in the control of HPA axis and the secretion of cortisol in individuals with ASD. Mothers of autistic children share a profile of the HPA axis hypoactivity with their children [8]. Several animal as well as human studies have shown that a wide variety of prenatal stressors increase the risk for a diverse range of adverse neurodevelopmental outcomes in the child. In animals, the morphology and function of the offspring's hippocampus is negatively affected by prenatal maternal stress [9]. Maternal stress during gestation is involved in abnormal behavior, mental, and cognition disorders in offspring [10-14]. In addition, physiological stress reactivity in experimental studies with autistic individuals has also been shown to be responsive to social support [15, 16].

CORTISOL AND ACTH LEVELS IN ASD

Early studies suggested that there might be a dysfunction in the HPA-axis of the poorly-developed autistic children [17]. In order to examine the function of HPA axis in autistic children, the diurnal rhythm of saliva cortisol was investigated using saliva samples. The plasma and saliva cortisol levels showed a positive correlation in normal healthy adults. Moreover, the saliva cortisol level exhibited a similar diurnal rhythm as did the plasma cortisol level [18]. Hoshino with co-workers found that some children with infantile autism showed an abnormal diurnal rhythm for saliva cortisol. Such abnormality was observed more frequently in poorly-developed cases than in highly-developed cases. These authors suggested that the negative feedback mechanism of the HPA-axis may be disturbed in autistic children (Fig. **2**). Lower blood cortisol and higher ACTH levels were reported in individuals with autism

[19, 20-22]. Corbett with co-workers [21] estimated circadian rhythms of cortisol in 22 children with and 22 children without autism via analysis of salivary samples collected in the morning, afternoon and evening over 6 separate days. Children with autism showed a decrease in cortisol in the morning over 6 days while maintaining higher evening values. Children with autism also showed more within-and between-subject variability in circadian rhythms. Children with autism thus showed dysregulation of the circadian rhythm evidenced by variability between groups, between children, and within individual child comparisons.

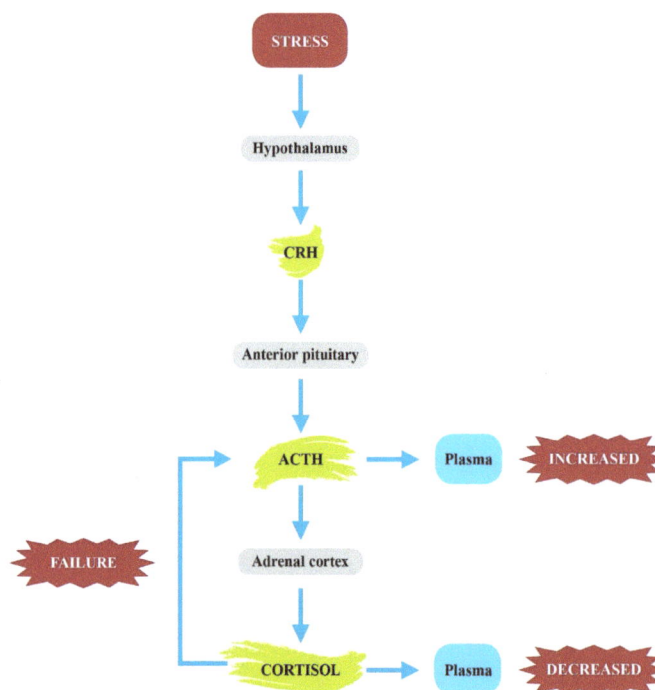

Figure 2: A flow chart showing the factors that regulate the secretion of glucocorticoids. Circulating glucocorticoids inhibit the secretion of CRH by the hypothalamus and that of ACTH by the anterior pituitary. Circulating ACTH also probably inhibits the secretion of CRH. The negative feedback mechanism of the HPA-axis is disturbed in autistic children. Lower blood cortisol and higher ACTH levels were reported in individuals with autism and their mothers.

Increased plasma ACTH levels were also found in adults with Asperger syndrome (AS) [23]. Twenty medication-free individuals with high-functioning AS were recruited from a clinic specialized in ASD and compared with ten age-matched healthy persons. However, plasma cortisol levels were similar in both groups. Brosnan with co-workers [24] examined whether the CAR was evident in 20 adolescent males with AS and 18 age-matched typically developing controls (aged 11-16). Whilst a significant CAR was evidenced in the control group, this was not the case for those with AS. A normal diurnal decrease in cortisol, however, was evident in both groups. The authors discussed the implication that individuals with AS may have an impaired response to change in their environment due to a refractory HPA axis. A failure to suppress corticol secretion in response to dexamethasone was reported in a group of high-functioning children with autism and matched controls [25]. There was a tendency towards cortisol hypersecretion during the day, predominantly in those autistic children who were integrated into the normal school system. While the temporal parameters of the cortisol circadian rhythm in these children with autism were probably normal, the tendency towards cortisol hypersecretion may indicate an environmental stress response in this group.

Corbett with coworkers [21] assessed HPA responsiveness by examining changes in salivary cortisol in response to a mock MRI. One-half of the children were re-exposed to the MRI environment. A statistically significant elevation in cortisol in response to the initial mock MRI was not observed. Rather, both groups showed heightened cortisol at the arrival to the second visit to the imaging centre, suggesting an anticipatory response to the re-exposure to the mock MRI.

Infants show cortisol response to receiving inoculations. Lewis and Thomas [26] observed that at 6 months the basal levels of cortisol also were influenced by an adult-like circadian rhythm; infants tested shortly after awakening had higher basal levels than those tested later in the day. These data provide strong evidence that studies of stress and cortisol release in infants must be sensitive to basal level, circadian rhythm, and behavioral effects, and that appropriate statistical procedures should be employed. Routine inoculation was used as a standard stressor to examine adrenocortical reactivity in infants [27-29]. Observation of cortisol and behavioral responses to routine inoculation was conducted at 2, 4, 6, and 18 months of age for infants in a longitudinal sample whose stress responses had been observed.

Infant cortisol and behavioral responses to receiving one versus two inoculations on one pediatric office visit were observed at 2 and 6 months of age. Cortisol level (pre- plus postinoculation level) decreased with age, whereas cortisol response (post- minus preinoculation level) did not vary with age when the data were aggregated over infants showing a pre- to postinoculation cortisol increase and those showing a decrease. Nonetheless, for those infants who showed a cortisol increase, cortisol level and response decreased with age. Infants quieted faster at the older age. There was a moderate relation between quieting behavior and cortisol response, at least for infants who showed a pre- to post-inoculation cortisol increase. These findings indicate a developmental trend for a decline over age in adrenocortical reactivity to inoculation for infants showing a cortisol release following the perturbation. Results were comparable whether infants received one or two inoculations. At 18 months, infants showed an increase in cortisol level over base to the perturbation. The magnitude of this response did not differ from the 6-month response. Moreover, level of cortisol response at 18 months was related to level of cortisol response at 6 months, but not at 2 or 4 months of age. These findings indicate that a developmental shift in adrenocortical functioning has occurred by 6 months of age.

The higher levels of the HPA axis hormones ACTH and β-endorphin were significantly higher in 48 autistic individuals than in 26 normal controls [19, 30]. During stress exposure peripheral β-endorphin is co-released with ACTH from the pituitary and does not cross the BBB. When measured in plasma, β-endorphin should be considered a stress hormone and not an indicator of central opioid functioning. Tordjman with coworkers suggested that elevated plasma levels of ACTH and β-endorphin found in individuals with autism reflect enhanced physiological and biological stress responses that are dissociated from observable emotional and behavioral reactions. The increased groups mean β-endorphin level in autism probably reflects an increased activation of the HPA axis, which appears to be associated with the stress of the blood drawing setting. These researchers thus suggested that individuals with severe autism have a heightened response to acute stressors rather than chronic hyper arousal or elevated basal stress response system functioning.

CHANGES IN CORTISOL LEVELS IN PARENTS OF AUTISTIC CHILDREN

The experience and some studies have indicated a higher level of stress in parents of children with autism. Behavioral problems of children correlated positively with parenting stress [31]. Mothers of children with autism scored higher than fathers in parental stress; no such differences were found in the group of parents of children with Down syndrome and typically developing children.

Recently, Seltzer with co-workers [8] reported the results from their longitudinal study with 406 recruited mothers of individuals with ASD. The data were collected between 1998 and 2005. The final set of analyses consisted of a sub-sample of 86 mothers. Importantly, mothers of adolescents or adults with ASD were found to have significantly lower level of cortisol throughout the day in comparison with a nationally representative group of mothers of similarly-aged unaffected children (n = 171). Their observations supported the view that mothers of individuals with ASD evidenced a profile of the HPA axis hypoactivity.

According to Seltzer *et al.* [8], the hypoactivity of HPA axis seems to be the possible physiological adaptation in the form of down-regulation of hormone activity of individuals experiencing chronic stress. The higher level of psychological distress and elevated levels of daily fatigue in these mothers of adolescents and adults with ASD are well known. However, these authors suggested and interesting alternative explanation for the observed between-group differences. Seltzer with co-workers suggested that there may be a pre-existing tendency for lower cortisol levels in mothers who have a child with ASD. In other words, the hypoactivation of HPA axis may be the result of a preexisting status rather than reflective of the reactive effects of the parenting a child with ASD.

These findings are in agreement with previous findings of Marinovic-Curin and co-workers [32], who conducted a study to test several elements of the HPA axis. Autistic subjects were studied as well as their parents. Cortisol circadian rhythm, cortisol daily secretion and its suppression response to dexamethasone had been measured from saliva or urine samples of the autistic children and their parents. The cortisol elevation after ACTH stimulation among the autistic individuals and their parents was slower than in healthy controls. No differences were found in salivary cortisol circadian rhythm or suppression response, as well as in cortisol daily excretion. The data of this study indicate that, compared to healthy subjects, autistic individuals have fine differences in cortisol response to ACTH stimulation.

The Impact of Prenatal Stress on Development of Offspring

A substantial number of human epidemiological data, as well as animal studies, suggest that adverse maternal stress during gestation is involved in abnormal behavior, mental, and cognition disorder in offspring. We mentioned that the hippocampus is the principal target site in the brain for corticosteroids, as it has the highest concentration of receptor sites for glucocorticoids [2]. Additionally, hippocampal corticosteroid receptor expression is critical in regulating glucocorticoid negative-feedback actions on HPA axis function and endogenous glucocorticoid release.

Animal Experiments

In animals, the morphology and function of the offspring's hippocampus is negatively affected by prenatal maternal stress. For example, Jia *et al.* [33] show that prenatal stress increases the glutamate level in hippocampus and causes apical dendritic atrophy of pyramidal neurons of hippocampal CA3 in offspring rats. The guinea-pig and pig have been used extensively to investigate the effects of prenatal glucocorticoids and stress on endocrine function and behavior in juvenile and adult offspring. The guinea pig, like the human, initiates the most rapid phase of brain growth during late fetal life. Adult male guinea pigs that were born to mothers exposed to a stressor during the phase of rapid fetal brain growth exhibit significantly increased basal plasma cortisol levels [9]. A significant reduction in glucocorticoid receptor mRNA in the CA3 region of the hippocampus was observed. In contrast, male guinea pig offspring, whose mothers were exposed to stress later in gestation exhibited a significantly higher plasma cortisol response to activation of the HPA axis. Prenatal exposure to synthetic glucocorticoids alters the expression of NMDA receptor subunits in the fetal and neonatal hippocampus of guinea pigs in a sex-specific manner [34]. Female offspring born to mothers treated with synthetic glucocorticoid betamethasone exhibited increased activity in an open-field and increased hippocampal NMDA receptors expression in early juvenile life. This hyperactivity was associated with reduced NR1 subunit mRNA levels in CA1/2 and CA3 regions of the hippocampus. Intriguingly, there were no effects of prenatal glucocorticoid exposure on NR1 mRNA levels in male fetuses. On the other hand, repeated prenatal glucocorticoid exposure has profound influences on HPA function and regulation in the juvenile guinea pig, and this involves altered regulation at the level of the pituitary and adrenal cortex. Furthermore, juvenile males appear to be more vulnerable to the effects of prenatal glucocorticoid exposure than females [35].

The increased basal salivary cortisol and a blunted ACTH response to exposure to the novel open-field enclosure in male and female guinea pig offspring born to chronically stressed mothers was reported by Emack *et al.* [36]. Prenatal chronic stress led to modification of growth trajectory, locomotor activity, and ACTH responses to stress in juvenile offspring. Importantly, males appear considerably more vulnerable to these effects than females.

Several studies demonstrate that piglet physiology and behavior can be affected when the mother has elevated cortisol concentrations during gestation.

The repeated social stress during pregnancy has long-lasting consequences on HPA axis and hippocampal neurotransmitter activity in the offspring of pigs [37, 38]. The prolonged oral administration of cortisol (using to model the prenatal stress) to pregnant sows resulted in elevated maternal plasma and salivary cortisol concentrations. This treatment induced elevated fetal basal and ACTH-induced plasma cortisol concentrations. Postnatally, it reduced birth weight of the piglets. In addition, it reduced the female offspring's salivary cortisol response to ACTH; moreover, the piglets were more aggressive in a social test. A repeated social stress applied to pregnant sows during late gestation can induce long-lasting effects on several parameters of the immune function of the offspring [39].

Human Studies

Recent human studies have shown that a wide variety of prenatal stressors increase the risk for a diverse range of adverse neurodevelopmental outcomes in the child.

Moreover, pregnant women in the developed world, which are at risk of preterm delivery, are prescribed antenatal glucocorticoid therapy. Many sick very low-birth-weight infants exhibit disproportionately low postnatal cortisol concentrations, which are associated with *e.g.* subsequent chronic lung disease. This is a known phenomenon in which corticosteroid supplementation reduces the mortality of hypocortisolemic patients. These observations have introduced the concept of early adrenal insufficiency, and in preliminary trials some infants indeed seem to benefit from low-dose early glucocorticoid replacement. Structural and functional alterations resulting from fetal glucocorticoid overexposure may persist throughout life and have been associated with diverse diseases [40]. Evidence is beginning to emerge indicating that children who were exposed to either prenatal synthetic glucocorticoid administration or endogenous glucocorticoid, because of maternal stress, may be at higher risk of emotional and behavioral abnormalities, such as postnatal aggressive/destructive behavior, increased distractibility, and hyperactivity. Children whose mothers experienced high anxiety during pregnancy are significantly more prone to developing attention deficit and hyperactivity disorder (ADHD) and other behavioral problems. There could be a direct effect of maternal mood on fetal brain development, which affects the behavioral development of the child [41, 42].

Ramsay *et al.* [43] used routine inoculation as a standard stressor to examine the effect of prenatal alcohol and/or cigarette exposure during prenatal development. They findings indicate that prenatal alcohol and/or cigarette exposure is associated with hyporeactivity of the adrenocortical system to stress at two months, but that this effect is no longer present by six months of age. Moreover, while there was a trend for a decrease with age in cortisol response in the nonexposed infants, there was a trend for an increase with age in cortisol response in the exposed infants. The authors suggested that there may be a continued long-term effect of prenatal alcohol and/or cigarette exposure on adrenocortical functioning that might have proved significant in a larger sample and would become more apparent at older ages. Children born to alcoholic mothers may show a profound mental retardation ranging to an apparent normality, and extending through epilepsy, attention deficit disorders with or without hyperactivity, autism and PDD, and different types of learning disorders. When adolescents, they may develop different kinds of personality disorders and substance abuse disorders. Finally, in adulthood, they may suffer from different types of affective and psychotic disorders [44]. It has been demonstrated in many animal studies that alcohol exposure induces neuronal and astroglial alterations in the hippocampal CA-1 area in offspring [45, 46]. Miles *et al.* [47] reported that children from high alcoholism families were more likely to have the onset of their autistic behavior occur with a loss of language (52.5% vs. 35.8%). This occurred primarily in families where the mother was alcoholic (80% vs. 40%), suggesting an association between maternal alcoholism and regressive onset autism.

Kajantie *et al.* [48] assessed whether human placental 11β-HSD2 activity is related to early adrenal insufficiency and postnatal clinical course in extremely low birth weight (<1000 g) infants. Impaired activity of this enzyme is common in intrauterine growth restriction and preeclampsia, conditions frequently associated with early preterm birth. Fetal glucocorticoid excess due to reduced 11β-HSD2 activity could make small preterm infants susceptible to early adrenal insufficiency when the maternal cortisol source is no longer sustained. Of all infants, those born severely preterm are probably most likely to be exposed to excess glucocorticoids. The 110 preterm infants born before 32 wk gestation were included into this study. The finding of a clear correlation between relative birth weight and reduced placental 11β-HSD2 activity rate is in agreement with similar previous findings.

Pretermed infants, treated with a prolonged tapering course of dexamethasone to decrease the risk and severity of chronic lung disease, were examined at one year of age [49]. Study participants were 118 very low birth weight infants. Twenty five children had cerebral palsy and 45% abnormal neurologic examination findings. Some recent studies are dealing with association between preterm birth and autism [13,14,50]. Hack *et al.* [14] studied the prevalence of behavioral problems and symptomatology suggestive of autism and Asperger's disorders at age 8 years among extremely low birth weight (ELBW) children, born 1992 through 1995. These authors compared parent reports of the behavior of 219 ELBW infants (mean birth weight, 810 g; gestational age 26 weeks) with 176 normal birth weight children of similar maternal sociodemographic status, sex, and age. ELBW children had significantly higher mean Symptom Severity Scores for the inattentive, hyperactive, and combined types of ADHD as well as higher scores for autistic and Asperger's disorders.

Gutteling with co-workers studied the effects of prenatal maternal stress on the development of their children [10-12]. This group of researchers analyzed prenatal stress at around 16 weeks of gestation through questionnaires and a cortisol day curve. Cortisol reactions were determined preceding and following the vaccination of children in the age between 3.11 and 5.9 years. Children of mothers who had higher concentrations of morning cortisol during pregnancy had higher concentrations of cortisol as compared to children of mothers who had lower concentrations of morning cortisol. Furthermore, more daily hassles and a higher level of fear of bearing a handicapped child during pregnancy were associated with higher concentrations of cortisol in the children. Gutteling *et al.* [11] examined whether pregnancy stress predicted HPA-axis reactions of children to the first day of school after the summer break. Children whose mothers had higher levels of morning cortisol during pregnancy, and more fear of bearing a handicapped child showed higher levels of cortisol on school days. In addition, the circadian rhythm of cortisol on school days appeared to have a steeper slope as compared to that of the circadian curve on a weekend day. The subsequent study investigated the influence of prenatal maternal stress on learning and memory of 112 children (50 boys, 62 girls, age 6.7 years). Results of hierarchical multivariate regression analyses showed that maternal life events measured during the first part of pregnancy were negatively associated with the child's attention/concentration index, while controlling for overall IQ, gender, and postnatal stress. No associations were found between prenatal maternal cortisol and the offspring's learning and memory.

NEW GENETIC AND BIOCHEMICAL LINKS TO HPA AXIS: IMPLICATION FOR ASD

Association between Autism and Polymorphisms of PITX1

Many authors report the dysregulation of the HPA axis in ASD; their conclusions are mostly based on measurement of ACTH and cortisol. Similar changes were reported in mothers of autistic children. Philippi and co-workers [51] suggested that paired-like homeodomain transcription factor 1(*PITX1)*, a key regulator of hormones within the HPA axis, such as ACTH, cortisol, and β-endorphin; may be implicated in the etiology of autism. The team of researchers investigated a total of 276 families from the Autism Genetic Resource Exchange (AGRE) repository composed of 1086 individuals including 530 affected children. They focused on chromosome 5q31 and identified three genes that they hypothesized could be involved in the development of autism: 1) paired-like homeodomain transcription factor 1 (*PITX1*); 2) histone family member *H2AFY*, which is involved in X-inactivation in females and therefore could be a positional candidate that could explain the 4:1 male:female gender distortion present in autism; and 3) neurogenin 1 (*NEUROG1*), which is a transcription factor involved in neurogenesis. Using single point association analyses and haplotype analyses, Philippi with co-workers found significant evidence for an association of autism with *PITX1* but not with *H2AFY* or *NEUROG1*. Since this study is of great importance and results of molecular biology cannot be freely interpreted, we bring some selected parts from article of Phillipi *et al.* [51]:

> "In this study, we have found a significant association between autism and polymorphisms of PITX1, a paired-like homeodomain transcription factor involved in hormonal regulation. Using a two step procedure we initially identified evidence for association for marker rs3805663 with autism. In the second step additional markers in the PITX1 gene were genotyped in an extended sample set of 276 families total. Although in this extended set marker rs3805663 did not reach a significant p-value any more, several additional markers showed highly significant results with the most significant result for rs6596189. Haplotype analyses yielded a couple of highly significant pairings, although none of these showed higher significance than rs6596189 alone. Individuals homozygous or heterozygous for the risk allele were 2.69 and 1.74 fold more likely to be autistic than individuals who were not carrying the allele, respectively.
>
> Linkage between chromosome 5q31-32 and autism is consistent with previously published studies. Both the IMGSAC (1998) genome-wide linkage scan and the screen performed by Risch and colleagues [52] found modestly elevated LOD-scores* in this region and exclusion mapping analyses did not clearly

exclude this region. Interestingly, this genomic region was recently identified as a potential locus for ADHD... In a genome-wide scan using large multi-generational pedigrees, Arcos-Burgos and colleagues [53] established an exclusion map for the region and defined a critical interval from 119 to 135 Mb, which encompasses the PITX1 gene region. Their most significant family-specific microsatellite marker D5S2117 at 133.4 Mb is less than 1 Mb from the PITX1 gene locus.

Although the mechanisms by which PITX1 may contribute to the susceptibility to autism are yet to be explored, the genetic association between PITX1 polymorphisms and autism, described here, could provide an explanation for the abnormal level of hormones of the HPA axis reported in the literature."

LOD-score: In genetics, a statistical estimate of whether two loci (the sites of genes) are likely to lie near each other on a chromosome and are therefore likely to be inherited together as a package.

These important genetic finding supports the view that the disturbance of the HPA axis observed in individuals with ASD and their mothers has important role in the etiology and pathogenesis of ASD.

Disorders of Cortisol Metabolism

The decreased blood/salivary level of cortisol under situation of increased ACTH level might be connected with some disorder of cortisol metabolism. Let us look on the activity of an endoplasmic reticulum-bound enzyme 11β-HSD1, which catalyzes the interconversion of hormonally active glucocorticoids (cortisol and corticosterone) to inactive glucocorticoids (cortisone in humans and 11-dehydrocorticosterone in rodents). For simplicity, we will further refer on cortisol only, but the interconversions of 11-dehydrocorticosterone proceeds by the same way. The reaction direction, which 11β-HSD1 catalyzes, is determined by the relative abundance of nicotinamide adenine dinucleotide phosphate (NADP) and/or reduced nicotinamide adenine dinucleotide phosphate (NADPH) [54].

In its native purified state, **11β-HSD1** acts as **a dehydrogenase inactivating cortisol to cortisone** (Fig. **3**). However, in the presence of reducing **NADPH**, generated through pentose phosphate shuttle, 11β-HSD1 switches to **reductase** with the **generation of active cortisol** in key tissues such as liver, adipose, and adrenal gland.

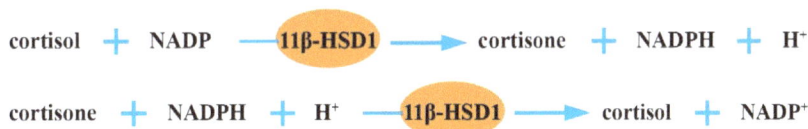

cortisol + NADP ——11β-HSD1—→ cortisone + NADPH + H⁺

cortisone + NADPH + H⁺ ——11β-HSD1—→ cortisol + NADP⁺

Figure 3: The activity of 11β-HSD1 under different redox states.

In hepatic microsomes from healthy animals or humans, reductase activity predominates and is significantly higher than dehydrogenase activity [54]. Thereby 11β-HSD1 plays a key role in the regulation of metabolic functions and in the adaptation of the organism to energy requiring situations. In humans, this implies that in low level of NADPH, 11β-HSD1 may be inactive or indeed switched from reductase to dehydrogenase activity.

The Effects of Redox Status on 11β-HSD1 Activity

NADPH is generated in the pentose phosphate pathway. The main function of NADP$^+$ is as a reducing agent in anabolism. Since NADPH is needed to drive redox reactions as a strong reducing agent, the NADP$^+$/NADPH ratio is kept very low. One of the best documented biochemical changes in ASD is a decrease in glutathione (GSH), the major intracellular antioxidant and an increase in oxidized glutathione (GSSG), resulting in a reduction in the ratio of reduced (active) GSH to (inactive) GSSG. Glutathione is pivotal for the maintenance of intracellular redox homeostasis and defense against oxidative damage in eukaryotic cells. There are no data on the intracellular ratio of NADP/NADPH in autistic individuals and no measurements of 11β-HSD1 activity. We can only speculate that the significantly reduced levels of GSH and thus reduced redox potential of the cell would favorite the conversion of active cortisol to inactive cortisone.

Studies of patients with genetic defects in 11β-HSD1 action show abnormal HPA axis responses with hyperandrogenism being a major consequence [55]. A lack of cortisol regeneration stimulates ACTH-mediated adrenal hyperandrogenism, with males manifesting in early life with precocious puberty and females presenting in midlife with hirsutism, oligoamenorrhea, and infertility. Elevated adrenal androgen levels are common in women with polycystic ovary syndrome or congenital adrenal hyperplasia [56]. Androgen excess in these women may be ovarian and/or adrenal in origin, but some authors proposed contributing mechanism is altered cortisol metabolism.

The association of low cortisol and higher androgen levels is evidently characteristic for some individuals with ASD. Some children with ASD have significant increases in their levels of plasma dehydroepiandrosterone (DHEA) and serum total testosterone [57]. David Geier and Mark Geier examined seventy consecutive patients with an ASD diagnosis who presented to the Genetic Centers of America for outpatient genetic/developmental evaluations from 2005-2007. Morning blood samples collected following an overnight fast, compared to the pertinent reference means, showed significantly increased relative mean levels for serum testosterone (158%), serum free testosterone (214%), percent free testosterone (121%), DHEA (192%), and androstenedione (173%), respectively. With respect to their age- and sex-specific reference ranges, females had significantly higher overall mean relative testosterone and relative free testosterone levels than males.

The significant increase in precocious puberty has been reported in autistic children [57, 58]. Geiers account the increase of DHEA level to decrease of reduced glutathione and the mutual cyclical interactions of androgen metabolites with methionine cycle transsulfuration pathways. According to their explanation, the decreased glutathione blocks the conversion of DHEA to DHEA-S (catalyzed by DHEA sulfotransferase) and therefore raises androgens, which in turn further lower glutathione levels. Elevated DHEA and DHEA-S levels have been also implicated in polycystic ovary syndrome.

The knowledge of the regulation of 11β-HSD1 activity could contribute to the further understanding of interactions between methionine-homocysteine cycle and HPA axis in pathophysiology of ASD.

Prenatal glucocorticoid administration, selectively during late gestation, results in early and persistent elevations in 11β-HSD1 mRNA expression and activity in the liver, pancreas, and subcutaneous fat. Niyrenda *et al.* [59] coined a "fetal programming" paradigm where brief antenatal exposure to glucocorticoids leads to the metabolic syndrome in the offspring. Polymorphisms in *HSD11B1,* the gene encoding 11β-HSD1, have been associated with metabolic phenotype in humans, including type 2 diabetes and hypertension [60]. Compared with the common G allele, the A allele of rs13306421, a polymorphism located two nucleotides 5' to the translation initiation site, gave higher 11β-HSD1 expression and activity *in vitro* and was translated at higher levels in *in vitro* translation reactions. The observed polymorphism may have direct functional consequences on levels of 11β-HSD1 enzyme activity *in vivo*. However, the rs13306421 A sequence variant originally reported in other ethnic groups (Pima Indians) may be of low prevalence because it was not detected in a population of 600 European Caucasian women.

Interestingly, Lewis *et al.* [27] observed the behavioral and cortisol responses of Japanese infants and Caucasian American infants, 4 months of age, during and following routine inoculation. The Caucasian American group showed a more intense initial affective response and a longer latency to quiet than the Japanese group; the Japanese group showed a greater cortisol response. Infants in the Caucasian American group were more likely to fall in the high behavior-low cortisol group, while infants in the Japanese group were more likely to fall in the low behavior-high cortisol group.

SOCIAL SUPPORT AND ENDOCRINE RESPONSE

Children with ASD exhibit social, communicative, and behavioral deficits. The presence of social support has been associated with decreased stress responsiveness and with attenuated free cortisol concentrations in saliva [15]. Although men in the partner support condition showed significant attenuation of cortisol responses compared with unsupported men, women showed a tendency toward increased cortisol responses when supported by their boyfriends.

Lopata *et al.* [16] examined the effect of social familiarity on salivary cortisol and social anxiety/stress for a sample of children with high-functioning ASD. Participants interacted with a familiar peer on one occasion and an unfamiliar peer on another occasion. Data were collected using salivary cortisol and a scale measuring subjective stress. A mild-moderate correlation was found between self-reported distress and salivary cortisol within each condition.

The Effect of Oxytocin on HPA Axis

Cortisol level might be affected by oxytocin. Oxytocin is implicated in stress reduction as well as in social behavior. It inhibits the stress-induced activity of the HPA axis responsiveness. Oxytocin is involved in social affiliation, sexual and maternal-infant binding, anxiety, mood, feeding control and memory (see the Chapter 11). Oxytocin receptors were identified in the adenohypophysis, where oxytocin might participate on regulation of ACTH and CRH. Windle with co-workers [61, 62] reported that oxytocin attenuates stress-induced HPA activity and anxiety behavior in female rats. The response to noise stress was significantly and dose-dependently decreased by oxytocin. The behavioral responses show that oxytocin exerts a central anxiolytic-like effect on both endocrine and behavioral systems and could play a role in moderating behavioral and physiological responses to stress. Importantly, their experiments show that central oxytocin attenuates both the stress-induced neuroendocrine and molecular responses of the HPA axis and that the dorsal hippocampus, ventrolateral septum, and the hypothalamic paraventricular nucleus constitute an oxytocin-sensitive forebrain stress circuit. Several studies document that oxytocin significantly reduced salivary cortisol levels after intranasal application [63, 64].

The Effects of Service Dogs

Recently, Viau and co-workers [65] studied the physiological impact of service dogs on children with ASD. It has been already demonstrated that human interaction with dog results in a decrease of cortisol levels in healthy adults. Introducing service dog to children with ASD is an attractive idea that has received growing attention in recent decades. However, no study has measured the physiological impact of service dogs on these children. Viau *et al.* [65] assessed the effects of service dogs on the basal salivary cortisol secretion of children with ASD. They measured the salivary cortisol levels of 42 children with ASD in three experimental conditions; prior to and during the introduction of a service dog to their family, and after a short period during which the dog was removed from their family. They compared average cortisol levels and CAR before and during the introduction of the dog to the family and after its withdrawal. The introduction of service dogs diminished CAR. Before the introduction of service dogs, they measured a 58% increase in morning cortisol after awakening, which diminished to 10% when service dogs were present. The increase in morning cortisol jumped back to 48% once the dogs were removed from the families. However, service dogs did not have an effect on the children's average diurnal cortisol levels. These results show that the CAR of children with ASD is sensitive to the presence of service dogs, which lends support to the potential behavioral benefits of service dogs for children with autism.

CONCLUSION

Kendall and Reichstein, who independently isolated and synthesised cortisol and then ACTH, were awarded the Nobel Prize for Medicine and Physiology in 1950. Today, physiology and biochemistry offer the detailed understanding of the multiple roles of glucocorticoids in human body and their key role in the response to stress. Glucocorticoids are among the most widely used drugs in the world and are effective in many inflammatory and immune diseases. The predominant effect of corticosteroids is to switch off multiple inflammatory genes encoding cytokines, chemokines, inflammatory enzymes, receptors, and proteins that have been activated during the chronic inflammatory process. Moreover, glucocorticoid therapy is prescribed to pregnant women in the developed world, which are at risk of preterm delivery.

Several studies have demonstrated the adverse effects of prenatal stress, alcohol and smoking, as well as the excess of synthetic glucocorticoids on the neurodevelopment in offsprings. The biochemical and genetic studies highlight the importance of the redox and 11β-HSD1 control of cortisol metabolism in regulating activity of HPA axis. It has been well documented that pathology of ASD is connected with hypoactivity of HPA axis and decreased plasma level of cortisol. Moreover, ASD pathology is connected with chronic inflammatory processes. The questions thus arise: Is autism a stress disorder? Are glucocorticoids beneficial or harmful for autistic individuals? Is the chronic inflammation of the brain and gut in many autistic individuals associated with the low cortisol levels? Our review shows that scientific research has not yet provided the satisfying responses.

This work was supported by VZ MSM 0021620806.

REFERENCES

[1]   Pocock G, Richards CD. Human Physiology. The Basis of Medicine. Oxford, New York: Oxford University Press 1999.

[2]   Sapolsky RM, Krey LC, McEwen BS. Glucocorticoid-sensitive hippocampal neurons are involved in terminating the adrenocortical stress response. Proc Natl Acad Sci U S A 1984; 81: 6174-7.

[3]   Kremen WS, O'Brien RC, Panizzon MS, *et al.* Salivary cortisol and prefrontal cortical thickness in middle-aged men: A twin study. Neuroimage 2010; Feb 13. [Epub ahead of print]

[4]   Hinkelmann K, Moritz S, Botzenhardt J, *et al.* Cognitive impairment in major depression: association with salivary cortisol. Biol Psychiatry 2009; 66: 879-85.

[5]   Klein JF. Adverse psychiatric effects of systemic glucocorticoid therapy. Am Fam Physician 1992; 46: 1469-74.

[6]   Lightman SL, Windle RJ, Julian MD, *et al.* Significance of pulsatility in the HPA axis. Novartis Found Symp 2000; 227: 244-57; discussion 57-60.

[7]   Lightman SL, Windle RJ, Ma XM, *et al.* Hypothalamic-pituitary-adrenal function. Arch Physiol Biochem 2002; 110: 90-3.

[8]   Seltzer MM, Greenberg JS, Hong J, *et al.* Maternal Cortisol Levels and Behavior Problems in Adolescents and Adults with ASD. J Autism Dev Disord 2010; 40: 457-69.

[9]   Kapoor A, Leen J, Matthews SG. Molecular regulation of the hypothalamic-pituitary-adrenal axis in adult male guinea pigs after prenatal stress at different stages of gestation. J Physiol 2008; 586: 4317-26.

[10]   Gutteling BM, de Weerth C, Buitelaar JK. Maternal prenatal stress and 4-6 year old children's salivary cortisol concentrations pre- and post-vaccination. Stress 2004; 7: 257-60.

[11]   Gutteling BM, de Weerth C, Buitelaar JK. Prenatal stress and children's cortisol reaction to the first day of school. Psychoneuroendocrinology 2005; 30: 541-9.

[12]   Gutteling BM, de Weerth C, Zandbelt N, *et al.* Does maternal prenatal stress adversely affect the child's learning and memory at age six? J Abnorm Child Psychol 2006; 34: 789-98.

[13]   Buchmayer S, Johansson S, Johansson A, *et al.* Can association between preterm birth and autism be explained by maternal or neonatal morbidity? Pediatrics 2009; 124: e817-25.

[14]   Hack M, Taylor HG, Schluchter M, *et al.* Behavioral outcomes of extremely low birth weight children at age 8 years. J Dev Behav Pediatr 2009; 30: 122-30.

[15]   Kirschbaum C, Klauer T, Filipp SH, Hellhammer DH. Sex-specific effects of social support on cortisol and subjective responses to acute psychological stress. Psychosom Med 1995; 57: 23-31.

[16]   Lopata C, Volker MA, Putnam SK, Thomeer ML, Nida RE. Effect of social familiarity on salivary cortisol and self-reports of social anxiety and stress in children with high functioning autism spectrum disorders. J Autism Dev Disord 2008; 38: 1866-77.

[17]   Hoshino Y, Ohno Y, Murata S, *et al.* Dexamethasone suppression test in autistic children. Folia Psychiatr Neurol Jpn 1984; 38: 445-9.

[18]   Hoshino Y, Yokoyama F, Watanabe M, *et al.* The diurnal variation and response to dexamethasone suppression test of saliva cortisol level in autistic children. Jpn J Psychiatry Neurol 1987; 41: 227-35.

[19]   Tordjman S, Anderson GM, McBride PA, *et al.* Plasma beta-endorphin, adrenocorticotropin hormone, and cortisol in autism. J Child Psychol Psychiatry 1997; 38: 705-15.

[20]   Curin JM, Terzic J, Petkovic ZB, *et al.* Lower cortisol and higher ACTH levels in individuals with autism. J Autism Dev Disord 2003; 33: 443-8.

[21]   Corbett BA, Mendoza S, Wegelin JA, Carmean V, Levine S. Variable cortisol circadian rhythms in children with autism and anticipatory stress. J Psychiatry Neurosci 2008; 33: 227-34.

[22]   Corbett BA, Schupp CW, Levine S, Mendoza S. Comparing cortisol, stress, and sensory sensitivity in children with autism. Autism Res 2009; 2: 39-49.

[23]   Tani P, Lindberg N, Matto V, *et al.* Higher plasma ACTH levels in adults with Asperger syndrome. J Psychosom Res 2005; 58: 533-6.

[24]   Brosnan M, Turner-Cobb J, Munro-Naan Z, Jessop D. Absence of a normal cortisol awakening response (CAR) in adolescent males with Asperger syndrome (AS). Psychoneuroendocrinology 2009; 34: 1095-100.

[25]   Richdale AL, Prior MR. Urinary cortisol circadian rhythm in a group of high-functioning children with autism. J Autism Dev Disord 1992; 22: 433-47.

[26]   Lewis M, Thomas D. Cortisol release in infants in response to inoculation. Child Dev 1990; 61: 50-9.

[27]   Lewis M, Ramsay DS, Kawakami K. Differences between Japanese infants and Caucasian American infants in behavioral and cortisol response to inoculation. Child Dev 1993; 64: 1722-31.

[28]　Ramsay DS, Lewis M. Developmental change in infant cortisol and behavioral response to inoculation. Child Dev 1994; 65: 1491-502.

[29]　Lewis M, Ramsay D. Stability and change in cortisol and behavioral response to stress during the first 18 months of life. Dev Psychobiol 1995; 28: 419-28.

[30]　Tordjman S, Anderson GM, Botbol M, *et al.* Pain reactivity and plasma beta-endorphin in children and adolescents with autistic disorder. PLoS One 2009; 4: e5289.

[31]　Wulffaert J, Scholte EM, Dijkxhoorn YM, *et al.* Parenting Stress in CHARGE Syndrome and the Relationship with Child Characteristics. J Dev Phys Disabil 2009; 21: 301-13.

[32]　Marinovic-Curin J, Marinovic-Terzic I, Bujas-Petkovic Z, *et al.* Slower cortisol response during ACTH stimulation test in autistic children. Eur Child Adolesc Psychiatry 2008; 17: 39-43.

[33]　Jia N, Yang K, Sun Q, *et al.* Prenatal stress causes dendritic atrophy of pyramidal neurons in hippocampal CA3 region by glutamate in offspring rats. Dev Neurobiol 2010; 70: 114-25.

[34]　Owen D, Matthews SG. Repeated maternal glucocorticoid treatment affects activity and hippocampal NMDA receptor expression in juvenile guinea pigs. J Physiol 2007; 578: 249-57.

[35]　Owen D, Matthews SG. Prenatal glucocorticoid exposure alters hypothalamic-pituitary-adrenal function in juvenile guinea pigs. J Neuroendocrinol 2007; 19: 172-80.

[36]　Emack J, Kostaki A, Walker CD, Matthews SG. Chronic maternal stress affects growth, behaviour and hypothalamo-pituitary-adrenal function in juvenile offspring. Horm Behav 2008; 54: 514-20.

[37]　Kranendonk G, Mulder EJ, Parvizi N, Taverne MA. Prenatal stress in pigs: experimental approaches and field observations. Exp Clin Endocrinol Diabetes 2008; 116: 413-22.

[38]　Otten W, Kanitz E, Couret D, *et al.* Maternal social stress during late pregnancy affects hypothalamic-pituitary-adrenal function and brain neurotransmitter systems in pig offspring. Domest Anim Endocrinol 2010; 38: 146-56.

[39]　Couret D, Jamin A, Kuntz-Simon G, Prunier A, Merlot E. Maternal stress during late gestation has moderate but long-lasting effects on the immune system of the piglets. Vet Immunol Immunopathol 2009; 131: 17-24.

[40]　Tegethoff M, Pryce C, Meinlschmidt G. Effects of intrauterine exposure to synthetic glucocorticoids on fetal, newborn, and infant hypothalamic-pituitary-adrenal axis function in humans: a systematic review. Endocr Rev 2009; 30: 753-89.

[41]　O'Connor TG, Heron J, Golding J, Beveridge M, Glover V. Maternal antenatal anxiety and children's behavioural/emotional problems at 4 years. Report from the Avon Longitudinal Study of Parents and Children. Br J Psychiatry 2002; 180: 502-8.

[42]　O'Connor TG, Heron J, Golding J, Glover V. Maternal antenatal anxiety and behavioural/emotional problems in children: a test of a programming hypothesis. J Child Psychol Psychiatry 2003; 44: 1025-36.

[43]　Ramsay DS, Bendersky MI, Lewis M. Effect of prenatal alcohol and cigarette exposure on two- and six-month-old infants' adrenocortical reactivity to stress. J Pediatr Psychol 1996; 21: 833-40.

[44]　Evrard SG. [Diagnostic criteria for fetal alcohol syndrome and fetal alcohol spectrum disorders]. Arch Argent Pediatr 2010; 108: 61-7.

[45]　Ramos AJ, Evrard SG, Tagliaferro P, Tricarico MV, Brusco A. Effects of chronic maternal ethanol exposure on hippocampal and striatal morphology in offspring. Ann N Y Acad Sci 2002; 965: 343-53.

[46]　Tagliaferro P, Vega MD, Evrard SG, Ramos AJ, Brusco A. Alcohol exposure during adulthood induces neuronal and astroglial alterations in the hippocampal CA-1 area. Ann N Y Acad Sci 2002; 965: 334-42.

[47]　Miles JH, Takahashi TN, Haber A, Hadden L. Autism families with a high incidence of alcoholism. J Autism Dev Disord 2003; 33: 403-15.

[48]　Kajantie E, Dunkel L, Turpeinen U, Stenman UH, Andersson S. Placental 11beta-HSD2 activity, early postnatal clinical course, and adrenal function in extremely low birth weight infants. Pediatr Res 2006; 59: 575-8.

[49]　O'Shea TM, Kothadia JM, Klinepeter KL, *et al.* Randomized placebo-controlled trial of a 42-day tapering course of dexamethasone to reduce the duration of ventilator dependency in very low birth weight infants: outcome of study participants at 1-year adjusted age. Pediatrics 1999; 104: 15-21.

[50]　Boulet SL, Schieve LA, Boyle CA. Birth Weight and Health and Developmental Outcomes in US Children, 1997-2005. Matern Child Health J 2009;

[51]　Philippi A, Tores F, Carayol J, *et al.* Association of autism with polymorphisms in the paired-like homeodomain transcription factor 1 (PITX1) on chromosome 5q31: a candidate gene analysis. BMC Med Genet 2007; 8: 74.

[52]　Risch N, Spiker D, Lotspeich L, *et al.* A genomic screen of autism: evidence for a multilocus etiology. Am J Hum Genet 1999; 65: 493-507.

[53]　Arcos-Burgos M, Castellanos FX, Pineda D, *et al.* Attention-deficit/hyperactivity disorder in a population isolate: linkage to loci at 4q13.2, 5q33.3, 11q22, and 17p11. Am J Hum Genet 2004; 75: 998-1014.

[54] Walker EA, Ahmed A, Lavery GG, *et al.* 11beta-Hydroxysteroid Dehydrogenase Type 1 Regulation by Intracellular Glucose 6-Phosphate Provides Evidence for a Novel Link between Glucose Metabolism and Hypothalamo-Pituitary-Adrenal Axis Function. J Biol Chem 2007; 282: 27030-6.

[55] Cooper MS, Stewart PM. 11Beta-hydroxysteroid dehydrogenase type 1 and its role in the hypothalamus-pituitary-adrenal axis, metabolic syndrome, and inflammation. J Clin Endocrinol Metab 2009; 94: 4645-54.

[56] Gambineri A, Vicennati V, Genghini S, *et al.* Genetic variation in 11beta-hydroxysteroid dehydrogenase type 1 predicts adrenal hyperandrogenism among lean women with polycystic ovary syndrome. J Clin Endocrinol Metab 2006; 91: 2295-302.

[57] Geier DA, Geier MR. A prospective assessment of androgen levels in patients with autistic spectrum disorders: biochemical underpinnings and suggested therapies. Neuro Endocrinol Lett 2007; 28: 565-73.

[58] Knickmeyer RC, Wheelwright S, Hoekstra R, Baron-Cohen S. Age of menarche in females with autism spectrum conditions. Dev Med Child Neurol 2006; 48: 1007-8.

[59] Nyirenda MJ, Carter R, Tang JI, *et al.* Prenatal programming of metabolic syndrome in the common marmoset is associated with increased expression of 11beta-hydroxysteroid dehydrogenase type 1. Diabetes 2009; 58: 2873-9.

[60] Malavasi EL, Kelly V, Nath N, *et al.* Functional effects of polymorphisms in the human gene encoding 11 beta-hydroxysteroid dehydrogenase type 1 (11 beta-HSD1): a sequence variant at the translation start of 11 beta-HSD1 alters enzyme levels. Endocrinology 2010; 151: 195-202.

[61] Windle RJ, Shanks N, Lightman SL, Ingram CD. Central oxytocin administration reduces stress-induced corticosterone release and anxiety behavior in rats. Endocrinology 1997; 138: 2829-34.

[62] Windle RJ, Kershaw YM, Shanks N, *et al.* Oxytocin attenuates stress-induced c-fos mRNA expression in specific forebrain regions associated with modulation of hypothalamo-pituitary-adrenal activity. J Neurosci 2004; 24: 2974-82.

[63] Meinlschmidt G, Heim C. Sensitivity to intranasal oxytocin in adult men with early parental separation. Biol Psychiatry 2007; 61: 1109-11.

[64] Ditzen B, Schaer M, Gabriel B, *et al.* Intranasal oxytocin increases positive communication and reduces cortisol levels during couple conflict. Biol Psychiatry 2009; 65: 728-31.

[65] Viau R, Arsenault-Lapierre G, Fecteau S, *et al.* Effect of service dogs on salivary cortisol secretion in autistic children. Psychoneuroendocrinology 2010; Feb 26. [Epub ahead of print]

Reproductive Hormones and Autism Spectrum Disorders

Russell L. Blaylock, MD

Theoretical Neurosciences Research, LLC, and Visiting Professor of Biology, Belhaven University, Ridgeland, MS 39157, USA

Abstract: There is some evidence that autistic individuals have elevated exposure to androgenic hormones, even female autistics, and that autistic behaviors may, in part, be explained by influences of testosterone on neurodevelopment and higher-order brain function. Some studies indicate early exposure to maternal levels of testosterone *in utero* may play a predominate role. Similarities of ASD disorders to childhood schizophrenia are strengthened by a similar relationship between high levels of androgen exposure early in life and this disorder. The fact that estrogens play a major role in neuroprotection may explain the male preponderance of ASD. Recent studies have shown that testosterone can enhance excitotoxicity and estrogen can reduce excitotoxicity. Rather than a direct sex hormonal effect on brain function, I propose that the sex hormones are playing a modulating role on immunoexcitotoxicity.

INTRODUCTION

Several studies have shown an androgen elevation in autistics as compared to age-matched controls. Geier and Geier compared 70 consecutive patients, aged 6 years or greater, with an ASD diagnosis meeting DSM-IV criteria, in which they measured total serum free testosterone, percentage of free testosterone, serum/plasma dehydroepiandrosterone (DHEA), androstenedione and follicle-stimulating hormone (FSH) [1]. They found a significant elevation in several measures, including an elevation in serum total testosterone (158%), an increase in free testosterone (214%), an elevation in percent free testosterone (123%), an elevation in DHEA (192%) and in androstenedione (173%). Interestingly, they found a significant decrease in FSH levels (51%) in the ASD subjects. Also of interest, they found that females with ASD had elevated testosterone levels as well.

Tordjman and co-workers, in a French study, failed to find elevated androgens in ASD subjects [2]. In their study, plasma levels of testosterone and DHEA-S were measured in male autistic subjects (31 prepubertal and 8 postpubertal), mentally retarded subjects (MR12 prepubertal), and normal controls (NC 10 prepubertal and 11 postpubertal). The mean plasma testosterone levels were similar in both the postpubertal autistic subjects and the postpubertal NC group. No differences were seen for testosterone or DHEA-S in the prepubertal autistic, MR or NC individuals.

Other studies indicate that androgens may be elevated either *in utero* during development or within the affected child. For example, a study from the Galton Laboratory at the University College London, found that pooled data of reading disability, ASD and Attention Deficit Hyperactivity Disorders (ADHD) suggested a link to higher intrauterine levels of testosterone collectively in these disorders [3]. Yet, when examined alone, ASD did not correlate with high intrauterine testosterone exposures. The weakness of this study was that it was based on the assumption that higher maternal testosterone levels would increase the number of male siblings, which is a weak association.

Another indirect, but more convincing study, was conducted by Igudomnukul and co-workers at the Autism Research Centre, Department of Psychiatry at the University of Cambridge, in which they compared control mothers of autistic children with mothers of normally developing children [4]. Compared to the controls, mothers with ASD children had a far greater number of disorders associated with elevated androgen levels, such as hirsutism, bisexuality or asexuality, irregular menstrual cycles, dysmenorrhea, polycystic ovarian syndrome, severe acne, tomboyism and a family history of ovarian, uterine and prostate cancers. For some reason, actual testosterone levels during pregnancy were not measured.

*Address correspondence to: **Russell L. Blaylock,** Theoretical Neurosciences Research, LLC, and Visiting Professor of Biology, Belhaven University, Ridgeland, MS 39157, USA; E-mail: Blay6307@bellsouth.net

Because a balance between androgens and estrogens is essential for normal differential sexual development and behavioral function, we must explore further the possibility that such an imbalance in sexual hormones is playing a role in ASD.

TESTOSTERONE, BEHAVIOR AND BRAIN DEVELOPMENT

Most studies on testosterone's effects on the developing CNS are concerned with the dimorphic nuclei in the hypothalamus and sexual difference in the male and female brain. They are also involved in reproduction as suggested by the presence of numerous androgen receptors seen within the hippocampus, preoptic area, amygdala and medial hypothalamic area.

On a neuronal level, androgens play a significant role in calcium homeostasis, especially during the most intense period of brain formation. It has been shown that migrations of developing neurons and glia depend on the induction of calcium waves by androgens [5, 6]. Therefore, in conjunction with glutamate and cytokines, androgens play an important role in the architectonic development of the CNS.

Baron-Cohen in 2002 proposed his "extreme male" theory of autism, which is based on the observation that males are better at "systemizing" and females are better with "empathizing" [7]. He attributed this to an overexposure of developing babies to testosterone during critical periods of brain development. A subsequent study they examined fetal testosterone exposures by testing amniotic fluid in the mothers of 38 children (24 males and 14 females) [8]. These children were studied 4 years later, using a social relationship test not involving faces, but rather interacting cartoon triangles. He found that normal females used more mental and affective terms to describe the cartoons than did males. Fetal testosterone was not associated with the frequency of mental or affective state terms, but was negatively correlated with the frequency of intentional propositions. Males used more neutral propositions than females, suggesting that exposure to higher levels of testosterone may lead to a suppression of empathic emotions.

Further evidence comes from another study by the same group, in which they again measured amnionic fluid testosterone levels (fT) in mothers of 193 children (100 males and 93 females) [9]. When the children were 6-8 years of age, they measured their child's Empathy Quotient (EQ-C) and completed the "Reading the Mind in the Eyes Task" (Eyes-C). They found a significant negative correlation between fT levels and scores on both measures.

Other studies have shown that fT is also inversely correlated with other social behaviors, such as eye contact in infancy, peer relations in preschoolers and mentalistic interpretations of animate motion. Most agree that ASD individuals have profound impairments in interpersonal social function. A study by Lombardo and co-workers, in which they studied 30 adults aged 19-45 with Asperger syndrome or high-functioning autism and 30 age, sex and IQ-matched controls, looking specifically for abnormalities in self-related testing and found abnormalities in a self-reference effect (SRE) paradigm. That is, they had broad impairments in both self-referential cognition and empathy [10].

Further analysis of a relationship between (fT), cognition and behavior disclosed a link to higher scores on the Childhood Autism Spectrum Test (CAST) and Child Autism Spectrum Quotient (AQ-Child) [11]. In testing 235 children, ages 6-10 years, they found a positive association between higher CAST and AQ-Child scores and fT. No correlation was found with IQ. The effect was present with each sex and when sexes were combined, demonstrating it was related to fT and not sexual differences itself.

Chura and co-workers studied the effects of amniotic fT levels measured during the second trimester of pregnancy as correlated with corpus callosum symmetry in 38 boys, aged 8 to 11 years, using a high-resolution structural MRI and found that there was a significant relationship with a rightward asymmetry of a posterior subsection of the corpus callosum; which projects mainly to the parietal and superior temporal areas, both of which are involved in ASD [12].

Based on the idea that elevated testosterone levels early in development can affect dimorphic neurological development, researchers began to look at a condition in which such levels existed early in life—congenital adrenal hyperplasia (CAH). Knickmeyer and co-workers examined this question in a study in which they looked at 60 individuals having CAH (34 females, 26 males) and found that the females scored significantly higher than normal females on the AQ [13]. It was the subscales measuring social skills and imagination that mostly affected the test scores. Other studies have linked increased aggressive behavior with early exposure to higher levels of testosterone [14].

Barbeau *et al.* have questioned these observations as to their relation to ASD behavior. In an article appearing in the British Journal of Psychology in 2009, they question in particular the notion that reduced empathy and increased systemizing are in fact male behaviors [15]. They also note that the cerebral hemisphere laterality patterns of individuals with ASD do not coincide with those of typical males, all valid points.

ANDROGENS, ESTROGENS, AND IMMUNOEXCITOTOXICITY

Beyond the effect of sex steroids on neurodevelopment, there are a number of effects on the pathophysiology of the brain itself. If we look at studies utilizing experimental strokes and other neurological injuries, we see that testosterone can act as a survival factor. For example it has been shown to be a survival factor for axotomized motoneurons and in fact, promotes axon regeneration [16].

Interestingly, stroke studies have shown that chronic testosterone replacement increased and castration and chronic 17ß-estradiol treatment decreased ischemic damage in middle cerebral artery occlusion models in male rats [17]. Immunoexcitotoxicity plays a prominent role in stroke pathophysiology. Some studies have shown that testosterone can increase glutamate excitotoxicity in a murine hippocampal cell line and an *in vitro* stroke model [18]. In the cell culture, 10 µM of testosterone significantly added to glutamate toxicity and 10 µM of 17ß-estradiol ameliorated the glutamate toxicity almost by half. Using an implanted testosterone pellet to minimize the stress of injections, they found that animals with the testosterone pellet had significantly larger stroke volumes as compared to the sham operated animals following a middle cerebral artery occlusion (MCAO).

It has been shown that testosterone levels decline rapidly in response to both physical and psychological stress and with strokes [19, 20]. Testosterone also declined in this study following MCAO. Some studies suggest that testosterone is neuroprotective, but a number of new studies indicate that only when testosterone is converted by aromatase to estrogens is it protective [21]. The same has been found for the protective abilities of pregnenolone and DHEA [22]. Some have assumed that only a small population of neurons secretes aromatase and these are located within the hypothalamus. More recent studies have shown that with injury, stroke or stress, reactive astrocytes can upregulate aromatase, thus increasing the conversion of testosterone to neuroprotective estrogens in males and females [23]. Under non-stressed conditions, astrocytes do not express aromatase.

Anabolic steroid abuse studies indicate that high levels of androgens can induce behavioral hyperexcitability, aggressiveness and suicidal tendencies, indicating a physiological effect in the developed brain [24].

The neurotoxic effects of high levels of testosterone appear to be related to sustained elevations of intraneuronal calcium. Low levels, 10 to 100 nM, produce physiological calcium oscillations, whereas higher levels, 1µM and higher, produce sustained calcium elevations in the cytosol, with triggering of cell death-signaling pathways [25]. Specific neurons in the CNS contain 5α-reductase, which can increase local dihydrotestosterone levels that are significantly higher than plasma levels. Dihydrotestosterone is not metabolized by aromatase. One sees a differential expression of androgen receptors in the CNS, which can produce variable sensitivity to testosterone. In the rat, numerous androgen receptors (AR) are found in the hippocampus, with the highest concentration in the CA1 area [26].

In the hypothalamus and limbic connections, glutamate is the primary neurotransmitter [27]. Carbone and co-workers found that GluR transmission was the major regulator of gonadotropin-releasing hormone (GnRH) release, with a differential effect controlling luteinising hormone (LH) and FSH release and surge [28]. *N*-methyl-D-aspartic acid (NMDA) receptors regulated LH release while kainate controlled FSH release. Other studies have shown, using female rats, that central administration of α-amino-3-hydroxy-5-methyl-4-isoxazole propionic acid (AMPA) stimulated LH release in an estrogen-primed ovariectomized adult rat, and potently stimulates GnRH release *in vitro* from mediobasal hypothalamic (MBH) fragments [29]. The effect of the AMPA on LH secretion was steroid dependent, that is, it inhibited LH release in non-estrogen primed ovarectomized rats. AMPA stimulated GnRH release equally well from estrogen-primed and non-primed MBH fragments.

GluRs are found in a number of hypothalamic nuclei, including the arcuate nucleus, the suprachiasmatic nucleus, the suproptic nucleus, the paraventricular nucleus and the preoptic area [30]. The main site of NMDA action is the preoptic area, a site in which GnRH cell bodies reside. AMPA and kainate acts mainly at the arcuate/median eminence, which is the site of GnRH neural terminals. They point out that NMDA receptors also act via noradrenergic neurons in the locus coeruleus, an area that influences GnRH release from the hypothalamus.

Glutamate neurotransmission also plays a major role in hypothalamic regulation of adrenocorticotropin hormone (ACTH), growth hormone, prolactin, oxytocin, vasopressin as well as gonadotropins. One sees extensive and reciprocal connections between the limbic systems and diencephalic areas with mesencephalic, lower brainstem and spinal cord areas [31]. These areas control various physiological and behavioral processes related to higher cognitive function. Approximately 50% of medial amygdala, bed nucleus and medial preoptic ER/AR-containing neurons express AMPA GluR, with glutamate being the dominate transmitter. The septal area, especially the lateral septum, is rich in glutamate and aspartate because of its massive innervation by excitatory amino acid-containing limbic cortical efferents. A substantial population of neurons in the region is a direct target of circulating gonadal steroids, including estrogen and testosterone. It is concluded that by altering expression of AMPA receptors, androgen and estrogen readily can influence excitatory neurotransmission in a sexually dimorphic manner. As we have seen, inflammatory cytokines, such as TNF-α and less so IL-1ß can enhance the excitability of AMPA receptors as well as NMDA receptors and increase extraneuronal glutamate concentrations. This will have a substantial effect, not only on hypothalamic function, but also higher cortical function.

Estrogen seems to possess a considerable ability to promote synaptogenesis and dendritic growth. For example, estrogen has been shown to increase dendritic spine density, pre and post-synaptic, and synaptic connectivity of the female hypothalamus [32]. The increased spinogenesis is mediated by estrogen effects on NMDA receptor function in this same CA1 area of the hippocampus [33]. In addition, estrogens enhance acetylcholine production in the forebrain in males [34]. In this study, repeated, low-dose estrogens increased the number of choline acetyltransferase (ChAT)-like immunoreactive cells in the medial septum. Higher doses significantly increased the number of ChAT cells in the nucleus basalis magnocellularis. The effect was not sustained over time and the effect was lost with higher doses of estrogen.

Weiland found that estradiol treatment in rats increased GAD nRNA levels in the CA1 region of the hippocampus. This would favor an anti-excitotoxic balance by stimulating GABAergic activity and may, in part, offer an explanation for the observed higher incidence of ASD in males. Overall, estrogen enhances memory, learning and reduces seizure risk by its positive effects on synaptogenesis, dendritic extension, anti-inflammatory cholinergic effects and tilting the balance toward GABAergic activity.

Dihydrotestosterone (DHT) on the other hand was found to increase NMDA receptor binding in CA1 with resulting increased hippocampal excitability, independent of cholinergic stimulation, which would shift towards an excitotoxic balance without an associated anti-inflammatory cholinergic response, as seen with estrogens [35].

Because the reactions of androgens and estrogens on neuronal excitability are so rapid one must assume the effects are via membrane receptors and not genomic. The strongest link to testosterone-enhanced excitotoxicity is its effects on Ca^{2+} oscillations, which has been shown to be dependent on extraneuronal calcium. Testosterone has been shown to increase intracellular Ca^{2+} with activation of inositol 1,4,5-trisphosphate and diacylglycerol. Careful studies have shown that androgen receptors (AR) exist throughout the cell, with the highest concentrations in the nucleus, but also in some cells the highest expression was seen in the tip of the neurite [6]. The calcium originates from both extracellular sites and intracellular stores. Testosterone was found to be acting via G protein coupled receptors (GPCR), again suggesting a link to aluminofluoride toxic effects [36]. Calcium has also been shown to increase binding of androgen to its receptor, at least in platelets [37].

It is of interest that extensive AR have been demonstrated on axons and dendrites themselves, with numerous collections on axons from the anterior commisure, cingulum and dorsal external capsule [38]. The most abundant collections of axonal AR were in the cortex, especially in cortical layers I, II and III perpendicular to the cortical surface and layer VI of no particular orientation. This would mean that excess androgen concentrations in the brain could increase excitotoxicity in neurons, dendrites and axons, again suggesting a link between males and autism.

ESTROGENS AND NEUROPROTECTION

In explaining the male preponderance of ASD, one may look at the powerful neuroprotective effects of estrogens, especially against inflammation. A recent study by Chiapetta and co-workers, found that estrogens played a major role in reducing damage produced by experimental middle cerebral artery occlusion (MCAO) by inhibiting IL-1ß release caused by the infarction [39]. In the study, they gave 17β-estradiol one hour before male rats were subjected to transient (2h) MCAO. In controls not given estrogen, the MCAO followed by 2 hours of reperfusion resulted in a 3-fold increase in IL-1ß levels. Those given 17β-estradiol demonstrated an attenuation of cytokine elevation and a significant reduction in infarct volume. Using Western Blotting they also found the estrogen treatment reduced cytochrome c translocation to the cytosol in the striatum and cortex. Others have shown that estrogen's neuroprotective effect may involve its influence of endocannabinoid signaling [40].

It appears that both membrane estrogen receptors (ER) and classical nuclear receptors are involved in this protection. Several recent studies have examined the role of the various ER subtypes in estrogen neuroprotection. Dubal and co-workers, using a MCAO model, found that there is a temporal expression of both receptor subtypes with ERα induction occurring early and ERβ modulation occurring later [41, 42]. It is also interesting to note that estrogen had no effect on early neuronal death with MCAO, but profound effects on delayed neuronal death. This might indicate that it was having a significant protective effect against immunoexcitotoxicity, the primary delayed pathophysiologic response.

Simpkins and Dykens found that estrogen analogs, including steroidal phenols, having significantly less hormone potency, were just as neuroprotective as 17β-estradiol [43, 44]. They found that estrogens and its analogs stabilized mitochondrial membranes during calcium loading, which would normally produce a collapse of membrane potential. Excitotoxicity is greatly enhanced by mitochondrial energy failure.

Aromatase, by its ability to convert testosterone into estrogen, is also of considerable importance in neurodevelopment as well as neuroprotection under conditions of immunoexcitotoxicity. Immunohistochemical studies of the rhesus monkey brain has shown that aromatase is widely distributed in a large population of CA1-3 pyramidal neurons, in granule cells of dentate gyrus and some interneurons co-expressed with caldindin, calretinin and parvalbumin calcium binding proteins [45]. The greatest density of aromatase immunoreactivity was in the pyramidal cells of the cortex and only a small collection of interneurons. Estrogen appears to play a major role in synaptic plasticity in both the hippocampus and cortex. They also found that the pattern of aromatase enzyme activity was not dependent on plasma estradiol. A newer study, looking at the normal, as well as sclerotic human hippocampus in epileptic subjects, found that aromatase distribution was present in both types of hippocampal CA1-3 neurons and granule cells, and as with the rhesus monkey study, was co-expressed in interneurons with the calcium-binding proteins [46].

More direct evidence of the protective effect of estrogen against immunoexcitotoxicity come from a study in which researchers found that trauma and hemorrhage up-regulated microglial inflammatory responses, including dramatic elevation in brain TNF-α [47]. Treatment with 17β-estradiol after the event prevented these inflammatory responses and provided protection.

Taken together, estrogens appear to play a major role in neurodevelopment by regulating calcium waves and modulating chronic immunoexcitotoxicity. This is particularly important when considering the chronic inflammatory state of the ASD brain and may explain, in part, the dramatic male preponderance of ASD in males. No studies have been done to determine aromatase levels in the ASD male brain, which would be important to know, since dysfunctional aromatase or reduced levels of aromatase in the male brain would put them at a higher risk of damage by immunoexcitotoxicity. Chakrabarti and co-workers have studied some 68 candidate genes in ASD cases and found several related to sex steroid function to be involved [48].

REFERENCES

[1]   Geier DA, Geier MR. A prospective assessment of androgen levels in patients with autistic spectrum disorders: biochemical underpinnings and suggested therapies. Neuro Endocrinol Lett 2007; 28: 565-73.

[2]   Tordjman S, Anderson GM, McBride PA, *et al.* Plasma androgens in autism. J Autism Dev Disord 1995; 25: 295-304.

[3]   James WH. Further evidence that some male-based neurodevelopmental disorders are associated with high intrauterine testosterone concentrations. Dev Med Child Neurol 2008; 50: 15-8.

[4]   Ingudomnukul E, Baron-Cohen S, Wheelwright S, Knickmeyer R. Elevated rates of testosterone-related disorders in women with autism spectrum conditions. Horm Behav 2007; 51: 597-604.

[5]   Estrada M, Espinosa A, Gibson CJ, Uhlen P, Jaimovich E. Capacitative calcium entry in testosterone-induced intracellular calcium oscillations in myotubes. J Endocrinol 2005; 184: 371-9.

[6]   Estrada M, Uhlen P, Ehrlich BE. Ca2+ oscillations induced by testosterone enhance neurite outgrowth. J Cell Sci 2006; 119: 733-43.

[7]   Baron-Cohen S. The extreme male brain theory of autism. Trends Cogn Sci 2002; 6: 248-54.

[8]   Knickmeyer R, Baron-Cohen S, Raggatt P, Taylor K, Hackett G. Fetal testosterone and empathy. Horm Behav 2006; 49: 282-92.

[9]   Chapman E, Baron-Cohen S, Auyeung B, *et al.* Fetal testosterone and empathy: evidence from the empathy quotient (EQ) and the "reading the mind in the eyes" test. Soc Neurosci 2006; 1: 135-48.

[10]   Lombardo MV, Barnes JL, Wheelwright SJ, Baron-Cohen S. Self-referential cognition and empathy in autism. PLoS One 2007; 2: e883.

[11]   Auyeung B, Baron-Cohen S, Ashwin E, *et al.* Fetal testosterone predicts sexually differentiated childhood behavior in girls and in boys. Psychol Sci 2009; 20: 144-8.

[12]   Chura LR, Lombardo MV, Ashwin E, *et al.* Organizational effects of fetal testosterone on human corpus callosum size and asymmetry. Psychoneuroendocrinology 2010; 35: 122-32.

[13]   Knickmeyer R, Baron-Cohen S, Fane BA, *et al.* Androgens and autistic traits: A study of individuals with congenital adrenal hyperplasia. Horm Behav 2006; 50: 148-53.

[14]   Berenbaum SA, Resnick SM. Early androgen effects on aggression in children and adults with congenital adrenal hyperplasia. Psychoneuroendocrinology 1997; 22: 505-15.

[15]   Barbeau EB, Mendrek A, Mottron L. Are autistic traits autistic? Br J Psychol 2009; 100: 23-8.

[16]   Kujawa KA, Kinderman NB, Jones KJ. Testosterone-induced acceleration of recovery from facial paralysis following crush axotomy of the facial nerve in male hamsters. Exp Neurol 1989; 105: 80-5.

[17]   Hawk T, Zhang YQ, Rajakumar G, Day AL, Simpkins JW. Testosterone increases and estradiol decreases middle cerebral artery occlusion lesion size in male rats. Brain Res 1998; 796: 296-8.

[18]   Yang SH, Perez E, Cutright J, *et al.* Testosterone increases neurotoxicity of glutamate *in vitro* and ischemia-reperfusion injury in an animal model. J Appl Physiol 2002; 92: 195-201.

[19]   Dash RJ, Sethi BK, Nalini K, Singh S. Circulating testosterone in pure motor stroke. Funct Neurol 1991; 6: 29-34.

[20]   Elman I, Breier A. Effect of acute metabolic stress on plasma progesterone and testosterone in male subjects: relationship to pituitary-adrenocortical axis activation. Life Sci 1997; 61: 1705-12.

[21]   Azcoitia I, Sierra A, Veiga S, *et al.* Brain aromatase is neuroprotective. J Neurobiol 2001; 47: 318-29.

[22]   Veiga S, Garcia-Segura LM, Azcoitia I. Neuroprotection by the steroids pregnenolone and dehydroepiandrosterone is mediated by the enzyme aromatase. J Neurobiol 2003; 56: 398-406.

[23]   Garcia-Segura LM, Wozniak A, Azcoitia I, *et al.* Aromatase expression by astrocytes after brain injury: implications for local estrogen formation in brain repair. Neuroscience 1999; 89: 567-78.

[24]   Thiblin I, Lindquist O, Rajs J. Cause and manner of death among users of anabolic androgenic steroids. J Forensic Sci 2000; 45: 16-23.

[25]   Estrada M, Varshney A, Ehrlich BE. Elevated testosterone induces apoptosis in neuronal cells. J Biol Chem 2006; 281: 25492-501.

[26]   Kerr JE, Allore RJ, Beck SG, Handa RJ. Distribution and hormonal regulation of androgen receptor (AR) and AR messenger ribonucleic acid in the rat hippocampus. Endocrinology 1995; 136: 3213-21.

[27]   Brann DW. Glutamate: a major excitatory transmitter in neuroendocrine regulation. Neuroendocrinology 1995; 61: 213-25.

[28]   Carbone S, Szwarcfarb B, Rondina D, Feleder C, Moguilevsky JA. Differential effects of the N-methyl-D-aspartate and non-N-methyl-D-aspartate receptors of the excitatory amino acids system on LH and FSH secretion. Its effects on the hypothalamic luteinizing hormone releasing hormone during maturation in male rats. Brain Res 1996; 707: 139-45.

[29]   Ping L, Mahesh VB, Bhat GK, Brann DW. Regulation of gonadotropin-releasing hormone and luteinizing hormone secretion by AMPA receptors. Evidence for a physiological role of AMPA receptors in the steroid-induced luteinizing hormone surge. Neuroendocrinology 1997; 66: 246-53.

[30]   Brann DW, Mahesh VB. Excitatory amino acids: function and significance in reproduction and neuroendocrine regulation. Front Neuroendocrinol 1994; 15: 3-49.

[31]   Diano S, Naftolin F, Horvath TL. Gonadal steroids target AMPA glutamate receptor-containing neurons in the rat hypothalamus, septum and amygdala: a morphological and biochemical study. Endocrinology 1997; 138: 778-89.

[32]   Leranth C, Petnehazy O, MacLusky NJ. Gonadal hormones affect spine synaptic density in the CA1 hippocampal subfield of male rats. J Neurosci 2003; 23: 1588-92.

[33]   Weiland NG. Estradiol selectively regulates agonist binding sites on the N-methyl-D-aspartate receptor complex in the CA1 region of the hippocampus. Endocrinology 1992; 131: 662-8.

[34]   Gibbs RB. Effects of estrogen on basal forebrain cholinergic neurons vary as a function of dose and duration of treatment. Brain Res 1997; 757: 10-6.

[35]   Romeo RD, Staub D, Jasnow AM, *et al.* Dihydrotestosterone increases hippocampal N-methyl-D-aspartate binding but does not affect choline acetyltransferase cell number in the forebrain or choline transporter levels in the CA1 region of adult male rats. Endocrinology 2005; 146: 2091-7.

[36]   Benten WP, Lieberherr M, Giese G, *et al.* Functional testosterone receptors in plasma membranes of T cells. Faseb J 1999; 13: 123-33.

[37]   Cabeza M, Flores M, Bratoeff E, *et al.* Intracellular Ca2+ stimulates the binding to androgen receptors in platelets. Steroids 2004; 69: 767-72.

[38]   DonCarlos LL, Garcia-Ovejero D, Sarkey S, Garcia-Segura LM, Azcoitia I. Androgen receptor immunoreactivity in forebrain axons and dendrites in the rat. Endocrinology 2003; 144: 3632-8.

[39]   Chiappetta O, Gliozzi M, Siviglia E, *et al.* Evidence to implicate early modulation of interleukin-1beta expression in the neuroprotection afforded by 17beta-estradiol in male rats undergone transient middle cerebral artery occlusion. Int Rev Neurobiol 2007; 82: 357-72.

[40]   Amantea D, Spagnuolo P, Bari M, *et al.* Modulation of the endocannabinoid system by focal brain ischemia in the rat is involved in neuroprotection afforded by 17beta-estradiol. Febs J 2007; 274: 4464-775.

[41]   Miller NR, Jover T, Cohen HW, Zukin RS, Etgen AM. Estrogen can act via estrogen receptor alpha and beta to protect hippocampal neurons against global ischemia-induced cell death. Endocrinology 2005; 146: 3070-9.

[42]   Dubal DB, Rau SW, Shughrue PJ, *et al.* Differential modulation of estrogen receptors (ERs) in ischemic brain injury: a role for ERalpha in estradiol-mediated protection against delayed cell death. Endocrinology 2006; 147: 3076-84.

[43]   Simpkins JW, Dykens JA. Mitochondrial mechanisms of estrogen neuroprotection. Brain Res Rev 2008; 57: 421-30.

[44]   Simpkins JW, Yi KD, Yang SH, Dykens JA. Mitochondrial mechanisms of estrogen neuroprotection. Biochim Biophys Acta 2010;

[45]   Yague JG, Wang AC, Janssen WG, *et al.* Aromatase distribution in the monkey temporal neocortex and hippocampus. Brain Res 2008; 1209: 115-27.

[46]   Yague JG, Azcoitia I, DeFelipe J, Garcia-Segura LM, Munoz A. Aromatase expression in the normal and epileptic human hippocampus. Brain Res 2010; 1315: 41-52.

[47]   Akabori H, Moeinpour F, Bland KI, Chaudry IH. Mechanism of the anti-inflammatory effect Of 17beta-estradiol on brain following trauma-hemorrhage. Shock 2010; 33: 43-8.

[48]   Chakrabarti B, Dudbridge F, Kent L, *et al.* Genes related to sex steroids, neural growth, and social-emotional behavior are associated with autistic traits, empathy, and Asperger syndrome. Autism Res 2009; 2: 157-77.

Addendum. Autism: Is It All in the Head?

Mark A. Hyman, MD

Chairman, Institute for Functional Medicine. Volunteer, Partners in Health Founder and Medical Director The UltraWellness Center, Lenox, MA 01240, USA

Abstract: Mark A. Hyman, MD, is Chairman of the Institute for Functional Medicine. Anna Strunecka asked him to provide his Editorial [1] as an addendum of this eBook. Autism is described as a hologram for chronic disease; an extreme manifestation of disruptions in normal biology that exist in varying degrees in most chronic illness. Autism is a complex, multi-system disorder rooted in a series of toxic, infectious, and allergic insults. Through the story of one boy, the author looked carefully at the few biological systems manifesting as the clinical features of autism: gut and immune dysfunction, nutritional deficiencies, toxicity and impaired detoxification, mitochondrial dysfunction and oxidative stress, and genetic polymorphisms that set the stage for biochemical train wrecks. In autistic children, the results of testing often reveal results that show deviations orders of magnitude higher than in other chronic illness, but nonetheless, the same patterns exist. The lessons learned from the dissection of the functional causes and mechanisms of autism can illuminate the path for whole system medicine and clinical research and the potential for it to address the global crisis of chronic disease.

INTRODUCTION

A hologram is a 3 dimensional photographic image. But it is much more. If you take a glass plate that stores any holographic image and break it into a thousand pieces, each fragment when illuminated with a laser, will recreate the entire image. Autism is a hologram for chronic disease. In it is reflected all the causes and cures for chronic disease. Autism is an extreme manifestation of disruptions in normal biology that exist in varying degrees in most chronic illness. Shining a light deep into the biology of autism will illuminate not only the mysteries of "brain disorders" such as Alzheimer's, attention deficit disorder, and depression, but also heart disease, autoimmune disease, digestive disorders, cancer, obesity, chronic fatigue, and more.

The discoveries that have led to the picture of autism as a reversible systemic disorder that is influenced by genetics and that affects the brain, rather than a genetically determined fixed brain-based disorder, emerged from a unique process in the history of medicine—the mining of the collective intelligence of scientists, clinicians, and parents of children with autism. What they have discovered is this: the broken brain of autism is caused by a broken body. Fix the body, and the brain can recover. Out of their experience emerged a road map that can be generalized to nearly all chronic illness because the roots of the biochemical disasters and metabolic dysfunction are the same—genetic predispositions (rather than determinants), a toxic environment, and a nutrient-deficient diet. In the case of autism, the effects of these insults are magnified by the overuse of medications such as antibiotics and vaccinations, which increase susceptibility to infections and promote allergy and autoimmunity.

What has emerged is the extraordinary insight that autism is a complex, multi-system disorder rooted in a series of toxic, infectious, and allergic insults. Inflammation, disruptions in normal energy metabolism and ATP production resulting in mitochondrial dysfunction, and impairment in critical regulation of oxidative stress and detoxification through a breakdown in the twin interconnected cycles of methylation (B_6, folate and B_{12} dependent transfer of methyl groups or CH_3) and sulfation (which produces glutathione) produce a metabolic encephalopathy.

Remove the dates of birth from the laboratory results and remove the diagnostic labels from a patient with autism and a patient with Alzheimer's and you will discover the same biological forces at work—inflammation, oxidative stress, impaired methylation and detoxification, mitochondrial dysfunction, and even the genetic polymorphisms. What are we to make of this observation? Is it coincidence, or does it reflect deeper patterns hidden in biological systems?

*****Address correspondence to: Dr. Mark A. Hyman, MD.** The UltraWellness Center. 45 Walker Street, Lenox, MA 01240, USA; E-mail: mark@drhyman.com

The increase in mood, developmental, and neurodegenerative disorders in the 21st century makes it imperative to learn from the autism experience. The central insight of systems biology that holds the key to solving the puzzle of chronic disease is this: the plethora of diseases of modern life (and codified in the ICD-9 classification system) can be explained by a few general biological laws.

The 18th century physicist Pierre LaPlace (1749-1827) observed this principle, which applies not only to biological systems, but also must be applied to the diagnosis and treatment of chronic disease in the 21st century [1]. He said, "The simplicity of nature is not to be measured by that of our conceptions. Infinitely varied in its effects, nature is simple only in its causes, and its economy consists in producing a great number of phenomena, often very complicated, by means of a small number of general laws."

The media and scientific literature present a jumble of confusing information and apparently disconnected data points—measles, vaccines, mercury, genetics, toxic insults, food allergies, gut inflammation, brain trauma and more, leaving scientists, clinicians, parents, and policy makers bewildered and misguided. It is all of these things and none of them. Through the story of one boy, the current paradigm of medicine is cracked open, illustrating the collapse of the medical system, the failure of medical care, and the end of medicine as we know it. His story is one of the thousands with autism and the millions with chronic disease and paints a picture of the future of medicine through the lenses of the philosophy of science, epidemiology, toxicology, biochemistry, genetics, and systems theory.

Is Recovery Possible? Patterns and Systems, Not Diagnoses and Symptoms

A desperate mother came to see me because her 2½-year-old son, "Sam," had just been diagnosed with autism. He was born bright and happy, breast-fed, had the best medical care available (including all the vaccinations he could possible have). He talked, walked, loved, and played normally—until his measles, mumps, and rubella vaccination at 22 months.

He was vaccinated for diphtheria, tetanus, whooping cough, measles, mumps and rubella, chicken pox, hepatitis A and B, influenza, pneumonia, hemophilous, and meningitis—all before the age of 2 years.

After this string of vaccinations, he lost his language, became detached, withdrawn, less interactive, and was unable to relate in normal ways with his parents and other children—all signs of autism. He also developed foul-smelling, sticky stools, dark circles under his eyes, and itchy ears. How could a normal boy be transformed so quickly?

He was taken to the best doctors in New York and "pronounced" as having autism (as if it were his fate), and told that there was nothing to be done except arduous, painful, and minimally effective behavioral and occupational therapy. The doctor told his mother the progress would be slow and she should keep her expectations low.

Devastated, his mother sought other options and found her way to me. When I first saw this little boy he was deep in the inner wordless world of autism—watching him was like watching someone on a psychedelic drug trip. We dug into his biochemistry and genetics and found many things to account for the problems he was having.

We looked carefully at the few biological systems that go awry, manifesting as the clinical features of autism: gut and immune dysfunction, nutritional deficiencies, toxicity and impaired detoxification, mitochondrial dysfunction and oxidative stress, and genetic polymorphisms that set the stage for biochemical train wrecks. In autistic children, the results of testing often reveals results that show deviations orders of magnitude higher than in other chronic illness, but nonetheless, the same patterns exist. By unraveling the tangled roots of his distress, we were able to address the systemic causes of his broken brain. Let's examine each of these areas of dysfunction.

GENETIC POLYMORPHISMS (PREDISPOSITIONS)

Impaired Glutathione Metabolism

- *Homozygous for 2 glutathione S-transferase P1 (GSTP1 ++) genes (I105V and A114V) [2, 3]. This reduces the ability to biotransform toxicants such as toxic metals, xenobiotics, solvents, pesticides, herbicides, and polycyclic aromatic hydrocarbons. Point mutations in the gene coding for glutathione s-transferase enzymes have been associated with increased risk for autism.*

Impaired Methylation

- *Methylenetetrahydrofolate reductase (MTHFR 677C > T and 1298A > C) heterozygous polymorphism. MTHFR is the enzyme involved in the final methylation step of folic acid, producing 5-methyltetrahydrofolate from 5,10 methylenetetrahydrofolate. Impairment of this enzyme usually results in an elevated homocysteine, as 5-methyltetrahydrofolate is required to recycle homocysteine back to methionine. However, in autistic children, increased oxidative stress results in shunting of homocysteine to provide cysteine for glutathione production, resulting in the low levels of homocysteine [4]. James demonstrated in 2004 that treating ASD children with methyl donors including B12, folic acid and trimethylglycine normalized glutathione and homocysteine levels [5]. This patient had a series of genetic predispositions and insults that led to accumulation of toxins and increased oxidative stress, triggering the vicious cycle of impaired methylation and glutathione production. This was manifested by his low homocysteine of 3 mmol/L (nL 6-8 mmol/L).*

- *Catechol-O-methyltransferase (COMT 472G > A) was heterozygous, which tends to slow the detoxification of the neurotransmitters needed for attention, focus, and cognitive skills, such as dopamine, epinephrine, and norepinephrine. COMT polymorphisms have been noted to occur with increased incidence in autistic children.*

Allergy and Autoimmunity

- *Elevated IgG anti-gliadin antibodies of 91 units (nL < 20), indicating an autoimmune response to gluten.*

- *Elevated total IgG antibodies to not only wheat but to 28 foods, including dairy, eggs, yeast, and soy, indicating disrupted intestinal permeability.*

Digestive Function

- *Stool analysis cultured 3 species of yeast and a deficiency of beneficial flora, including Lactobacillus and Bifidobacteria.*

- *Elevated stool markers in intestinal inflammation consistent with allergy, infection, or inflammatory bowel disease (eosinophil protein X 18.2 mg/g (nL < 7 mg/g) and calprotectin 46 mg/g (nL < 40 mg/g).*

- *The urinary organic acids revealed very high levels of D-lactate—an indicator of overgrowth of bacteria in the small intestine resulting in intestinal fermentation of carbohydrates.*

- *Urinary peptide analysis revealed very elevated IAG (indoyl-acyloylglycine 161 µg/mg creatinine (nL < 9.5), a toxic phenylalanine metabolite derived from dysbiotic bacterial metabolism, deltorphins, and enkephalins. These are neuroactive peptides, which disrupt cognitive function and have been associated with autism spectrum disorder [6,7].*

Nutritional Deficiencies

- *Low amino acids reflect inadequate protein intake (unlikely) or maldigestion, malabsorption, or over-utilization in phase 2 detoxification pathways because of toxicity. For example, methionine and threonine are metabolized to cysteine and glycine respectively, which when combined with glutamate form the glutathione tripeptide.*

- *Mineral deficiencies: zinc (important for immune function, activation of digestive enzymes, as a cofactor for metallothionein, necessary for intracellular detoxification of heavy metals and 200+ other enzymes), magnesium (a natural N-D-methyl aspartate or NMDA receptor antagonist that reduces brain excitotoxicity and is a cofactor for 300 + enzymes), and manganese (a cofactor for super oxide dismutase, a critical mitochondrial antioxidant enzyme).*

- *Impaired methylation: elevated urinary methylmalonic acid, indicating B12 deficiency, and low homocysteine.*

- *Vitamin A and vitamin D deficiencies.*

- *Essential fatty acid deficiencies: eicosapentaneoic acid (EPA) deficiency and elevated AA/EPA ratio (an excess of inflammatory to anti-inflammatory fatty acids) are associated with autism spectrum disorder [8,9].*

Mitochondrial Dysfunction and Oxidative Stress

- *Organic acid analysis revealed widespread impairment in fatty acid, carbohydrate, and citric acid metabolism (elevated lactate, citrate, isocitrate, succinate, malate), indicating mitochondrial dysfunction resulting in energy deficits linked to cognitive dysfunction and demonstrated in 70% of autistic children [10].*

- *Elevated suberate indicated impaired fatty acid transport into the mitochondria from carnitine deficiency.*

- *Very elevated lactic acid (L-Lactate) 101mg/mg creatinine (nL < 22), indicating cellular acidosis and coenzyme Q10 deficiency.*

- *Increased oxidative stress, indicated by elevated DNA adducts or a 8-Hydroxy-2' deoxyguanosine of 11.5 mg/mg creatinine (nL < 5).*

Toxicity and Impaired Detoxification

- *Elevated red blood cell aluminum and lead.*

- *Elevated hair antimony and arsenic, but low levels of mercury because of impaired ability to excrete mercury.*

- *Elevated post 2,3-dimercapto-1-propane sulfonate (DMPS) provocation urinary mercury 14 mg/gram creatinine (nL < 3).*

- *Markers of impaired sulfur (low sulfate) and glutathione status (elevated pyroglutamate) on urinary organic acids. Pyroglutamate elevation is indicative of glutathione wasting via a number of possible mechanisms, including cysteine and glycine insufficiency, and a low urinary sulfate is a functional marker for total body sulfur stores.*

- *Elevated urinary porphyrins indicate enzymatic disruption of normal porpryhin metabolism from heavy metals, which has been documented in autism.*

TREATING THE WHOLE SYSTEM

Is it possible to point to any one gene, biomarker, or biological dysfunction, and say "Aha! This is the cause of autism and this is what we should treat!" The answer is an unequivocal no. None of these is the cause of autism. All of them exist in varying degrees and patterns in each individual. The key to unraveling the tangle of molecules and metabolic disruption is seeing patterns and working systematically with those patterns in the right order. Gathering the information is the first step. Understanding how to navigate is critical. Rather than a focus on just one thing, Sam's treatment involved a simple concept. Identify impediments to optimal function (toxins, infections, allergens) and remove them, and identify the ingredients necessary for optimal function and provide them—a simple idea, but perhaps the most powerful in medicine today.

Sam's treatment started with repairing his gut and immune system. We then added nutrients needed for optimal function and removed heavy metals after we had optimized his nutritional status, methylation, and transsulfuration, the highways of detoxification and the quenchers of oxidative stress. Specifically, his treatment consisted of the following.

Step 1: Correct Digestive Imbalances and Remove Food Allergens and Sensitivities

- *Eliminate gluten and IgG food sensitivities.*

- *Treat small bowel bacterial overgrowth with non-absorbed antibiotic (rifaximin).*

- *Treat intestinal yeast with antifungals (fluconazole, itraconazole).*

- *Re-inoculate with beneficial bacteria (broad-spectrum probiotic) and Saccharomyces boulardii.*

- *Correct maldigestion with plant-based enzymes, including dipeptidyl peptidase IV (DPP-IV).*

Step 2: Correct Nutritional Deficiencies and Optimize Nutritional Status

- *Multivitamin, topical zinc, magnesium, methyl folate, methylcobalamin, pyridoxine (B6).*
- *Cod liver oil for EPA/DHA and vitamins A and D.*
- *Coenzyme Q10 to correct mitochondrial dysfunction (elevated L-lactate).*

Step 3: Enhance Detoxification and Treat Oxidative Stress

- *High dose intramuscular methylcobalamin B12 necessary to enhance methylation and overcome toxic injury to methionine synthase (part of the methylation cycle) and activate dopamine receptors.*
- *Topical glutathione to support detoxification.*
- *A chelating medication dimercaptosuccinic acid (DMSA) to remove mercury and lead.*

TREATMENT RESULTS

After 3 weeks on a gluten-free diet, Sam showed dramatic and remarkable improvements. He began to talk again and showed much more connection and relatedness to people.

After 4 months, he was more focused, used more words, and was able to enter a more mainstream school.

Ten months into treatment he was retested. The gut inflammation resolved (normal eosinophil protein X and calprotectin), the small bowel bacterial overgrowth resolved (normal D-lactate), but a mild yeast overgrowth (elevated arabinitol) persisted. Urinary peptide markers (IAG, enkephalins) dramatically improved. His methylmalonic acid was still elevated but improved.

The glutathione deficiency markers improved (pyroglutamate and sulfur), and urinary porphyrins were improved but still elevated. Urinary organic acid testing revealed normalization of his mitochondrial function and a 50% reduction in L-lactate. The oxidative stress marker (8OHDG) was normal.

Most importantly, he went from non-verbal to verbally fluent and no longer qualified for a special school or special services because he "lost" his diagnosis of autism. And his bowel movements normalized.

Sam now has a wonderful sense of humor (typically completely absent in autistic children), and engages in spontaneous play and hugs with friends and family.

This result was not random but the result of a deliberate application of a few simple biological laws that explain the diverse phenomena observed in Sam's clinical presentation and laboratory evaluation. It is the clinical application of systems biology without which the puzzle of chronic disease cannot be solved. Now let us take a journey into the frontiers of medicine and science and see how that puzzle intersects with Sam's story.

A PARADIGM SHIFT: A REVERSIBLE METABOLISM ENCEPHALOPATHY, NOT A FIXED BRAIN DISORDER

Martha Herbert is an assistant professor of neurology at Harvard Medical School and the director of TRANSCEND (Treatment, Research and Neuroscience Evaluation of Neurodevelopmental Disorders). She has put together a remarkable story of autism, which is like a hologram through which we can see the systemic nature of most illness. Her landmark paper, "Autism: A Brain Disorder, or A Disorder of the Brain?" will change forever our thinking about mental and brain illness [11].

Rather than ignore the almost universal physical complaints found in autistic children, most of which have been described in the scientific literature since the 1940s, she explains how they could be at the root of the behavioral symptoms found in autistic children. She explains how the incoherent brain connections those show up as the inability to talk, connect with other people, or produce odd repetitive behaviors have their root not in the brain but in problems with the digestive system and the immune system, environmental toxins, mitochondrial dysfunction, and oxidative stress. These breakdowns in the body, which lead to behavioral problems, occur because of genetic susceptibilities, which are amplified by environmental stresses and toxins.

Why, she asks, do 95% to 100% of autistic children have gastrointestinal dysfunction? Why do 70% of them have immune system abnormalities? It has also been noted that autistic children have frequent infections and allergies and often have had multiple courses of antibiotics. After examining all the accumulated research on autism, including her own work on brain imaging and the structure and function of the brain in autistic children, she concludes that autism is not a brain disorder, but a systemic disorder that affects the brain.

Dr Herbert noticed that brains of autistic children on MRIs are bigger, swollen perhaps [12]. At the same time Dr Carlos Pardo and his group from Johns Hopkins examined on autopsy the brains of 11 autistic children who had died [13]. They also looked at the spinal fluid of living autistic children. By examining and comparing these factors, they found the children's brains to be swollen and inflamed. The swollen brains are filled with activated immune cells and inflammatory molecules.

This brought up another question: Why were their brains inflamed to begin with? The short answer is allergens, toxins, infections, and nutritional deficiencies.

Where do problems like these come from and how do they affect the brain? Are they in the brain to begin with? Or are we looking in the wrong place?

According to Dr Herbert, this gut, immune, and toxicity problems are integrally related to and often the cause of what happens in the brain. In fact, she suggests that autism is a systemic metabolic disorder that changes brain function. The brain and body function as a whole system, and multiple chronic, insidious triggers can throw the brain into chaos.

This is a 180-degree turn from conventional thinking. If an altered response to a microbe or bug in the body by the immune system can affect brain function, or if a molecule made in the gut can change behavior or perception, then of course the brain is in communication with the rest of the body.

We insult our digestive tract every day. We do everything to harm it and hardly anything to help it work as it was designed to work. We eat food that is low in fiber, high in sugar, and full of antibiotics, pesticides, and hormones; we drink alcohol and caffeine; we take antibiotics, acid blocking medication, anti-inflammatory drugs, hormones, and steroids (all of which can inhibit proper gut function); we are under constant stress; and we are exposed to thousands of environmental toxins, all of which damage the gut. These factors trigger widespread inflammation because our gut-immune system reacts to all the foreign proteins in food and the myriad of bugs and becomes "angry" and inflamed. So if inflammation starts in the belly (many autistic children have swollen bellies), then it spreads to the brain, it can literally lead to a swollen brain.

Imagine the extraordinary beauty and dance of the brain where everything is exquisitely regulated. The timing and coordination of nerve cell firing and the amplitude (or volume) of the message has to be just right. Filters that modulate our sensory inputs must let in only the information needed. The activity of the brain must be perfectly synchronized for us to be awake, alert, receptive, interactive, communicative, flexible, and happy.

But what if the signals start misfiring and the coordination and synchronization break down because of multiple metabolic disturbances such as ineffective enzymes or cell function due to insults from toxins, microbes, allergens, or nutritional deficiencies? This is the net effect of inflammation, whatever the original cause. It triggers a runaway cascade of damaging effects. All mental processing slows, neurotransmitters can't do their job, cell membranes don't function the way they were designed to, cellular enzymes get hijacked or derailed, cells get trigged into a death spiral called apoptosis, and the delicate network of cellular connections and communications is interrupted and/or altered.

How does that show up? As autism or Alzheimer's disease or depression. It depends on the individual's unique genetic makeup and environment.

Scientists are now asking why. Why do we find more mercury in autistic children, and what is the effect of mercury on the brain? Why do these children have altered immune function? Why do they have more viral infections? Why do we find measles virus in the intestinal lining of these children and in their spinal fluid?

And there are other questions. What is the effect of giving babies 9 immunizations at one doctor's visit, or more than 27 vaccinations by the age of 2 years? What is the effect of being born with an average of 287 toxins already in the bloodstream? How do all these trigger inflammation, and how does this cause autism? These questions force us to ask how genes, biology, brain, and behavior are connected

THERE IS NO GENE FOR AUTISM: LOOKING AT ALL THE CAUSES

Researchers searched for the one autism gene, or the one location in the brain that is damaged that leads to autism. Such an approach implies that the cause of autism (or other "brain disorders") is genetically hard-wired and therefore treatment is hopeless. It is now clear that there is no one gene to be found. The search for dozens of genes continues; however, the effects of the environment and metabolism remain neglected.

The same metabolic and environmental problems hold true for the 1 in 6 children with some type of developmental problem, the 1 in 10 with ADHD, and the 1 in 150 with autism. Each of these may be a problem related to underlying metabolic disorders, and not the result of genetically hard-wired diseases or damaged brains. It is all the same problem; it just shows up slightly differently in different kids.

Jill James, PhD, and her group from the Department of Pediatrics, University of Arkansas for Medical Sciences, has done studies showing that children with autism have common patterns of "abnormal" genes that affect methylation and sulfation [2]. Dr James examined the blood for signs of abnormal methylation and sulfation in 80 autistic and 73 normal children. She also examined the genes of 360 autistic children and 205 normal children finding common patterns. In another key study, Dr James and her colleagues were able to fix this biochemical disruption of methylation and sulfation through nutritional supplementation with methyl donors (B_{12}, folate, B_6, and trimethylglycine) [5].

Dr Herbert suggests that many different causes can lead to the same symptoms—namely the lack of language and social connection and the rigidity and inflexibility of behavior seen in children with autism, as well as many "behavioral" problems in children, such as those with ADHD. A few common pathways result in the same symptoms from a host of different insults. In fact, Dr Herbert challenges the idea that there is only one kind of autism. Researchers are realizing that there may be many "autisms" because each child has unique genetic and environmental factors that can lead to very similar symptoms and behaviors.

Rather than hunting for drugs that target the brain to treat autism, the better path may be to study treatments that target the inflammation, toxins, allergens, infections, biochemistry (like problems with methylation and sulfation), and digestive dysfunction, which alter brain function in the first place.

Treating the gut, replacing B_{12}, B_6, and folate, omega 3 fats, vitamins A and D or magnesium and zinc, eliminating gluten and casein from the diet, or detoxifying mercury and lead from their little bodies may be the best way to get autistic children's brain connections working again.

The experience of thousands of children, parents, scientists, and doctors who are part of a unique collaborative effort called Defeat Autism Now! confirms this approach and helps children recover—some slightly, some miraculously—from a disorder that was thought to be incurable.

In the treatment of psychiatric and neurological disorders, we must look at the body. We need to look for the connections, patterns, and final common pathways, which have enormous implications for so many "fixed" diseases. If recovery and improvement are documented in autism, what does that mean for Alzheimer's, chronic depression, bipolar disease, psychosis, eating disorders, or violent sociopathic behavior?

These problems, it seems, are not hard-wired into the brain, as we believed, but the result of a few common systemic problems that completely disrupt the fine dance and coordination of the brain—problems than can be fixed metabolically and systemically.

THE GUT-IMMUNE SYSTEM AND THE BRAIN

Many medical discoveries are made by accident. An open inquiry of observed phenomena sometimes reveals unexpected clues. Many doctors and scientists have ignored the fact that up to 95% of autistic children have intestinal problems, such as altered bowel function and abdominal distention. How can their intestinal problems affect their brains, interrupt their language, and lock them in their own private world?

Dr Wakefield asked how. He happened to notice inflammation (or lymphoid nodular hyperplasia) in the bowels of some children with autism. Could this observation be brushed off as coincidence? In a study of 148 children with autism compared to 30 normal controls (children without autism), 90% of autistic children showed inflamed bowels on biopsy compared to only 30% of controls (although 30% is a lot! This makes me wonder if many non-autistic children have bowel inflammation from poor diet and allergies as well) [14]. He also noticed the inflammation was much more severe in autistic children. Food allergens, bacteria, viruses, and toxins (such as mercury) could all be the cause.

In addition to all the potential digestive problems that autistic children face, it also seems they are more susceptible to allergy and gut inflammation triggered by certain foods, such as gluten and casein [15].The extreme inflammation in the guts of autistic children contributes to their inflamed brains.

The bottom line is that the guts of these genetically susceptible autistic children are damaged for many reasons—live measles vaccinations, toxic metals, over use of antibiotics, abnormal gut flora, and food allergies. And some children have different combinations of these triggers than others.

The net result is that their digestion breaks down. Digestive enzymes don't work properly. Food particles (especially from gluten and dairy) are partially digested and become brain-fogging toxic compounds (like the peptides mentioned above). Toxins, viruses, bacteria, and food allergens leading to brain inflammation activate the immune system in the gut. Toxic bacteria and yeasts take over, releasing compounds that change normal brain operations. All this overwhelms the system and creates chaos between the brain, and the gut-immune system.

Through the extreme example of autism we can see one end of a spectrum of disorders that affects millions in small and large ways, from full on psychosis and dementia, to mild anxiety and a little depression. Addressing gut inflammation is a back door into healing the brain.

Making Sense of the Measles Vaccine Controversy

Other studies have linked the live measles virus from vaccinations to the inflamed gut. Living measles viruses have been identified in some people with inflamed guts. Vaccines, even an "inactivated" live virus, stimulate the body's humoral immune system to produce protective antibodies. But sometimes, as in the case of autistic children, a weakened immune system can't manage this "inactivated" live virus. Then the live attenuated virus persists, producing low-grade inflammation—in both the gut and the brain.

A study of children with developmental delay found that 75 of 91 patients with autism and inflamed bowels had live measles virus detected in samples of their intestinal tissues. Only 7 of the 70 control patients were found to have the measles virus in their gut [16]. In another study, DNA analysis was performed on the measles strains found in autistic children and compared to that of strains found in non-autistic children with inflamed bowels. The DNA of the measles virus in autistic children came from vaccine strains of measles, not wild types [17]. A more recent study did not find persistent measles virus in children with autism, but even this does not rule out the possibility that the virus did its harm in a "hit and run" fashion [18].

This doesn't mean that ALL children who are vaccinated have problems, but for some reason autistic children are unable to handle the live measles viruses used in immunizations, and it triggers an inflammatory response in the gut and the brain. These children can't handle the vaccine (maybe because mercury suppresses their immune system) and then the normally benign live measles virus in the vaccine takes root in the body and sends these kids into an even deeper spiral of brain dysfunction.

More alarming is that vaccine strains of measles virus seem to migrate into the brains of some children with autism. That means it may not be only gut-related inflammation that is causing the problems, but the measles virus may take root in the brain itself. How this happens is not clear, but the trail from vaccine to the gut to the brain is smoking hot. Vaccine measles strains have been isolated from the spinal fluid of autistic children [19].

Large-scale population studies show no connection between the measles, mumps, and rubella (MMR) vaccine and autism [20]. That is because in such large populations, the effect on children who are susceptible to MMR is "washed out." If you study large groups of people, you won't pick up small effects on genetically or biochemically unique individuals. It's also possible that this is only one of a variety of autism triggers. Looking at the problem using this kind of statistical analysis is unhelpful for treating individual patients.

Oddly, in the major study "disproving" this connection, the authors noted an increase in autism diagnosis in the 6 months after the MMR vaccine but dismissed it as unimportant because it "appeared to be an artifact related to the difficulty of defining precisely the onset of symptoms in this disorder" [20]. If your child had autism or autistic behaviors you would you know it, and you would know when it started! This is yet another example of conventional science seeing what it believes instead of believing what it sees.

The vaccine probably only affects a few genetically susceptible children who are biochemical and immunological train wrecks because of toxic overload. The methods of these larger population studies are not designed to ferret out the uniqueness of individuals. If they looked at all the genetic subgroups in the population, then there could be meaningful data. If they did intestinal biopsies and spinal fluid examinations on affected and unaffected kids, as Dr Wakefield did, then they might obtain some meaningful information.

Roger Williams [21], the author of *Biochemical Individuality* said that, "Nutrition [and medicine] is for real people. Statistical humans are of little interest. People are unique. We must treat people with respect to their biochemical uniqueness."

MIND AND METALS: A DANGEROUS COMBINATION

A group of dedicated parents, scientists, and doctors led by Sidney Baker, MD, John Pangborn, PhD, and the late Bernard Rimland, have created a map for this new territory where we now find ourselves. Through their organization Defeat Autism Now!, this "think tank" has made clear that many children on the autism spectrum, and those with ADHD and even learning difficulties, are toxic [22].

Autistic children have low levels of glutathione, the major detoxification compound in the body, so they cannot excrete metals. Their hair shows low levels of mercury because genetically they can't excrete it [23], but they have higher levels in their baby teeth [24]. Chelation Challenge test using DMSA or DMPS show that autistic kids have more mercury and other metals than normal children.

Thimerosal and Autism

Until recently, mercury, in the form of thimerosal, was the most common disinfectant placed in vaccines (most flu vaccines still contain it) and contact lens fluid. A recent study "proved" that thimerosal has no link to autism or ADHD [25]. Or does it?

A study in *The New England Journal of Medicine* was apparently designed to show no link. Here is how the vaccine industry–funded scientists designed the study, which could not accurately answer the question of the effect of thimerosal in autism:

- They excluded all children with ADHD and autism. These are the children with the genetic susceptibilities to problems. These are the children who cannot detoxify. This is like doing a study to see if peanuts cause allergies, but excluding all kids with an allergy to peanuts from the study. It's just plain bad science.

- They did not measure mercury in hair, urine, or blood or the total body burden of metals in the children—just their exposure. So if some children were good detoxifiers, they would be able to excrete mercury. They should have measured the total body load of mercury in these children and then noted how that correlates to any neurologic or other effects. They also should have measured the genes involved in detoxification of mercury, such as apoE4, GSTM1, and MTHFR. Again, this is just plain bad science.

- Manufacturers who put mercury in vaccines in the first place employed or funded the authors of the study and its accompanying editorial. That's like putting tobacco companies in charge of studies on the risks of smoking. This is documented in the disclosures section of the paper.

- They didn't explain how it could be safe that babies received 187.5 g of mercury by the time they were 6 months old when the safe level is 0.5 g/kg of mercury at any one time, according to the EPA.

If thimerosal is as safe as studies like this attempt to suggest, why was it removed from use after 50 years from all childhood and adult vaccines in 2001 (except, of course, the ones we export to the third world)?

Metals Stuck in Our Chemistry

Mercury and other heavy metals block many metabolic pathways, including those related to building new hemoglobin molecules. Porphyrin metabolism is disrupted by mercury. Studies show that markers of abnormal porphyrins can be found in the urine of patients with heavy metal toxicity [26]. Genetic polymorphisms in porphyrin metabolism are linked to the development of neurotoxic and neurobehavioral effects from mercury exposure [27].

In addition, polymorphisms for brain derived neurotrophic factor (BDNF) is important in helping the brain repair, and neuroplasticity significantly increases the risk of mood, cognitive, and motor problems even at very low levels of mercury exposure [28,29].

This explains variable susceptibility to heavy metal poisoning and why studies of large populations often show no harmful effects from toxins. If 95 of 100 children can detoxify metals and are not affected, they are 100% fine. The problem, however, becomes significant and the effect severe for the 5% of children who cannot detoxify well.

It is not the porphyrin, BDNF, GST, or MTFHR genes that are the problem. It is the unique combination of all genes with the toxic environment and immune challenges that triggers brain dysfunction and illness.

CONCLUSION

Autism is a hologram. Through it a 3-dimensional picture of the failure of our current medical paradigm and the promise of a new one has formed. The lessons learned from the dissection of the functional causes and mechanisms of autism can illuminate the path for whole systems medicine and clinical research and the potential for it to address the global crisis of chronic disease. All we have to do is shine the light through the fragments of the hologram scattered at our feet. Then we have to pay attention and act collaboratively socially, politically, environmentally, and personally.

REFERENCES

[1]   Hyman MA. Autism: Is it all in the head? Altern Ther Health Med 2008; 14: 12-5.
[2]   James SJ, Melnyk S, Jernigan S, *et al.* Metabolic endophenotype and related genotypes are associated with oxidative stress in children with autism. Am J Med Genet B Neuropsychiatr Genet 2006; 141B: 947-56.
[3]   Williams TA, Mars AE, Buyske SG, *et al.* Risk of autistic disorder in affected offspring of mothers with a glutathione S-transferase P1 haplotype. Arch Pediatr Adolesc Med 2007; 161: 356-61.
[4]   Reddy MN. Reference ranges for total homocysteine in children. Clin Chim Acta 1997; 262: 153-5.
[5]   James SJ, Cutler P, Melnyk S, *et al.* Metabolic biomarkers of increased oxidative stress and impaired methylation capacity in children with autism. Am J Clin Nutr 2004; 80: 1611-7.
[6]   Bull G, Shattock P, Whiteley P, *et al.* Indolyl-3-acryloylglycine (IAG) is a putative diagnostic urinary marker for autism spectrum disorders. Med Sci Monit 2003; 9: CR422-5.
[7]   Wright B, Brzozowski AM, Calvert E, *et al.* Is the presence of urinary indolyl-3-acryloylglycine associated with autism spectrum disorder? Dev Med Child Neurol 2005; 47: 190-2.

[8]   Johnson SM, Hollander E. Evidence that eicosapentaenoic acid is effective in treating autism. J Clin Psychiatry 2003; 64: 848-9.

[9]   Amminger GP, Berger GE, Schafer MR, *et al.* Omega-3 fatty acids supplementation in children with autism: a double-blind randomized, placebo-controlled pilot study. Biol Psychiatry 2007; 61: 551-3.

[10]   Poling JS, Frye RE, Shoffner J, Zimmerman AW. Developmental regression and mitochondrial dysfunction in a child with autism. J Child Neurol 2006; 21: 170-2.

[11]   Herbert M. Autism: A brain disorder or a disorder of the brain? Clin Neuropsychiatry 2005; 2: 354-79.

[12]   Herbert MR. Large brains in autism: the challenge of pervasive abnormality. Neuroscientist 2005; 11: 417-40.

[13]   Vargas DL, Nascimbene C, Krishnan C, Zimmerman AW, Pardo CA. Neuroglial activation and neuroinflammation in the brain of patients with autism. Ann Neurol 2005; 57: 67-81.

[14]   Wakefield AJ, Ashwood P, Limb K, Anthony A. The significance of ileo-colonic lymphoid nodular hyperplasia in children with autistic spectrum disorder. Eur J Gastroenterol Hepatol 2005; 17: 827-36.

[15]   Millward C, Ferriter M, Calver S, Connell-Jones G. Gluten- and casein-free diets for autistic spectrum disorder. Cochrane Database Syst Rev 2004; CD003498.

[16]   Uhlmann V, Martin CM, Sheils O, *et al.* Potential viral pathogenic mechanism for new variant inflammatory bowel disease. Mol Pathol 2002; 55: 84-90.

[17]   Kawashima H, Mori T, Kashiwagi Y, *et al.* Detection and sequencing of measles virus from peripheral mononuclear cells from patients with inflammatory bowel disease and autism. Dig Dis Sci 2000; 45: 723-9.

[18]   Hornig M, Briese T, Buie T, *et al.* Lack of association between measles virus vaccine and autism with enteropathy: a case-control study. PLoS One 2008; 3: e3140.

[19]   Bradstreet J, Dahr JE, Anthony A, Kartzinel J, Wakefield A. Detection of measles virus genomic RNA in cerebrospinal fluid of children with regressive autism: a report of three cases. J Am Phys Surgeons 2004; 9: 38-45.

[20]   Taylor B, Miller E, Farrington CP, *et al.* Autism and measles, mumps, and rubella vaccine: no epidemiological evidence for a causal association. Lancet 1999; 353: 2026-9.

[21]   Williams R. Biochemical Individuality. New York: McGraw Hill; 1998.Pages.

[22]   Autism Research Initiative. Treatment Options for Mercury/metal Toxicity in Autism and Related Developmental Disabilities: Consensus Position Paper. San Diego, CA: Autism Research Initiative; 2005. Available at: http://www.autism.com/triggers/vaccine/heavymetals.pdf. Accessed September 17, 2008.

[23]   Holmes AS, Blaxill MF, Haley BE. Reduced levels of mercury in first baby haircuts of autistic children. Int J Toxicol 2003; 22: 277-85.

[24]   Adams JB, Romdalvik J, Ramanujam VM, Legator MS. Mercury, lead, and zinc in baby teeth of children with autism versus controls. J Toxicol Environ Health A 2007; 70: 1046-51.

[25]   Thompson WW, Price C, Goodson B, *et al.* Early thimerosal exposure and neuropsychological outcomes at 7 to 10 years. N Engl J Med 2007; 357: 1281-92.

[26]   Geier DA, Geier MR. A prospective study of mercury toxicity biomarkers in autistic spectrum disorders. J Toxicol Environ Health A 2007; 70: 1723-30.

[27]   Echeverria D, Woods JS, Heyer NJ, *et al.* The association between a genetic polymorphism of coproporphyrinogen oxidase, dental mercury exposure and neurobehavioral response in humans. Neurotoxicol Teratol 2006; 28: 39-48.

[28]   Heyer NJ, Echeverria D, Bittner AC, Jr., *et al.* Chronic low-level mercury exposure, BDNF polymorphism, and associations with self-reported symptoms and mood. Toxicol Sci 2004; 81: 354-63.

[29]   Echeverria D, Woods JS, Heyer NJ, *et al.* Chronic low-level mercury exposure, BDNF polymorphism, and associations with cognitive and motor function. Neurotoxicol Teratol 2005; 27: 781-96.

INDEX

A

Acrolein 23, 53
Adenosine triphosphate (ATP) 54, 93, 103, 104, 107, 110, 152, 206
Adenylyl cyclase 34, 51, 152, 165, 179
Adhesion molecules 56
Adrenal gland 165, 186, 193
Adrenocorticotropin hormone (ACTH) 186-195, 202
Aggressiveness 2, 4, 11, 93, 201
Alanine transaminase 101-104
Allergy 82, 123, 149, 206, 208, 213, 214
Alpha7-nicotinic acetylcholine receptors 25, 57
Aluminofluoride complexes (ALFx) 75, 148-158
Aluminum 23, 24, 38, 59, 64, 75, 116, 148-158, 162, 170, 209
Alzheimer's disease 61, 108, 115, 151, 152, 158, 206, 211, 212
Amelioration 35, 83, 96, 100, 157, 162, 170
AMPA 24, 32, 33, 36, 39, 41, 42, 47, 51-53, 56, 63, 77, 153, 201, 202
Androgens 42, 50, 199-205
Antibiotics 13, 87, 88, 105, 206, 209, 211, 213
Antifungals 13, 209
Antigen presenting cells 58
Antimycotics 13
Antioxidants 12, 13, 63, 140, 141, 157, 158, 165
Arcuate nucleus 47, 202
Area postrema 22, 57
Aripiprazol 12
Ascorbic acid 157
Aspartate 23, 26, 32, 33, 35, 38-42, 53, 76, 101, 113, 114, 140, 156, 162, 167-170, 202
Aspartate aminotransferase 101, 102
Asperger syndrome 1, 7, 8, 10, 40, 88, 91, 103, 176, 178, 179, 188, 200
Astrocytes 22, 23, 26, 33-37, 47, 48, 51, 53, 54, 56, 59, 60, 63, 76, 78, 108, 135, 137-141, 166, 169, 201
Attention Deficit Hyperactivity Disorders (ADHD) 2, 121, 133, 191-3, 199, 212, 214
Autism Spectrum Quotient 200
Autoimmune disease 73, 75, 135, 206
Autoimmunity 23, 73, 74, 76, 78, 84, 86, 135, 137, 153, 156, 206, 208

B

Basic fibroblast growth factor (BFGF) 50
Bcl-2 anti-apoptotic protein 25
BDE (tetrabrominated biphenyl) 77, 78
Bipolar disorder 2, 4, 11, 36, 39
Blood 3, 7, 37-41, 48, 50, 57, 61, 76, 90, 91, 93, 95,100-112, 123-5, 127, 129-138, 148-152, 154, 162, 163, 165, 167, 169, 173, 176, 178, 182, 186-189, 193, 194, 212, 214
Blood-brain barrier (BBB) 37, 41, 48, 56, 57, 82, 90, 91, 127, 134, 138, 149, 153, 155, 166, 169, 176, 189
Bowel disease 56, 82, 83, 85, 86, 165, 208
Brain-derived neurotrophic factor (BDNF) 50, 78, 215

C

Ca^{2+} 25, 32, 33, 34, 37, 63, 93, 107, 113, 115, 135, 140, 141, 151, 153, 157, 158, 165, 169, 202
Calcium 34, 51, 52, 55, 63, 77, 104, 114, 115, 135, 140, 148, 154, 156, 158, 169, 200-203
Candida 62, 78, 88, 91, 102
Candidate gene 38, 39, 83, 91, 181, 182, 203

Carnitine 100-105, 209
Carnosine 13
Casein 13, 73, 86, 88-91, 95, 212, 213
Casomorphin 90-91
Central nervous system (CNS) 3, 23, 32, 36, 38, 41, 47, 48, 52, 57, 62, 63, 73, 74, 78, 82, 86, 88, 90, 94-96, 101, 102, 104, 107, 115, 127, 134, 140, 165, 186, 200, 201
Cerebellum 17-25, 35, 36, 39, 42, 52, 55, 60, 76, 77, 83, 91, 93, 104, 106, 114, 139, 153
Cerebrospinal fluid (CSF) 38, 50, 103, 114, 179
Chelation 13, 141, 158, 214
Chemokines 26, 49, 50, 54, 56, 74, 76, 114, 140, 153, 195
Childhood autism rating scale (CARS) 112
Childhood autism spectrum test (CAST) 200
Choline acetyltransferase 202
Cholinergic anti-inflammatory pathway 57
Circumventricular organs (CVO) 22, 41, 48, 56, 57, 74, 169,
Clostridium 88
Coeliac disease 62, 89
Colitis 61, 62, 79, 84-86, 165, 169
Colon 62, 84, 85, 87, 165, 169
Complement 50, 53, 60, 86
Congenital adrenal hyperplasia 194, 200
Connexin 32 hemichannel 55
Constipation 74, 82-84, 89
Corticotropin- releasing hormone (CRH) 186, 188, 195
Cortisol 108, 176, 177, 186-195
Cortisol Awakening Response (CAR) 187
Cosmetics 123, 148
Creatine kinase (CK) 102, 104
Cyclic adenosine monophosphate (cAMP) 34, 51, 52, 115, 153, 154, 165, 168, 180
Cytokines 23, 24, 26, 32, 48-58, 61, 64, 73-78, 89, 114, 140, 148, 153, 166, 176, 195, 200, 202

D

Dehydroepiandrosterone (DHEA) 194, 199, 201
Diacylglycerol (DAG) 34, 51, 154
Diarrhea 82, 84, 85, 87, 89
Digestion 90, 91, 165, 208, 209, 213
Dihydrotestosterone 201, 202
Diphtheria-tetanus 133, 152, 207
Docosahexaenoic acid (DHA) 125, 210
Dogs 195
Dorsal vagal complex 56, 74
DSM-IV 1, 5-10, 101, 106, 112, 178, 199

E

Elastase 50
Electron transport chain 100,101
Embryogenesis 24, 56, 77, 93, 95
Endocrine 95, 108, 153, 162-169, 174, 177, 178, 190,194,195
Enteric nervous system 84, 94, 95
Enzyme immunoassay (EIA) 177, 178
Epilepsy 2, 3, 12, 22, 25, 33, 38, 60, 101, 104, 115, 121, 168, 191
Estrogen 50, 141, 199-203
Ethylmercury (EtHg) 121-142
Etiology 3, 23, 32, 38, 40, 41, 74, 83, 92, 114, 156, 173-181, 186, 192, 193
Excitatory amino acid transporters (EAAT 1-5) 35, 39, 51, 53, 141

Excitotoxic factors 2, 32, 75, 170
Excitotoxicity 23-26, 32, 36-41, 107, 108, 113-116, 135,137-140, 148, 153, 157-159, 166, 199, 201, 203, 208
Excitotoxin 23, 41, 47, 49, 54, 58, 64, 75, 76, 100, 107, 113, 140, 148, 158
Extracellular signal-regulated kinase (ERK) 53, 56, 93

F

Fcy receptors 49,50
FDG PET [^{18}F]-2-fluoro-2-deoxy-D-glucose positron emission tomography 105-108
Feedback 21, 52, 153, 167, 186-188, 190
Fluoride 24, 38, 64, 75, 107, 116, 148-159, 162, 167, 170, 202
Fluoxetine 11, 39
Folic acid (folate) 13, 100, 109, 110, 116, 206, 208, 210, 212
Follicle-stimulating hormone (FSH) 199, 201
Food allergies 61, 156, 207, 209, 213
Food intolerance 61, 78, 82, 84
Formula 59, 64, 150, 155
Fragile-X syndrome 52

G

GABA (gama-amino butyric acid) 3, 17, 25-26, 35, 36, 38, 39, 50-53, 55, 56, 77, 115, 141, 202
Gastroesophageal reflux disease (GERD) 83, 84
Gastrointestinal (GI) 12, 13, 62, 73, 82-96, 211
Gastrointestinal tract (GIT) 82-96, 123, 149, 162
Gliadin 62, 73, 87, 89-91
Glial fibrillary acidic protein (GFAP) 37
Gliosis 3, 17, 22, 38, 42
Glucocorticoids 186-195
Glutamate aspartate transporter (GLAST) 35, 36, 141
Glutamate transport proteins (EAAT) 35, 39, 51, 53, 141
Glutamate transporter 1 (GLT-1) 35, 36, 141, 169
Glutamate/cystine X_c- antiporter 53
Glutamatergic neurotransmission 32-46, 47, 55
Glutamic acid (= glutamate) decarboxylase (GAD) 25, 35-40, 42, 51, 53, 54, 56, 77, 115, 139, 141, 202
Glutamic acid (= glutamate) dehydrogenase (GDH) 17, 35, 36, 51, 59, 77, 141
Glutamine 3, 35, 36, 40, 42, 51, 55, 90, 139, 141, 173
Glutamine synthetase (GS) 35-37, 48, 51, 53, 54, 59, 77, 138, 139, 141
Glutathione (GSH) 35, 36, 53, 60, 63, 100, 109-112, 114, 127, 128, 132, 137, 138, 141, 157, 158, 166, 193, 194, 206-210, 214
Glutathione oxidized (disulfide) (GSSG) 100, 110-112, 166, 193
Glutathione peroxidase (GPx) 109, 110, 166
Gluten 62, 73, 86, 89-91, 95, 208-210, 212, 213
Gluten free (GF) diet 13, 89, 95
Gluten free-casein free (GFCF) diet 13, 89-91, 95
Gonadotropin-releasing hormone (GnRH) 201
Granulocyte macrophage colony-stimulating factor (GM-CSF) 78
Gut microflora 87, 96

H

Haloperidol 11, 39, 103
Hepatocyte growth factor/scatter factor HGF/SF 93
High mobility box-1 (HMGB-1) 57
Hippocampus 3, 23, 36, 37, 39, 42, 56, 60, 104, 115, 132, 138, 141, 153, 165, 186, 187, 190, 195, 200-203
HLA 73
Homocysteine 23, 26, 100, 109-114, 116, 194, 208

HSD1 = 11β-hydroxysteroid dehydrogenase type 1 (11β-HSD1) 186, 193-195
HSD2 (11β- HSD2) 191
Hydroxyindole o-methyltransferase (HIOMT) 163, 168, 169
Hyperactivity 2, 4, 11, 23, 39, 41, 75, 76, 78, 133, 153, 190, 191
Hypophysis 195
Hypothalamic-pituitary-adrenal axis (HPA) 186-195
Hypothalamus 18, 21, 23, 41, 47, 56, 165, 167, 174, 177, 186, 188, 200-202

I

Ibotenate 24, 51
ICD-10 1, 4-10
IgG 22, 23, 62, 74, 86, 208, 209
Immune system 47, 61, 73-76, 78, 79, 86, 88, 93, 94, 135, 136, 142, 154, 162, 163, 165, 166, 169, 209, 211, 213
Immunity 73, 87, 166, 170,
Immunoexcitotoxicity 17, 22-24, 38, 47-64, 73-79, 83, 142, 148, 170, 199, 201-203
Immunoglobulin 23, 85
Indolamine-2,3-dioxygenase (IDO) 49
Infantile autism 2, 3, 24, 89, 101, 106, 108, 156, 187
Infants 4, 21, 40, 41, 88, 105, 121-137, 149, 150, 152, 155, 156, 169, 176-178, 189, 191-195
Inflammation 24, 26, 34, 47-49, 60-62, 64, 75, 76, 82, 84-87, 95, 96, 105, 107-109, 114, 115, 148, 164, 195, 203, 206-214
Inositol 1,4,5-trisphosphate (Ins(1,4,5)P$_3$) 34, 51, 202
Insomnia 42, 127, 128, 168, 169
Interferon 24, 26, 49, 50, 53, 54, 57, 76, 135, 166
Interleukin 23, 24, 48-62, 73-78, 135, 166
Intestine 84, 85, 90, 114, 208
Iodine 153
Iodine deficiency disorders (IDD) 153
Ionotropic glutamate receptors (GluRs) 33, 34, 39, 52
IQ 4, 8-11, 18, 125, 129, 133, 134, 136, 152, 153, 158, 176, 192, 200

K

Kainate 32, 33, 51-53, 63, 115, 201, 202
Kynurenine 49

L

Lamellipodia 48
Language 1, 3-9, 12, 19, 20, 25, 42, 55, 85, 109, 132, 133, 135-137, 142, 176, 191, 207, 212, 213
Leaky gut 78, 88, 90
Lipid peroxidation products (LPP) 23, 26, 34, 50, 53, 54, 60, 64, 76, 77, 79, 107, 140, 141, 148, 153, 156
Lipopolysaccharide (LPS) 24, 48, 51, 53, 56-58, 61, 63, 75-77
Long-chain n-3 polyunsaturated fatty acids (n-3 LC PUFA) 125
Long-term depression (LTD) 19
Luteinising hormone (LH) 201
Lymphoblastoid cells (LCLs) 110
Lymphoid nodular hyperplasia (LNH) 84-86, 213

M

Macrophage 22, 24, 48, 49, 56, 57, 73, 75, 77, 78
Macrophage antigen complex (MAC1) 49
Macrophage inflammatory protein (MIP-1) 49, 50, 56, 77, 78
Magnesium 13, 33, 41, 140, 208, 210, 212
Magnetic resonance imaging (MRI) 2, 4, 17, 18, 20, 57, 151, 175-177, 188, 200, 211

Magnetic resonance spectroscopy (MRS) 104, 106
Maneb 59, 63
Measles virus 60, 75, 85, 86, 212, 213, 214
Measles-mumps-rubella (MMR) 61, 83, 85, 86, 207, 214
Melatonin 11-13, 52, 53, 90, 154, 156, 158, 162-170
Mercury 23, 24, 38, 59, 60, 64, 75, 100, 107, 109, 116, 121-142, 152, 168-170, 207-215
Mesenchymal epithelial transition factor (MET) 83, 91-94
Metabotropic glutamate receptors (mGluRs) 17, 32-34, 37, 39, 51-53, 113, 156, 158, 167, 168
Metalloproteases 50
Metallothionein 60, 100, 137, 208
Methyl-4-phenyl-1,2,3,6-tetrahydropyridine (MPTP) 58, 59
Methyl-4-phenylpyridinium (MPP+) 61
Methylfenidate 11
Methylmercury 59, 121-142, 153
Microglia 22-24, 26, 32-34, 37, 47, 48-64, 73-79, 108-110, 137-142, 151, 157, 158, 203
Microglial priming 58-62, 73, 75, 140, 157
Mitochondria 23, 34, 35, 39, 52, 53, 63, 77, 100-5, 107, 110, 113, 115, 116, 140, 141, 166, 203, 206-11
Mitogen-activated kinase (MAPK) 48, 53, 93
Modified Checklist for Autism in Toddlers (M-CHAT) 4
Monocytes 22, 48, 50, 57, 75-77, 166
Monosodium glutamate (MSG) 25, 41
Mood 4, 11, 58, 93, 121, 164, 179, 186, 191, 195, 207, 215
Mother 3, 22-4, 39, 40, 59, 64, 73, 74, 77, 111, 112, 114-16, 123, 124, 127, 128, 131, 135-7, 141, 151, 155, 156, 176, 177, 186-93, 199, 200, 207
MTHFR 114, 208, 215

N

N-acetyltransferase (NAT) 163, 164, 202
Natural killer cells 75, 135, 166
Nerve growth factor (NGF) 73, 135
Newborns 40, 41, 47, 63, 105, 130, 134, 136, 148, 155, 158, 167, 178
Nicotinamide adenine dinucleotide phosphate (NADP) 49, 64, 166, 193
Nicotinic acetylcholine receptors 25, 57
Nitric oxide 48, 76, 140, 153
Nitric oxide synthetase (NOS) 107, 153
N-methyl-D-aspartic acid (NMDA) 24, 25, 32-34, 37, 38, 41, 47, 48, 51-56, 60, 63, 106, 107, 113, 139, 140, 190, 201, 202, 208
Non-allergic food hypersensitivity 87
No-observed-adverse-effect levels (NOAEL) 130, 133

O

Obsessive-compulsive disorder (OCD) 8, 11, 75
Opioid excess theory 90, 91
Opioid peptides 91
Oxidative stress 23, 49, 53, 58, 63, 100, 102, 103, 107-15, 137, 158, 162, 165, 166, 169, 170, 176, 206-11
Oxytocin 12, 173-182

P

Pain 32, 82-5, 89, 127, 129, 151, 169,174
Paired-like homeodomain transcription factor 1 (PITX1) 186, 192-3
Paraquat 63
Parents 2, 4, 10-12, 40, 82, 85, 87-90, 95, 103, 112-116, 121, 122, 124, 126, 168, 177, 181, 186, 189, 190, 206, 207, 212, 214
Parturition 3, 173-6

Pathogen-associated molecular patterns (PAMPS) 48
Pattern recognition receptors (PRR) 48, 49
PBDE (polybrominated diphenyl ether) 78
PCR (polymerase chain reaction) 169, 181
Peripheral blood mononuclear cells (PBMC) 77, 78, 87, 89
Peroxynitrite 49, 53, 56, 63, 64, 107, 140
Pervasive developmental disorder (PDD) 1, 5, 8, 9, 83, 92, 162, 191
Pervasive developmental disorder-not-otherwise specified (PDD-NOS)
Pesticides 23, 24, 58, 59, 61, 63, 64, 77, 78, 207, 211
PET (positron emission tomography)105-8
Phosphatidylinositol (PI) 34, 93
Phosphatidylinositol-3-kinase (PI3K) 55, 93
Phosphatidylinositol-4, 5-bisphosphate (PIP$_2$) 34
Phospholipase C (PLC) 34, 51, 93, 165, 180
Pineal gland 153, 156, 162-170
Pinealocytes 167-169
Pituitary 123, 131, 173, 177, 186-190
Pivalic acid 105
Plasminogen activator 50
Platelets 3, 40, 167, 202
Prefrontal cortex 19, 21, 26, 36, 61, 74, 132, 165, 176, 186
pregnancy 3, 24, 50, 61, 63, 73, 74, 112, 121, 124, 125, 128, 135, 137, 167, 176, 177, 190-2, 199, 200
Prostaglandins 23, 26, 37, 50, 54, 76, 107, 114
Protein kinase C (PKC) 34, 115, 165
Provisional Tolerable Weekly Intakes (PTWI) 124
PTEN 73
Puberty 3, 42, 156, 167, 168, 170, 194
purinergic receptors 37, 50
Purkinje cells 17, 18, 22, 24-26, 38, 39, 42, 52, 62, 74, 77, 104, 113, 114, 141, 153
Pyridoxine 210 (see Vitamin B$_6$)
Pyruvate kinase (PK) 152

Q

Quinolinic acid (QUIN) 26, 49, 76, 140

R

RAGE receptor 49
Ramified microglia 57
Reactive nitrogen species (RNS) 23, 34, 49, 76, 107, 153, 164
Reactive oxygen species (ROS) 23, 34, 49, 76, 107, 153, 164, 165
Reelin 17, 25-26, 36, 39, 42, 56, 73
Regressive autism 22, 24, 75, 83-88, 102, 191
Retardation mental 2-5,11, 22, 38, 55, 104, 121, 141, 176, 191
Retardation motor 58, 104, 131, 136
Rett syndrome 1, 4, 6, 77, 105, 106, 108
Risperidone 11, 103

S

S-adenosylmethionine (SAM) 110, 111, 163
Saliva 162, 170, 178, 186-188, 190, 193-195
Scavenger 49, 110, 111, 157, 158, 162, 165, 167, 170
Secretin 13
Seizure 1, 3, 12, 17, 22-4, 37, 38, 42, 55, 62, 74, 101, 110, 113, 115, 121, 136, 141, 151, 152, 157, 168, 202
Serotonin 3, 11, 25, 50, 60, 73, 74, 154, 163, 164, 165, 169

Schizophrenia 2, 3, 6, 8, 9, 11, 20, 21, 23-5, 33, 36, 38, 39, 50, 59, 63, 74, 91, 110, 113, 121, 199
Siblings 3, 23, 39, 40, 82, 84, 87, 103, 106, 109, 152, 181, 199
Sickness behavior 56, 57, 61, 76
Single nucleotide polymorphisms (SNPS) 38-40, 181, 182
Sleep 2, 11, 12, 42, 52, 56-58, 82, 93, 105, 129, 156, 158, 162, 167-170, 187
Socialization 1, 42
Speech 6, 12, 20, 25, 104, 105, 128, 133, 151, 178, 179, 182
Stool 82, 84, 87, 89, 115, 129, 133, 134, 207, 208
Stress 23, 53, 62, 108, 137, 138, 145, 167, 174-6, 186-192, 194, 195, 211
Subfornical organ 57
Suicide 4, 11
Superoxide 49, 63, 64, 107, 109, 140, 158
Suprachiasmatic nucleus (SCN) 162, 202

T

Taurine 25, 47, 110, 112, 114-116
Tantrums 11
Testosterone 194, 199-204
Thimerosal 110, 121, 122, 125-126, 129-135, 137, 138, 141, 214 -215
Toddlers 4
Tumor growth factor (TGF) 23, 50, 57, 61, 75, 76, 78
T helper cells 89
toll-like receptors (TLR) 48, 77
Tumor necrosis factor α (TNF-α) 23, 24, 37, 48 – 56, 57, 60-63, 74, 76-78, 86, 87, 89, 202, 203
Tryptophan 3, 49, 50, 163, 164, 169, 170
Thyroid gland 153, 164
Thyroid-stimulating hormone (TSH) 153
Twins 3, 38

U

Umbilical 3, 125, 137
Urine 91, 100, 101, 103, 105, 106, 122-124, 127, 130, 134, 151, 152, 156, 158, 162, 167, 190, 214, 215
Uterus 173

V

Vaccines 38, 47, 59-61, 63, 75, 78, 85, 121-127, 131-138, 140,142, 149-152, 156, 207, 213-215
Very low-density lipoprotein (VLDL) 56
Vitamins 13, 100, 158, 210
Vitamin A 208, 210, 212
Vitamin B$_6$ 13, 40, 110, 116, 210, 212
Vitamin B$_{12}$ 110, 113, 116, 208, 210, 212
Vitamin C 13, 157, 158
Vitamin D 157, 158, 208, 210, 212
Vitamin E 157
Viral infection 24, 25, 47, 60, 61, 63, 75, 212

Y

Yeast 87, 88, 208-10, 213

Z

Zinc 208, 210, 212